Functional High Entropy Alloys and Compounds

Edited by

Yong Zhang
Fuyao University of Science and Technology & UST Beijing

Yonggang Yao
Huazhong University of Science and Technology

Haojie Liu
Huazhong University of Science and Technology

WILEY-VCH

Editors

Professor Yong Zhang
Uni of Science and Technology Beijing
No.30 Xueyuan Rd
Haidian District
Beijing
CH, 100083

Professor Yonggang Yao
Huazhong Uni of Science and Technology
1037 Luo Yu Road
Wuhan
CH, 430074

Haojie Liu
Huazhong Uni of Science and Technology
1037 Luo Yu Road
Wuhan
CH, 430074

Typesetting: Lumina Datamatics

Printing: CPI Group (UK) Ltd,
Croydon CR0 4YY

Library of Congress Card No.: applied for

British Library Cataloguing-in-Publication Data
A catalogue record for this book is available from the British Library.

Bibliographic information published by the Deutsche Nationalbibliothek
The Deutsche Nationalbibliothek lists this publication in the Deutsche Nationalbibliografie; detailed bibliographic data are available on the Internet at <http://dnb.d-nb.de>.

Print ISBN: 9783527354375
ePDF ISBN: 9783527849185
ePub ISBN: 9783527849178
oBook ISBN: 9783527849192

C9783527354375_030925

Contents

Contributors

Yong Zhang, Rongzhi Li, Hang Xiang, Yuxin Wen
School of Materials Science and Engineering, Fuyao University of Science and Technology & UST Beijing

Yonggang Yao, Haojie Liu
School of Materials Science and Engineering, Huazhong University of Science and Technology

Di Ouyang
Department of Metallic Engineering, Wuhan University of Technology

Cheng Zhang
School of Materials Science and Engineering, Huazhong University of Science and Technology

Bo Song, Yan Liu
School of Materials Science and Engineering, Huazhong University of Science and Technology

Yuhua Wang, Yifan Wan, Yitong Wang
College of Science, Wuhan University of Science and Technology

Pengfei Xie, Huibin Wang
College of Chemical and Biological Engineering, Zhejiang University

Dongshuang Wu, Bing Zhu
School of Materials Science and Engineering, Nanyang Technological University

Wei Luo, Yuwei Chen
School of Materials Science and Engineering, Tongji University

Foreword

The emergence of high-entropy materials (HEMs) has revolutionized the field of materials science and engineering, offering a bold departure from traditional alloy design paradigms. Unlike conventional materials, which typically revolve around one or two principal elements, HEMs combine multiple principal components in near-equimolar proportions. This unique approach yields materials with exceptional configurational entropy, stabilizing simple solid solutions and delivering extraordinary mechanical, thermal, and chemical properties.

Since their introduction in the early 2000s, HEAs and their derivatives, including high-entropy ceramics and oxides, have captured the attention of researchers worldwide. These materials have demonstrated outstanding performance in extreme environments, such as aerospace, nuclear, and energy applications, where traditional materials often fall short. Their resistance to high temperatures, corrosion, and radiation, coupled with their mechanical robustness, positions HEMs as pivotal for next-generation technologies.

This book represents a comprehensive exploration of HEMs, covering their fundamental principles, advanced characterization methods, and diverse applications. It provides an invaluable resource for both seasoned researchers and newcomers to the field, fostering a deeper understanding of the underlying mechanisms and inspiring innovative applications.

The editors have meticulously compiled insights from leading experts, offering a balanced perspective that bridges theoretical foundations and practical implementation. As you navigate through these chapters, you will discover the transformative potential of HEMs, which not only redefine material performance but also pave the way for sustainable, high-efficiency technologies.

We are witnessing the dawn of a new era in materials science, and this book serves as a beacon, guiding us toward the exciting possibilities that lie ahead. It is my privilege to introduce this seminal work, which I believe will significantly influence the scientific community and industry alike.

Yong Zhang, Yonggang Yao, and Haojie Liu

Preface

The concept of high-entropy materials emerged in the early 2000s as a transformative idea in materials science. Driven by the vision to overcome the limitations of traditional alloys, researchers pioneered the design of materials comprising multiple principal elements in near-equimolar ratios. This approach harnesses configurational entropy to stabilize simple solid solutions, unlocking properties that were previously unattainable in conventional materials.

This book is the culmination of years of research and innovation in the field of high-entropy materials. It aims to provide a comprehensive understanding of these materials, bridging fundamental principles with cutting-edge advancements and practical applications. The structure of the book reflects this holistic approach:

- The initial chapters delve into the theoretical underpinnings of HEMs, detailing the core effects of high entropy, severe lattice distortion, sluggish diffusion, and the cocktail effect.
- Subsequent sections focus on the synthesis, characterization, and tailored design of HEMs for specific applications, highlighting their performance in extreme environments, nanoscale synthesis, energy storage systems, and catalytic processes.
- The book concludes with a forward-looking perspective, addressing the challenges of scalability, cost, and environmental sustainability while envisioning the future of HEMs in emerging technologies.

This work would not have been possible without the contributions of our colleagues, collaborators, and students, whose dedication and expertise have enriched every chapter.

As the field of high-entropy materials continues to evolve, we hope this book serves as both a foundation and a catalyst for further exploration. Whether you are a researcher, engineer, or student, we invite you to join us in uncovering the vast potential of these extraordinary materials.

Yong Zhang
Fuyao University of Science and Technology & UST Beijing

Yonggang Yao & Haojie Liu
Huazhong University of Science and Technology
11 January 2025

1

High-Entropy Alloys and Compounds: Fundamentals

*Haojie Liu, Yonggang Yao**

School of Materials Science and Engineering, State Key Laboratory of Materials Processing and Die & Mould Technology, Huazhong University of Science and Technology, Wuhan, P.R. China

*Corresponding author: yaoyg@hust.edu.cn

1.1 Introduction (Historical Background and Development)

High-entropy alloys (HEAs) were first conceptualized and introduced in the early 2000s (Figure 1.1), pioneered by J.W. Yeh and his colleagues in 2004 [1]. Traditional alloy design typically revolves around a single principal element, supplemented with smaller amounts of secondary elements as dopants to achieve desired property enhancements. However, HEAs challenge this paradigm by combining five or more elements in nearly equal proportions, creating a fundamentally distinct multicomponent system characterized by high configurational entropy and complex microstructures or atomic coordination. The elevated entropy arising from the random distribution of multiple principal elements can stabilize simple solid-solution phases, such as face-centered cubic (FCC) or body-centered cubic (BCC) structures rather than the formation of more complex intermetallic compounds. This stabilization is a direct consequence of entropy-driven thermodynamics, where the high configurational entropy counteracts enthalpic contributions that typically drive phase separation.

The initial HEAs, often consisting of transition metals like Fe, Ni, Co, Cr, and Mn, exhibited outstanding properties [3–5], including high hardness, excellent wear resistance, and remarkable thermal stability. These properties stem from the synergistic combination of elements, which collectively enhance the material's performance under extreme conditions. The high configurational entropy in these alloys plays a key role in suppressing the formation of brittle intermetallic phases but instead promoting ductile and tough solid solutions [6].

Building on the foundational work of Yeh and others, the field of HEAs rapidly expanded in scope and application. Researchers began exploring diverse

Functional High Entropy Alloys and Compounds, First Edition. Edited by Yong Zhang, Yonggang Yao and Haojie Liu.

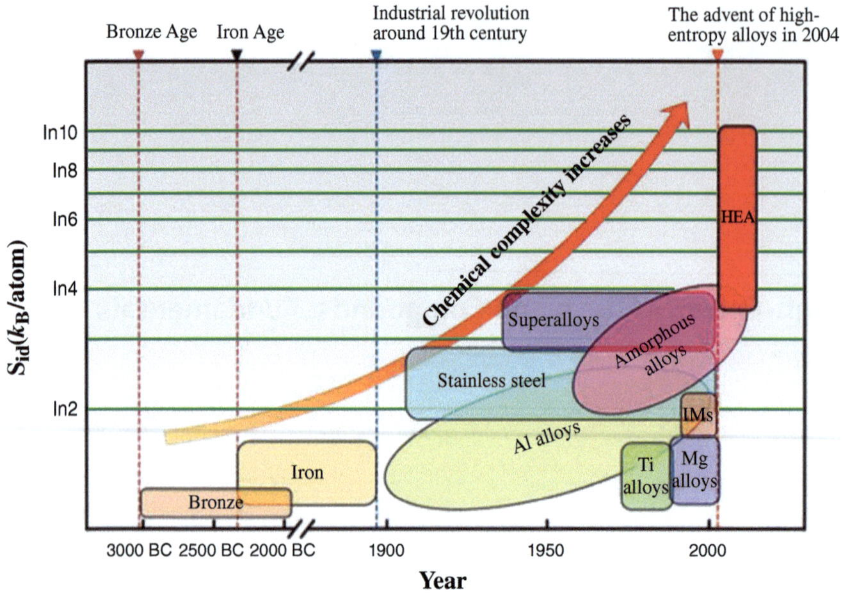

Figure 1.1 Rising trend of alloy chemical complexity versus time (IMs: intermetallics, HEA: high-entropy alloy) [2]. *Source*: [2] / with permission of Springer Nature.

compositional systems, incorporating elements from various groups of the periodic table, including refractory metals (e.g., Mo, W, Ta), lightweight elements (e.g., Al, Mg, Li), and even nonmetallic elements (e.g., Si, B, C) [7]. This diversity allowed for the tailoring of HEA properties to meet the specific demands of various applications. For instance, refractory HEAs demonstrated exceptional high-temperature strength and oxidation resistance, while lightweight HEAs became promising candidates for aerospace applications requiring materials with low density and high strength. HEAs have also shown remarkable potential in harsh environments, such as those encountered in aerospace, nuclear reactors, energy generation, and deep-sea exploration. These environments demand materials that can withstand extreme temperatures, radiation, corrosion, and oxidation. For example, corrosion-resistant HEAs have been developed for marine and chemical processing industries, while high-strength refractory HEAs are being evaluated for use in hypersonic vehicles and fusion reactors [8, 9].

The continued exploration of HEAs is fueled by advancements in computational modeling, materials processing techniques, and experimental characterization. Computational tools such as high-throughput simulations and machine learning algorithms are enabling the design and optimization of novel HEA compositions with tailored properties [10, 11]. These tools allow researchers to predict phase stability, mechanical performance, and environmental resistance with unprecedented accuracy, significantly reducing the time and cost associated with traditional trial-and-error approaches. In parallel, innovations in materials processing techniques, including additive manufacturing (AM), powder metallurgy, and advanced casting methods, are facilitating the scalable production of HEAs with

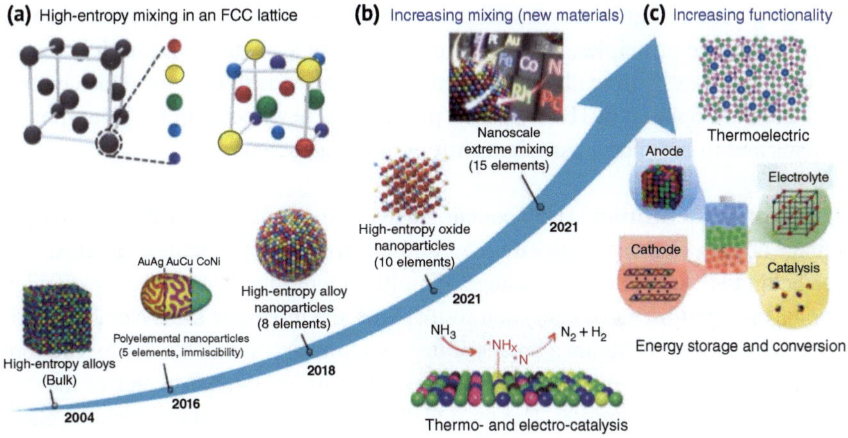

Figure 1.2 Schematic diagram of the development of high-entropy nanoparticles with multielemental composition and enhanced functionality [17]. *Source*: [17] / with permission of American Association for the Advancement of Science.

complex geometries and microstructures [12–14]. Moreover, state-of-the-art characterization tools, such as atom probe tomography (APT), transmission electron microscopy (TEM), and synchrotron-based techniques, provide deep insights into the atomic-scale structure and properties of HEAs, driving a better understanding of the underlying mechanisms that govern their exceptional performance [15, 16].

HEAs hold great promise for addressing critical challenges in materials science and engineering. Their unique combination of properties—including high strength, excellent thermal stability, superior corrosion resistance, and exceptional wear resistance—makes them highly attractive for applications in next-generation technologies. Furthermore, HEAs are poised to play a pivotal role in sustainable development by enabling materials with enhanced durability, reduced resource consumption, and improved energy efficiency. As research continues to advance, the integration of interdisciplinary approaches, combining theoretical modeling, experimental validation, and industrial-scale implementation, will be essential to unlocking the full potential of HEAs. The development of HEAs tailored to specific applications will not only expand the boundaries of materials science but also contribute to the creation of environmentally resilient and economically viable technologies. These advancements position HEAs as key enablers of innovation and sustainability across a broad range of industries (Figure 1.2).

1.2 Definition of HEAs and Compounds

The concept of high-entropy materials (HEMs), which are composed of at least five elements or components, represents a shift in how materials are designed and understood, leveraging entropy as a key factor in their creation. Typical examples are HEAs and high-entropy compounds (HECs), offering a wide array of unique properties derived from their multicomponent systems. While the effects of entropy have long been recognized in traditional materials and could be explained through thermodynamic principles, the impact of entropy in systems with multiple principal elements was historically overlooked or considered insignificant. This changed with the development of HEAs, which prompted a new focus on multicomponent systems and introduced a paradigm shift in materials science.

Composition Definition. Alloys containing five or more elements with an atomic percentage of each element greater than 5% and less than 35% are defined as HEAs. The idea behind HEAs is to take advantage of multielement mixing and the large configurational entropy (discussed later) that results from the mixing of these elements into a single-phase alloy. HEMs are a broader category of materials that, like HEAs, are composed of multiple elements in near-equal or equimolar proportions to replace the used one/few components position. The term "HEMs" extends beyond just metallic alloys, which also include nonmetallic systems, such as ceramics and polymers, that exhibit the same principle of high entropy. For example, high-entropy ceramics are ceramic materials wherein multiple metal or nonmetal elements are mixed to create a stable phase with desirable properties such as high hardness, thermal stability, and oxidation resistance, as in complex oxide ceramics of $(Ti, Zr, Hf, Nb, Ta)O_2$. In the later text, we will use HEA as the most important example in HEMs to illustrate those basic concepts.

Entropy Definition. The key defining characteristic of HEAs is their "high-entropy" composition, where the term "entropy" refers to the number of ways atoms can be arranged in a system. By mixing several elements in nearly equal amounts, the number of possible atomic arrangements increases, leading to a higher-entropy state, which can stabilize the solid-solution phase at room temperature, resulting in improved properties. This definition specifies the kinds and ratios of components and introduces the concept of high entropy, which aims to overcome the formation of intermetallic phases through high mixing entropy, thereby obtaining a single-phase solid-solution structure. According to Boltzmann formula, the entropy of a system is related to a given state as the following relationship:

$$S = k_B \ln \omega$$

where k_B is the Boltzmann constant and ω is the probability that a given state exists. The entropy of the system includes four parts: configurational entropy $\left(\Delta S_{mix}^{conf}\right)$, vibrational entropy $\left(\Delta S_{mix}^{vib}\right)$, magnetic entropy (ΔS_{mix}^{mag}), and electronic randomness entropy $\left(\Delta S_{mix}^{elec}\right)$. The total mixing entropy of HEAs is depicted by the equation:

$$\Delta S = \Delta S_{mix}^{conf} + \Delta S_{mix}^{vib} + \Delta S_{mix}^{mag} + \Delta S_{mix}^{elec}$$

But compared with vibrational entropy, magnetic entropy, and electronic entropy, configurational entropy plays a dominant role, in general. Therefore, to avoid the difficulties caused by the calculation of vibrational entropy, magnetic entropy, and electron entropy, the configuration entropy of the system is generally regarded as the mixing entropy ΔS_{mix} of the ideal solid solution, which can be expressed as:

$$\Delta S_{mix} = -R\sum x_i \ln(x_i)$$

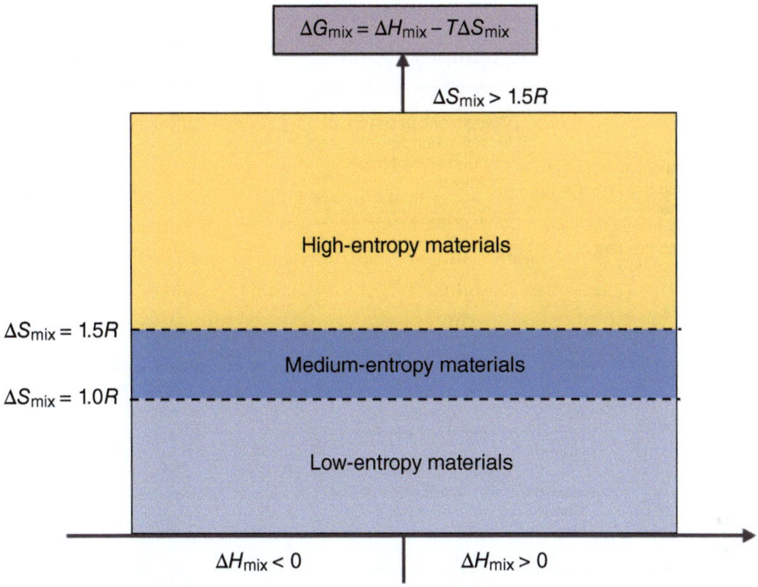

Figure 1.3 Basic definition of HEAs, in which ΔG_{mix}, ΔH_{mix}, ΔS_{mix}, and T are the Gibbs free energy of mixing, the enthalpy of mixing, the entropy of mixing, and the temperature, respectively. The increase in total entropy as the separate systems of different compositions are mixed without chemical reactions. ΔS_{conf} is the configurational entropy, a portion of a system's entropy related to discrete representative positions of its constituent particles.

where R is the molar gas constant ($R = 8.314\ \text{J} \cdot \text{mol} \cdot \text{K}^{-1}$), x_i is the mole fraction of each constituent element. The mixing entropy of the alloy can be maximized by designing the alloy with equimolar ratios of multiple elements, $x_i = 1/n$, that is:

$$\Delta S_{mix} = -R\left(\frac{1}{n}\ \ln\frac{1}{n} + \frac{1}{n}\ \ln\frac{1}{n} + \ldots\ \frac{1}{n}\ \ln\frac{1}{n}\right) = R\ \ln(n)$$

Based on $\Delta S_{mix} = R\ \ln(n)$, Figure 1.3 clearly shows the relationship between the number of equimolar elements (n) and the mixing entropy (s) of the system. When the number of HEA elements n is greater than or equal to 5, the system's mixing entropy s is greater than or equal to $1.61R$. Therefore, entropy ($1.61R$) can be used as a criterion for distinguishing HEAs/medium-entropy alloys.

Extended Definition. HEAs, initially defined as alloys with a single-phase solid-solution structure and containing more than five principal elements in near-equiatomic ratios, have been a focus of research for decades. Over time, this strict definition has evolved, breaking traditional constraints and paving the way for the development of more versatile materials. This progression has allowed researchers to shift from rigidly adhering to classic definitions to prioritizing tailored compositions that address specific functional needs. The so-called second generation of HEAs, often characterized by non-equimolar compositions and multiphase structures, exemplifies this shift (Figure 1.4): at least mixing of four elements would be good, and even multiple phases are allowed in the extended definition of HEAs. By

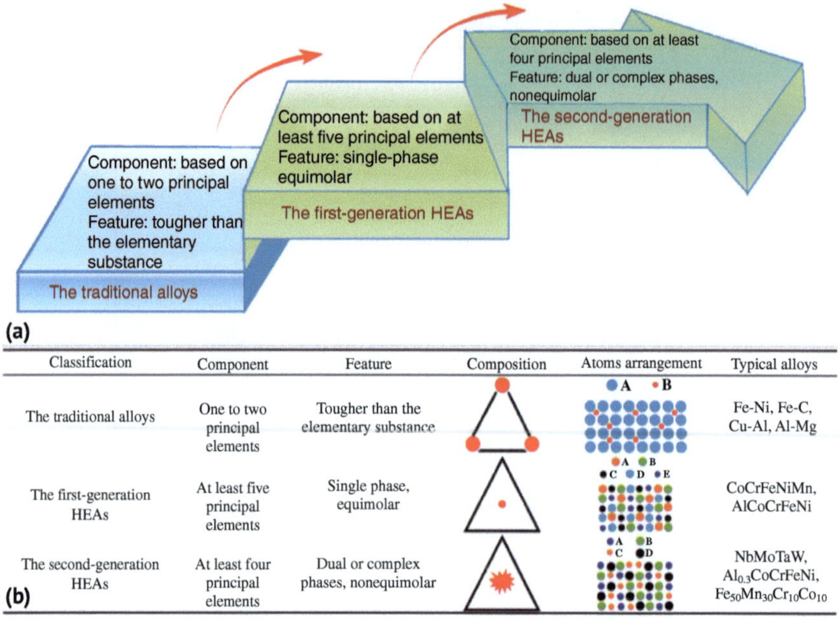

Figure 1.4 (a) The evolution of traditional alloys and HEAs [18]. (b) Characteristics for the traditional alloys and two generations of HEAs [18]. *Source*: [18] /with permission of Springer Nature.

carefully adjusting the proportions of components and introducing minor elements, these materials can be optimized to stabilize specific phases or enhance performance through mechanisms such as precipitation or composite strengthening. For example, alloys like $Fe_{50}Mn_{30}Cr_{10}Co_{10}$ demonstrate transformation-induced plasticity effects, while $(FeCoNiCr)_{94}Ti_2Al_4$ leverages precipitation hardening, and eutectic HEAs such as $AlCoCrFeNi_{2.1}$ combine exceptional phase stability with high mechanical strength.

This evolution reflects a broader trend in material design, where compositional or phase complexity is used as a tool to unlock novel properties, transcending the boundaries of traditional single-phase alloys. By embracing this expanded perspective, researchers are exploring not only alloys but also other high-entropy systems, creating opportunities for materials tailored to specific high-performance applications, from extreme environments to advanced structural and functional technologies. This holistic approach is driving the next wave of innovations, emphasizing adaptability and multifunctionality over rigid adherence to initial definitions.

1.3 Characteristics, Structures, Phase Transformations, and Microstructures

Expanding beyond alloys, the high-entropy concept has been applied to ceramics, carbides, nitrides, oxides, and intermetallic compounds, creating materials with superior mechanical, thermal, and particularly functional properties. The shared underlying principle of configurational entropy plays a critical role in stabilizing crystalline phases in both metallic alloys and compounds. This section explores the characteristics, crystal structures, phase transformations, and microstructures of these HEMs in detail.

1.3.1 Characteristics of HEAs and Compounds

The four core effects for HEAs are high entropy, severe lattice distortion, sluggish diffusion, and the cocktail effect—each corresponding to four key aspects of material behavior: thermodynamics, structure, kinetics, and performance. The high-entropy effect promotes the formation of stable multielement solid solutions, especially at high temperatures. Severe lattice distortion arises from varying atomic sizes, affecting structural properties and altering thermodynamics and kinetics. Sluggish diffusion slows down atomic movement, delaying phase transformations and enhancing stability. The cocktail effect refers to the synergistic improvement in properties like strength, toughness, and conductivity, surpassing predictions from traditional alloy theory. Together, these effects define the unique behavior and superior performance of HEAs compared to conventional alloys.

High-Entropy Effect. The concept of the high-entropy effect, initially introduced by Yeh, posits that an increase in entropy results in a reduction in the Gibbs free energy of solid-solution phases. High-entropy systems are driven by the concept of maximizing configurational entropy, which arises from the random distribution of multiple principal elements, as discussed in Section 1.2. In contrast to conventional materials where one or two elements dominate, the presence of multiple elements in nearly equal concentrations promotes the formation of simpler phases (e.g., solid solutions in alloys or basic crystal structures in compounds) rather than complex, brittle intermetallic or ordered phases [19, 20]. Rost et al. found an endothermic reaction upon cyclic heat treatment of rock salt (CoCuMgNiZnO high-entropy oxide [HEO]) [21]. They attributed this to the enthalpic penalties of Zn and Cu for the crystallization of a rock salt structure. To lower the barrier of formation of a single solid solution, these two elements were excluded from HEO. However, this was not the case; the exclusion of Zn and/or Cu led to the formation of multiphases. This was a strong indication that the role of mixing entropy was prominent since it decreased when one cation was removed. Besides, the overall degrees of order in HEAs are expected to decrease with the increase in temperature. Therefore, even as-cast alloys containing ordered phases can also be transformed into disordered solid-solution phases at elevated temperatures. However, it is noted that a single solid solution in HEAs is less likely to form as the number of constituent elements increases, which contradicts the high-entropy effect. As they have articulated, the configurational entropy rises slowly with the number of the increasing elements, whereas the probability of the formation of IMs increases more rapidly [22].

While the high-entropy effect is crucial for the formation of HEMs, it is not the sole determinant of phase formation in multicomponent alloys and ceramics. However, it remains a fundamental influence on HEM formation, minimizing the number of phases, as indicated by Gibbs phase rule, which allows for an increase in the number of equilibrium phases with the addition of components [23, 24] (Figure 1.5).

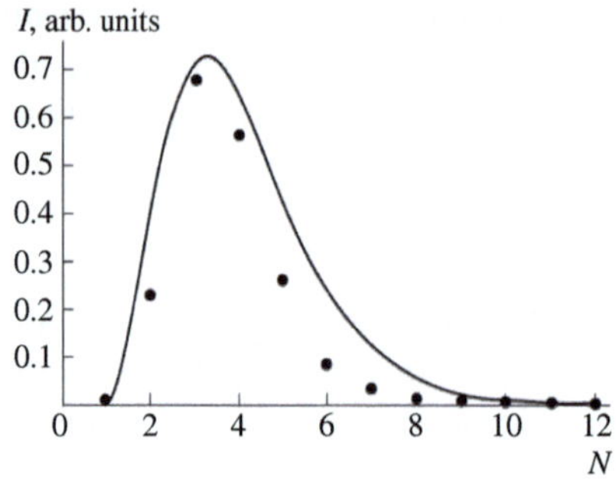

Figure 1.5 High-entropy effect in reducing the number of inorganic crystal structures. Points denote the numbers of inorganic crystal structures with 1, 2, 3, … N different elements. $N = 3$ has a maximum of 19,000 [24]. *Source*: [24] / with permission of Springer Nature.

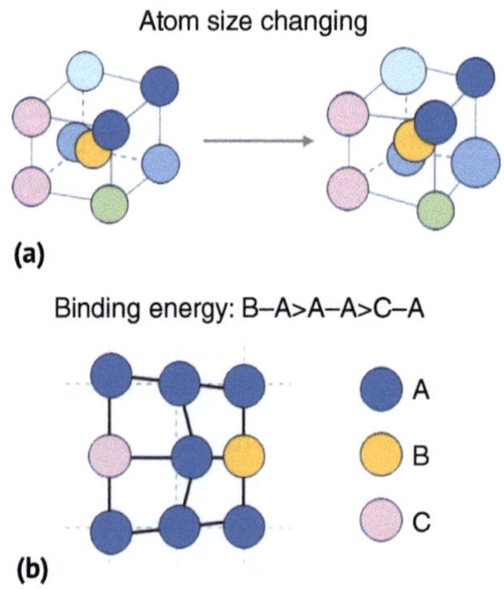

Figure 1.6 (a) Schematic representation of atomic size difference-induced distortion: introducing multiple different elements into the same crystal structures leads to severe lattice distortions. (b) Schematic representation of bonding energy difference-induced distortion. The bonding interaction between A and B is stronger than between A and A and between A and C, so the distance between A and B will be smaller than between A and C, which creates the lattice distortion.

Severe Lattice Distortion. Severe lattice distortion in HEAs is primarily caused by the presence of multiple principal elements with differing atomic sizes, bonding characteristics, and random distribution throughout the crystal lattice [25, 26] (Figure 1.6). The significant atomic size mismatch among these elements results in irregular stretching and compression of the lattice, as atoms of different sizes struggle to fit into the same crystallographic positions. This size disparity disrupts the regular arrangement of atoms, causing considerable strain on the lattice structure. Additionally, the bonding differences between the various elements, which have distinct electron configurations and bonding strengths, further contribute to this disruption. These nonuniform bonding forces lead to uneven interactions between atoms, creating additional strain and preventing the formation of a smooth, ordered lattice (Figure 1.6).

Moreover, the random atomic distribution of elements in HEAs, where each element is distributed randomly across lattice sites, increases the disorder [27]. Unlike traditional alloys, where a single principal element typically dominates, HEAs have near-equal concentrations of multiple elements, leading to greater structural complexity. This random placement of atoms of different sizes and bonding energies increases lattice distortion, as the structure cannot easily accommodate the varying atomic characteristics.

This lattice distortion has important effects on the properties of HEAs. The distorted lattice hinders the movement of dislocations, which are defects in the crystal structure that allow materials to deform. By impeding dislocation movement, the distorted lattice increases the strength and hardness of HEAs [28]. Additionally, the severe distortion enhances the alloy's resistance to deformation, improving its toughness and wear resistance. Overall, the severe lattice distortion in HEAs, caused by atomic size mismatch, bonding differences, and random distribution, is a key factor contributing to their superior mechanical properties [29].

Sluggish Diffusion. The random arrangement of atoms with varying sizes and bonding preferences in HEAs and HECs disrupts atomic mobility, leading to lower diffusion rates, particularly at high temperatures. This phenomenon of sluggish diffusion arises from two key factors [30].

First, each lattice site in HEAs and HECs is occupied by atoms with different neighboring atoms, creating unique local environments. When an atom moves into a vacant site, the surrounding atoms and bonding energies differ from its original location. Due to these varying local configurations and bonding energies, an atom that moves into a site with lower energy becomes "trapped," making it difficult to transition to a higher-energy site (Figure 1.7). Conversely, if an atom is

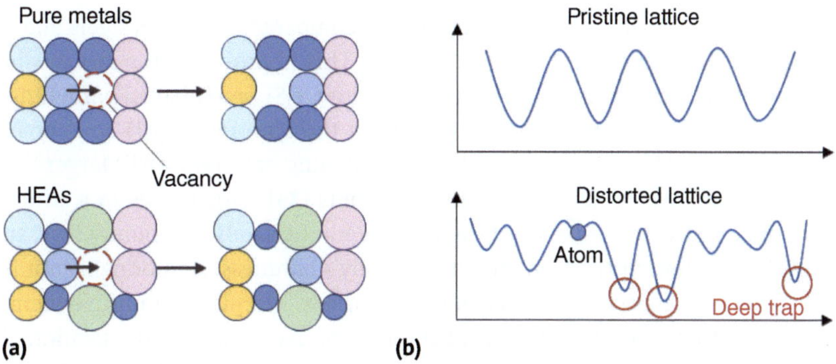

Figure 1.7 (a) Schematic representation of the neighbors an atom interacts with before and after jumping into a vacancy in different matrices. (b) Regular (top) and irregular (bottom) fluctuations of lattice potential energies along the diffusion path for an atom in pure element and HEA lattice. The potential energy of each lattice site in pure elements is nearly identical. By contrast, HEAs are present sites in which the configuration of bonding is preferable for the diffusing species to others; these sites act as traps, and their presence lowers the diffusion rate.

in a higher-energy site, it has a higher likelihood of returning to its original position. These energy differences slow the rate of diffusion. This sluggish behavior, combined with the high activation energy of diffusion in HEAs, is attributed to the significant variation in lattice potential energy (LPE) across different sites within the crystal lattice. Low LPE sites act as obstacles that impede atomic movement, contributing to what is known as the diffusion retardation effect (Figure 1.7b).

Second, the various components in HEAs diffuse at different rates. Some elements move less actively than others, making it more difficult for those components to fill vacancies compared to faster-moving elements. However, processes such as phase changes require the simultaneous diffusion of all components to achieve the desired composition. For example, forming a new phase or facilitating grain growth demands the collective diffusion of all elements to allow effective migration of grain boundaries. In such cases, the slow diffusion of certain components becomes a limiting factor, slowing down the overall transformation process.

Despite this, the sluggish diffusion phenomenon offers several advantages [31], which include the formation of supersaturated states, increased recrystallization temperature, delayed grain growth, reduced grain coarsening, and the formation of fine precipitates. These benefits enhance microstructure and property control, leading to improved material performance.

The Cocktail Effect. The "cocktail effect" refers to the synergistic interaction between different elements in an alloy, resulting in superior material performance. While this effect is also seen in conventional alloys, it is more significant in HEAs due to the complexity of their compositions. The cocktail effect describes the deviation from the simple rule of mixtures and encompasses all factors and mechanisms that contribute to this deviation. In HEMs, the cocktail effect can lead to unexpected and highly beneficial properties, such as increased strength, wear resistance, and stability at high temperatures.

For instance, a heat-treated HEA coating $\left(NiCo_{0.6}Fe_{0.2}Cr_xSiAlTi_y\right)$ demonstrates superior hardness and wear resistance compared to various steels and non-heat-treated HEA coatings [32] (Figure 1.8a). This improvement is due to silicon, which promotes the formation of wear-resistant Cr_3Si precipitates, hardening the coating. Another example involves increasing the aluminum (Al) content in an $Al_xCoCrCuFeNi$ HEA, leading to solution hardening because of Al's larger atomic radius and stronger bonding with other elements [33]. This results in a structural shift from an FCC to a mixed FCC and BCC and ultimately to a purely BCC structure as the Al content rises, enhancing the alloy's hardness. Additionally, multielement synergy in materials arises from the complex interplay of multiple elements rather than from the individual contributions of each element in isolation. This synergy allows the material to harness the diverse properties of the constituent elements, leading to enhanced overall performance. In the context of material design, electrochemically inert elements—such as Zn and Ti—are often chosen to stabilize the host structure, improving its structural integrity. These elements act as stabilizers, preventing phase degradation under operational stresses. On the other hand, electrochemically active elements like Co, Mn, Ni, Fe, and others serve as redox-active centers, facilitating charge transfer and improving the material's

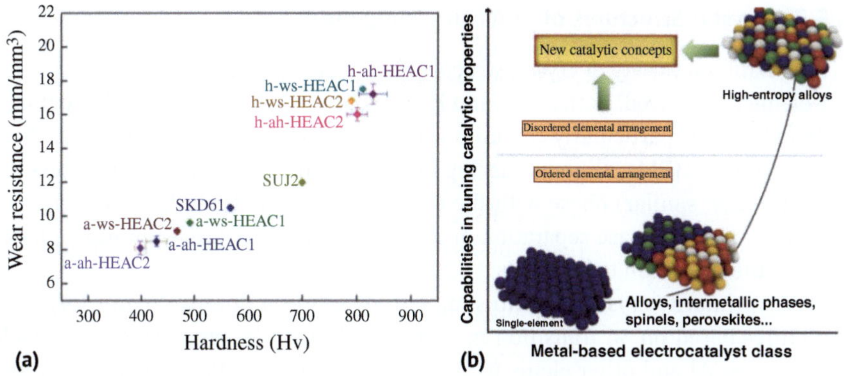

Figure 1.8 (a) The wear resistance of an as-prepared HEA, tool steel SKD61, bearing steel SUJ2, and a thermal-sprayed HEA coating is compared [32]. *Source*: [32] /with permission of IOP. (b) The cocktail effect increases the synergy effects of a vast number of different multielement active sites and expands the capabilities of tuning catalytic properties in HEMs [38]. *Source*: [38] /with permission of John Wiley & Sons.

electrochemical capacity [34–36]. An example of this synergistic approach can be found in the work of Wang et al. [37], who designed a novel O3-type cathode material, $NaNi_{0.2}Fe_{0.2}Mn_{0.35}Cu_{0.05}Zn_{0.1}Sn_{0.1}O_2$. In this material, Mn, Zn, and Sn are critical in mitigating interlayer slipping during the charge–discharge process, effectively reducing the structural degradation that can occur due to irreversible phase transitions. Meanwhile, Ni, Fe, and Cu contribute as redox-active elements, providing additional charge compensation during electrochemical reactions. As a result, this high-entropy cathode material exhibits significant improvements, including enhanced cycling stability (87% capacity retention after 500 cycles at 3C) and a superior discharge capacity of 128 mAh g^{-1} at 0.1C [38].

Importantly, the synergy effect in these materials is not just the simple summation of the individual properties of the elements involved. Instead, it results from a complex interaction where changes in element concentrations and types can dramatically influence the material's overall behavior. This highlights the considerable flexibility offered in customizing material performance, as adjusting the composition and proportions of the constituent elements can fine-tune the material's properties for specific applications, offering new pathways for innovation in electrochemical devices and other technologies (Figure 1.8b).

While some researchers argue that the cocktail effect is redundant, this is a misconception. The overall properties of an alloy are the result of the combined contributions of all related factors. Therefore, understanding the interactions at play when elements are mixed and processed is key to improving or developing new alloy designs. The cocktail effect influences many aspects, including composition, atomic interactions, crystal structure, and microstructure. For materials scientists, it is essential to comprehend the factors that affect a material's properties to select optimal compositions and processing techniques for creating new materials with enhanced characteristics. By minimizing negative factors and amplifying positive ones, better materials can be engineered.

1.3.2 Crystal Structures of HEAs and Compounds

HEMs exhibit a variety of crystal structures, depending on the type of material and the elements involved. Both HEAs and HECs tend to form simple crystal structures, although HEAs, particularly ceramics, can display more complex behaviors. From the concept of HEMs, although having multiple elements, they should possess the same (at least similar) phase with the parent structure but in a high-entropy configuration. Severe phase separation or a very complex mixture of multiple phases will fail the definition to be HEMs.

HEAs primarily crystallize into simple solid-solution phases (Figure 1.9). FCC structures based on 3d transition metals (Co, Cr, Cu, Fe, Mn, Ni), with possible additions of Al and other elements, are common, which offer good ductility and toughness, especially at low temperatures, and are suitable for applications requiring high toughness and moderate strength [40]. The arrangement of atoms in the FCC structure is highly symmetrical, with atoms positioned at each corner and the center of each face of the unit cell. This configuration allows for numerous slip systems, which enhances the material's ability to deform plastically, contributing to its toughness. Moreover, the high-entropy effect in FCC HEAs suppresses the formation of brittle intermetallic phases, stabilizing the alloy in a homogeneous solid-solution form, which leads to improved toughness and resistance to cracking.

In contrast, some HEAs, particularly those incorporating refractory metals such as Nb, Mo, Ta, and W, crystallize into the BCC structure. The BCC structure is typically harder and stronger than FCC but exhibits lower ductility, which makes BCC HEAs suitable for high-temperature applications that require high strength and stability [41]. BCC structures are characterized by a single atom at the center of the unit cell and additional atoms at the cube's corners. These structures tend to have fewer slip systems compared to FCC, leading to lower ductility but increased strength due to stronger atomic bonds and a more compact atomic arrangement. BCC HEAs can also exhibit enhanced thermal stability, which is vital for their use in environments that involve extreme temperatures, such as aerospace, power plants, and nuclear reactors. Hexagonal close-packed (HCP) structures can form in alloys containing elements such as Ti or Zr. HCP structures are often more brittle, but they can offer advantages in specific mechanical or corrosion-resistant applications [42]. The HCP structure consists of layers of atoms arranged in a repeating ABAB pattern, where each layer is tightly packed. The fewer slip systems in HCP structures make these materials more prone to brittle fracture under stress; however, the ordered atomic

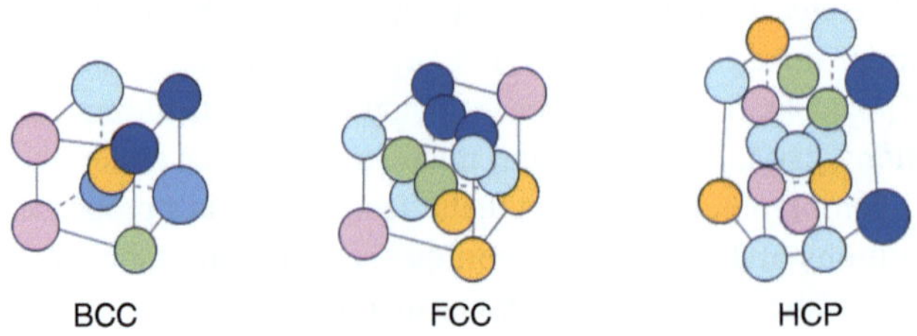

Figure 1.9 Graphic illustration displaying different lattice structures of HEAs.

arrangement can confer specific properties that are advantageous for applications that demand high wear resistance, such as cutting tools or coatings.

Besides, high-entropy intermetallic compounds are a main part of metallic materials, formed from combinations of elements that favor ordered atomic arrangements, that often crystallize in structures such as B2 (CsCl-type), L12 (Cu_3Au-type), or sigma phases. The B2 phase exhibits a simple cubic structure with alternating metal atoms occupying the corners and center of the unit cell. This structure tends to provide good mechanical strength and resistance to thermal degradation, though it is typically more brittle compared to FCC or BCC structures. The B2 structure is often favored in HEAs that involve elements with a large size difference, which helps stabilize the ordered arrangement of atoms. The L12 structure is characterized by an FCC lattice with a more complex atomic arrangement than the simple FCC structure. This phase is often found in HEAs that involve elements such as Ni, Al, or Ti, and it imparts superior strength and creep resistance at elevated temperatures. The L12 phase can enhance the high-temperature stability of alloys, making them suitable for applications in aerospace and high-performance engineering. The sigma phase is an ordered, high-temperature phase that typically forms in HEAs with certain combinations of transition metals. This phase exhibits complex crystal structures that not only often lead to increased brittleness but also provide exceptional resistance to oxidation and thermal damage [43, 44].

HEOs exhibit various crystal structures [45, 46], including rock salt, fluorite, perovskite, spinel, bixbyite, pyrochlore, and layered O3-type (Figure 1.10). Rock salt HEOs [47, 48], identified first, are the most extensively studied due to their facile formation. This structure features a single Wyckoff site for cations, facilitating a homogenous arrangement with superior chemical stability, thus enhancing their electrochemical application potential. Perovskite [49] and spinel-type HEOs are also of significant interest in energy conversion, possessing multiple cation sites that allow for complex atomic arrangements and greater compositional tunability. Other structures such as fluorite, bixbyite, pyrochlore, and O3-layered [50] have

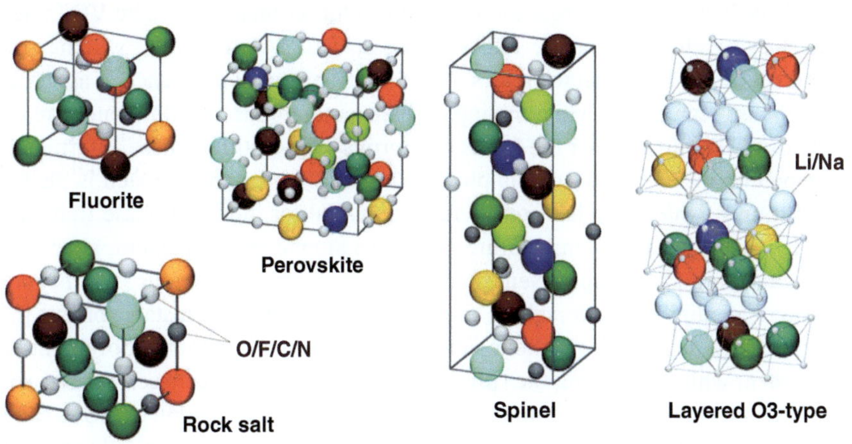

Figure 1.10 Graphic illustration displaying different lattice structures and compositions of HECs [58]. *Source:* [58] /with permission of The Royal Society of Chemistry.

been less explored in electrochemistry but exhibit excellent thermal stability. The phase-formation mechanisms for these HEOs vary significantly due to differences in crystal structures, elemental combinations, and compositions. Nevertheless, all demonstrate enhanced high-temperature stability, attributed to the random distribution of metals within the cation sublattice, leading to uniformly dispersed metal oxide solid solutions.

Bulk high-entropy borides [51], carbides [52], nitrides [53], carbonitrides [54], borocarbides [55], and silicides frequently exhibit enhanced hardness that exceeds predictions based on the rule of mixtures. This phenomenon can be attributed to solid-solution hardening or alterations in the availability of lattice slip systems. Similar to (MgCoNiCuZn)O, both carbides and nitrides generally crystallize in a rock salt structure. In contrast, borides predominantly adopt the AlB_2 hexagonal structure, characterized by alternating two-dimensional boride and high-entropy cation layers. Silicides, on the other hand, typically display relatively low-symmetry structures, with $(MoNbTaTiW)Si_2$ [56] adopting the C40 crystal structure and $(MoNbTaWZr)Si_2$ [57] exhibiting a hexagonal configuration.

1.3.3 Phase Transformations in HEAs and Compounds

As mentioned in Section 1.3.2, there are numerous HEA types and crystal structures, and, similarly, these structures are capable of transforming into each other under certain circumstances and external stimuli. Phase transformations in HEMs are complex and highly dependent on composition, temperature, and external conditions. These phase transformations influence the mechanical properties, microstructure, and stability of the material. Particularly, the phase structure transformation of HEA can be modified using two main approaches. The first approach is chemical tuning, which involves five key physical parameters: mixing entropy, mixing enthalpy, electronegativity difference, valence electron concentration (VEC), and atomic size difference. These parameters significantly influence the atomic arrangement and lattice parameters of HEAs [59]. High mixing entropy in HEAs facilitates the formation of a solid-solution phase [60]. The VEC affects the stacking character of the HEAs' phase structure, determining whether they exhibit FCC or BCC characteristics. The electronegativity difference is associated with phase segregation while mixing enthalpy and atomic size difference are crucial for predicting whether the HEAs will form a solid-solution phase, intermetallic compound phase, or amorphous phase [61, 62]. The second approach is physical field tuning, which also has considerable potential for inducing phase transitions in HEAs. For instance, temperature can cause atomic rearrangement, alter stacking fault energy, and help prevent phase separation. Strain can reduce the magnetic moment, modify dislocation glide, and trigger transitions to metastable phases [63]. Additionally, applying a magnetic field provides energy for atomic movement and alignment while decreasing the Gibbs free energy difference between phases, such as FCC and HCP [64]. In summary, effective property modification of HEAs can be achieved by tailoring their phase structures. This modification not only enhances performance but also offers significant opportunities to attain desired phases with predictable characteristics.

From Solid-Solution to Solid-Solution Phase. The transitions from one solid-solution phase to another demonstrate the complex interplay of many factors in determining the structural phases of HEAs. This ability to switch between various phases under different conditions reveals that solid-solution HEAs can exhibit a much broader range of structural behaviors than might be expected from the simple combination of their constituent elements. Wang et al. studied the $Al_{0.6}$CoCrFeNi HEA and observed several interesting phase transitions [65] (Figure 1.11). They found that the BCC phase transforms to an orthorhombic phase under pressure, starting at about 10.6 GPa and completing around 21.4 GPa. Upon pressure release, a body-centered tetragonal phase emerges alongside the orthorhombic phase, suggesting that the orthorhombic phase may be metastable at ambient conditions due to a significant energy barrier. Additionally, annealing at 1000 °C for 2 hours converts the initial BCC phase into a stable FCC phase. Under compression, the FCC phase can transform into an HCP phase, similar to Cantor's alloy. The study highlights that lattice distortion, influenced by pressure or temperature, plays a crucial role in the formation and transition of different phases in the HEA.

Form Solid-Solution Phase to Intermetallics. The transition from solid-solution alloys to intermetallic phases marks a significant change in atomic structure, profoundly affecting the material's mechanical and thermal properties. In solid

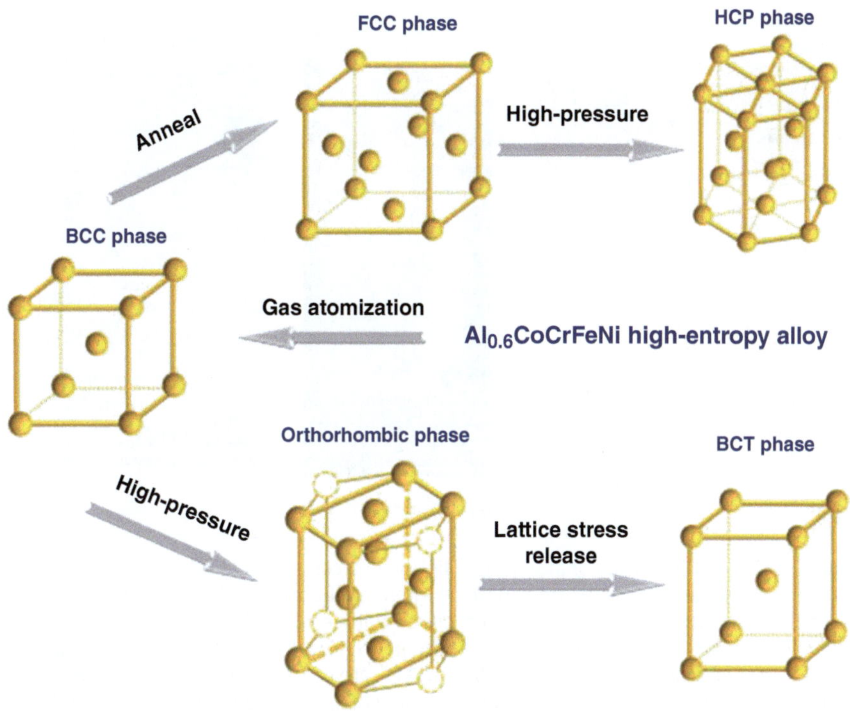

Figure 1.11 A schematic illustration of the atomic structure for five polymorphs observed in the $Al_{0.6}$CoCrFeNi HEA and the transition paths between them [66]. *Source:* [66] / with permission of Elsevier.

solutions, the atoms of different elements are randomly distributed across the crystal lattice, which provides a good balance of ductility and strength. In contrast, intermetallics are ordered phases where atoms of different elements occupy specific lattice positions, forming compounds with defined stoichiometric ratios and more complex crystal structures, such as B2 (CsCl-type), L12 (Cu_3Au-type), or sigma phases. These ordered structures are generally more chemically stable due to their strong atomic bonds but tend to be more brittle than solid solutions. The transformation to intermetallics is often triggered by thermal treatments, such as annealing or long-term aging, where atoms diffuse and segregate into ordered patterns (Figure 1.12a).

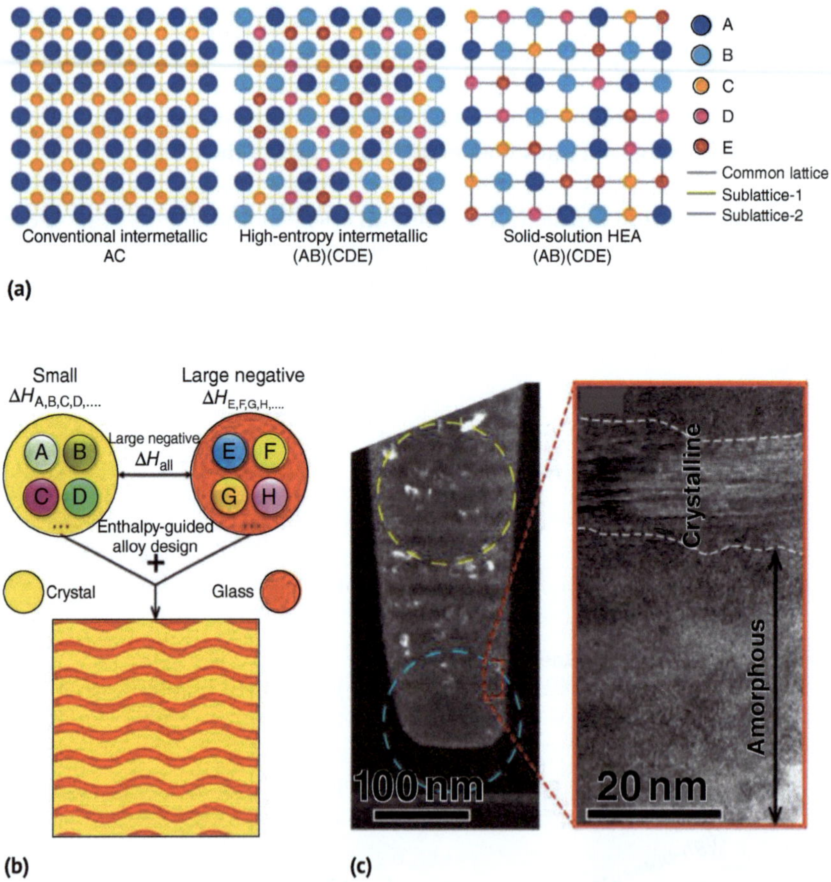

Figure 1.12 (a) Illustration of atomic distribution in a conventional B2-type binary intermetallic, a quinary high-entropy B2-type intermetallic and a quinary HEA in solid solution. It is assumed that A and B occupy one sublattice while C, D, and E occupy the other [71]. *Source*: [71] /with permission of Springer Nature. (b) Schematic presentation of the alloy design strategy. A, B, C, D ... and E, F, G, H ... are constituent elements of the crystal and the glass phases, having small mixing enthalpy $(\Delta H_{A,B,C,D...})$ and large negative mixing enthalpy $(\Delta H_{E,F,G,H...})$, respectively, while all the constituent elements of the two phases have large negative mixing enthalpy (ΔH_{all}) [70]. *Source*: [70] / Springer Nature / CC BY 4.0. (c) High-Resolution Transmission Electron Microscopy (HRTEM) images of the crystal and amorphous phase [70]. *Source*: Reproduced with permission from [70] / Springer Nature / CC BY 4.0.

This transition reduces configurational entropy but increases the overall chemical stability of the material.

Temperature plays a crucial role in this transformation. High temperatures provide the necessary kinetic energy for atoms in a solid solution to diffuse and segregate, leading to the formation of ordered intermetallic phases. During long-term annealing at elevated temperatures, this diffusion process accelerates, allowing phase transformations to occur more readily. For example, FCC solid solutions can transition into ordered structures like B2 or L12 [66–68]. As the temperature rises, atomic mobility increases, enabling atoms to rearrange into more energetically favorable configurations. Even at lower temperatures, given enough time through aging, intermetallics can still form, although the process occurs more slowly.

The composition of an alloy also strongly influences whether a solid solution will transform into an intermetallic phase. Certain elements, due to their atomic size differences and bonding characteristics, are more prone to forming intermetallics. For instance, aluminum (Al), titanium (Ti), and molybdenum (Mo) have a strong tendency to form intermetallic phases, particularly when combined with transition metals like nickel (Ni) or cobalt (Co). The specific ratio of these elements in the alloy can either stabilize the solid solution or encourage the formation of intermetallic compounds. Therefore, controlling the alloy's composition is critical for tailoring its properties to achieve a desired balance of strength, ductility, and thermal stability.

From Crystal to Amorphous Phases. The transformation from crystalline to amorphous phases in HEMs, including both alloys and compounds, signifies a profound change in atomic structure and material properties [69]. In crystalline phases, atoms are organized in a regular, repeating lattice structure, which provides stability and predictable properties. In contrast, amorphous phases lack this long-range atomic order, creating a disordered atomic arrangement similar to that found in glass. This shift from order to disorder has a marked impact on the material's mechanical, thermal, and chemical behaviors. The transition to an amorphous phase can enhance certain properties, such as hardness, corrosion resistance, and structural stability, though it often comes at the expense of reduced ductility and toughness.

One of the most common methods to induce amorphization in HEMs is through rapid cooling or quenching from the melt. When a material is cooled quickly, the atoms are prevented from rearranging into the stable, ordered lattice typical of a crystalline phase. In HEMs, which already benefit from high configurational entropy, this rapid cooling further delays the formation of ordered structures, making them highly susceptible to forming amorphous phases under these conditions. The complexity of the atomic interactions in HEMs, combined with the rapid solidification, leads to a disordered atomic state, which is often more stable under certain conditions due to the sluggish atomic diffusion characteristic of these materials.

Another method of inducing amorphization is through mechanical processes such as severe plastic deformation (SPD) [70]. When materials are subjected to extreme strain, significant dislocation accumulation and atomic disordering occur, pushing the material toward an amorphous state. In HEAs, this deformation-induced amorphization can happen under high strain rates, where dislocation movements are restricted, increasing the material's internal free energy. As dislocation density grows and atomic disorder worsens, the crystalline structure collapses, transforming into an amorphous phase. This phenomenon is particularly relevant in HEAs, where the high degree of atomic mixing and entropy promotes such transformations under mechanical stress (Figure 1.12c).

High pressure and shear also play a key role in transforming crystalline structures into amorphous ones. When subjected to high pressure, often in combination with shear forces, the crystalline lattice is forced to collapse, creating an energetically favorable amorphous structure. The applied stresses disrupt the ordered atomic arrangement, causing it to disintegrate into a disordered state. This process is commonly observed in high-entropy ceramics, where high-pressure torsion or ball milling can lead to amorphization. The inherent instability of the crystalline lattice under such conditions, combined with the atomic complexity of HEMs, makes amorphous transformation a likely outcome under high stress.

Composition control is another important factor in promoting the transformation from crystalline to amorphous phases in HEMs (Figure 1.12b). The choice of alloying elements and their concentrations can significantly influence the likelihood of amorphization. Certain elements, such as boron (B), phosphorus (P), and carbon (C), tend to disrupt the atomic ordering in a crystal lattice, thereby encouraging the formation of an amorphous phase. In high-entropy systems, the presence of multiple principal elements creates complex atomic interactions that further favor amorphization, especially under rapid cooling or mechanical stress. The high configurational entropy of these materials, which contributes to the stabilization of disordered phases, makes them more prone to forming amorphous structures when the right conditions are met.

In HEMs, the high configurational entropy further stabilizes the amorphous phase. The sluggish diffusion characteristic of HEMs plays a crucial role, as it inhibits the atomic movements necessary to form ordered crystalline structures during cooling or deformation. This sluggish diffusion delays the formation of stable, ordered phases, allowing the disordered, amorphous structure to persist. As a result, the unique combination of atomic complexity, slow diffusion, and high entropy in HEMs makes them particularly well-suited for stabilizing amorphous phases, which can be exploited in various applications that require high hardness, corrosion resistance, and thermal stability.

1.3.4 Microstructures of HEAs and Compounds

The microstructure of HEAs and HECs is central to their performance and the optimization of their mechanical, thermal, and chemical properties. Microstructural features [72] such as grain size, phase distribution, and defect structures play a pivotal role in determining the material's strength, toughness, hardness, and thermal stability. These microstructures are highly sensitive to processing conditions, including casting, sintering, annealing, mechanical deformation, and post-processing heat treatments. By carefully controlling these processing parameters, the microstructure of HEMs can be tailored to meet the specific demands of various applications (Figure 1.13).

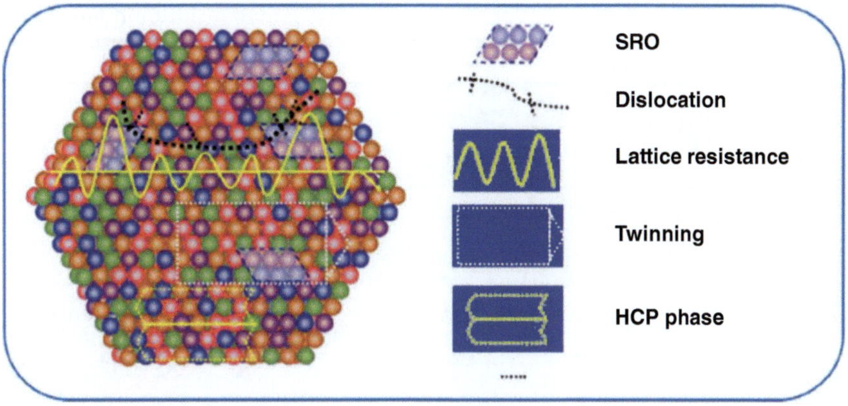

Figure 1.13 Schematic illustration of the interplay between short-range orderings (SROs) and different deformation carriers in a representative FCC HEA [72]. *Source*: [72] / with permission of Elsevier.

Grain Size and Morphology. One of the most influential factors in determining the mechanical properties of HEMs is the size and morphology of the grains. The arrangement and size of the grains significantly impact the material's strength, toughness, and fatigue resistance. Grain refinement can be achieved through various processing methods such as rapid cooling, high-temperature annealing, or SPD. Fine-grained microstructures generally offer enhanced mechanical properties, especially in terms of strength. Fine-grained structures are known to improve mechanical strength through the Hall–Petch effect, which describes the inverse relationship between grain size and yield strength. As the grain size decreases, the number of grain boundaries increases, and these boundaries act as obstacles to dislocation movement. Grain boundaries act as barriers to the motion of dislocations, thereby increasing the material's yield strength and hardness. This effect is particularly pronounced in HEAs, where the solid solution of multiple elements provides additional strength through a combination of solid solution strengthening and grain boundary strengthening [73]. For example, An et al. investigate a multimechanism approach to enhancing the mechanical properties of a CoCrFeMnNi HEA through a combination of non-equiatomic alloy design and innovative processing techniques. By deliberately altering the equiatomic composition of the alloy, the design achieves a single-phase FCC structure while lowering the stacking fault energy (SFE). This modification promotes the formation of deformation twins and stacking faults, contributing to the alloy's enhanced mechanical performance. The processing method applied in this study—rotationally accelerated shot peening (RASP)—was used to create a hierarchical grain size gradient within the material, resulting in a high density of nanotwins. These microstructural features significantly improve the mechanical properties of the HEA. As a result, the non-equiatomic CoCrFeMnNi alloy exhibited impressive mechanical properties, with a yield strength of 750 MPa, a tensile strength of 1050 MPa, and a tensile uniform elongation of 27.5%. Furthermore, the toughness of the alloy was measured at 2.53×10^{10} kJ/m^3, which is approximately twice that of the same alloy without the RASP treatment.

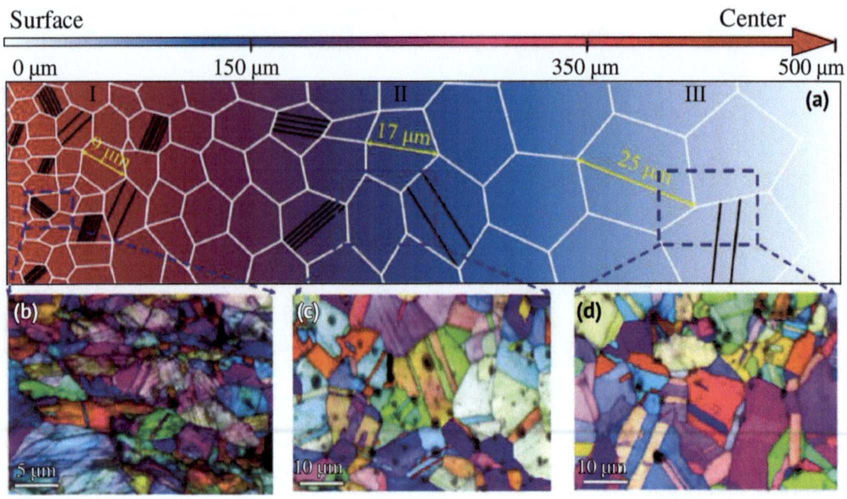

Figure 1.14 Gradient microstructure of the RASP treated $Co_{21.5}C_{r21.5}Fe_{21.5}Mn_{21.5}Ni_{14}$ plate [74]. (a) A schematic of the grain size gradient through the plate thickness. (b)–(d) Electron Backscatter Diffraction (EBSD) micrographs of the alloy at different depths from the surface. *Source*: Reproduced with permission from [74] / Elsevier.

The increase in strength is attributed to several reinforcing mechanisms within the material, including grain boundary strengthening, dislocation strengthening, twin strengthening, and hetero-deformation strengthening, all of which are linked to the heterogeneous microstructure. The simultaneous activation of multiple deformation mechanisms—such as dislocation movement, twinning, and microband formation—contributes to the alloy's remarkable strain hardening. This effect delays early necking and ensures a steady, continuous plastic deformation, which in turn enhances the material's toughness [74]. These findings underscore the potential of multimechanism design strategies for improving the performance of HEAs, offering a promising route for the development of high-strength, tough materials suitable for demanding applications (Figure 1.14).

Fine grains also contribute to better thermal stability and improved resistance to creep, as the grains limit the movement of dislocations at high temperatures. These refined microstructures allow HEAs to maintain their mechanical integrity under extreme thermal cycling and mechanical loading. On the other hand, coarse grains can improve toughness. Larger grains offer fewer grain boundaries, which allows for better dislocation motion, leading to enhanced energy absorption during plastic deformation. This mechanism is especially important in applications where the material needs to absorb impact energy or withstand sudden loading, as larger grains provide fewer barriers to dislocation movement and thus improve the material's ability to deform plastically before failure. While coarse grains may result in a slight reduction in strength, they can offer significant improvements in fracture toughness and ductility [75].

Phase Distribution and Multiphase Microstructures. The phase distribution in HEMs (HEAs and HECs) is a critical factor that influences a broad range of material properties, extending beyond mechanical strength to include catalytic efficiency, thermal stability, and chemical durability. The ability to tailor and control phase formation through various processing techniques enables the design of high-performance materials for demanding applications in energy systems, aerospace, automotive, and catalysis [76].

Single-phase HEAs, where the constituent elements are uniformly distributed within the lattice, typically exhibit good ductility and toughness, which are vital for mechanical applications. However, the absence of secondary phases does not preclude these materials from being highly effective in catalytic applications. The homogeneous distribution of elements and active sites within single-phase structures can enhance catalytic efficiency, particularly for reactions requiring redox stability or resistance to thermal degradation. For example, oxygen evolution reaction (OER) and hydrogen evolution reaction (HER), critical in water splitting and energy conversion technologies like fuel cells and electrolyzers, benefit from the well-distributed active sites in single-phase HEAs. The electronic configuration of each element within the single-phase structure can influence the binding energy of intermediates and facilitate optimal catalytic behavior. Additionally, the high resistance to corrosion and thermal stability offered by HEAs allow for long-term performance in high-temperature, high-oxidizing environments, such as those found in solar-to-fuel processes or electrochemical cells. For instance, in water electrolysis, a homogeneous solid solution of metals like Ni, Co, and Fe has demonstrated enhanced electrochemical stability and higher reaction rates due to the uniform distribution of cations, which promotes optimal catalytic pathways.

In contrast, multiphase microstructures can form when secondary phases, such as intermetallics or other compounds, coexist with the primary solid solution. These multiphase structures often provide increased hardness and wear resistance, as the presence of harder secondary phases reinforces the overall material. However, the trade-off is typically reduced ductility, as the additional phases create localized regions that are more prone to cracking. In HEOs and nitrides, multiphase systems offer a balance between strength and toughness, with different phases contributing to a combination of mechanical properties that are suitable for demanding applications.

The precipitation of secondary phases plays a significant role in the modification of mechanical and catalytic properties in HEMs. Secondary phases, such as carbides, oxides, or intermetallic compounds, form during cooling or aging, influencing both structural integrity and functional properties. In HEAs used in applications like aerospace and power generation, the precipitation of hard secondary phases like carbides or intermetallics can substantially increase high-temperature strength and oxidation resistance. These secondary phases obstruct dislocation movement and grain growth, creating a more robust and stable material. For instance, in Ni-based HEAs, the precipitation of carbides or Ni–Al intermetallics during aging can create a precipitate-hardening effect, which makes the alloy stronger at elevated temperatures and resistant to oxidation. This property is crucial in applications such as jet engine components, where the material must withstand both extreme temperatures and mechanical loading over extended periods. In catalytic applications, the precipitation of secondary phases can significantly influence active site formation and reaction kinetics. The controlled precipitation of phases can help tailor the electronic structure around catalytic sites, improving their reactivity and selectivity. For example, in CO_2 reduction reactions, where the conversion of CO_2 into useful chemicals is catalyzed, the presence of precipitates in HEA-based catalysts can enhance the electronic properties at the active site, thereby improving catalytic efficiency. The precipitation of phases like oxides or nitrides can also stabilize the active sites under harsh reaction conditions, ensuring consistent performance over long operational periods. In energy storage devices such as supercapacitors and batteries, the precipitation of secondary phases can also play an essential role in optimizing performance. Carbide and nitride precipitates can enhance the thermal stability and cycle life of materials used in these devices. The secondary phases can alter the electrode material's porosity, improving the ion diffusion rates and charge storage capacity. For instance, the precipitation of TiC or NbC in high-entropy carbide-based electrodes can increase the specific surface area, leading to better charge storage and cycle stability. Similarly, the precipitation of oxides in HEOs used in supercapacitors can improve electrical conductivity and charge retention during long-term cycling.

Defects and Dislocations in HEMs. The presence of defects and dislocations in the crystal structure of HEMs also contributes to their mechanical performance [77]. These materials are characterized by a high density of lattice defects, including vacancies, dislocations, and stacking faults (Figure 1.15). These defects arise due to severe lattice distortion caused by the incorporation of multiple elements with different atomic sizes. The high density of dislocations serves as a strengthening mechanism, as these defects impede further dislocation motion, improving both hardness and yield strength. In high-entropy ceramics, grain boundary strengthening mechanisms are further enhanced by the presence of these defects, while crack deflection mechanisms improve toughness. As cracks propagate through the material, they are often deflected at grain boundaries or by secondary phases, reducing the likelihood of catastrophic failure.

The high density of dislocations in HEMs provides additional dislocation hardening and increases strength. In electrocatalysts, dislocations can serve as effective

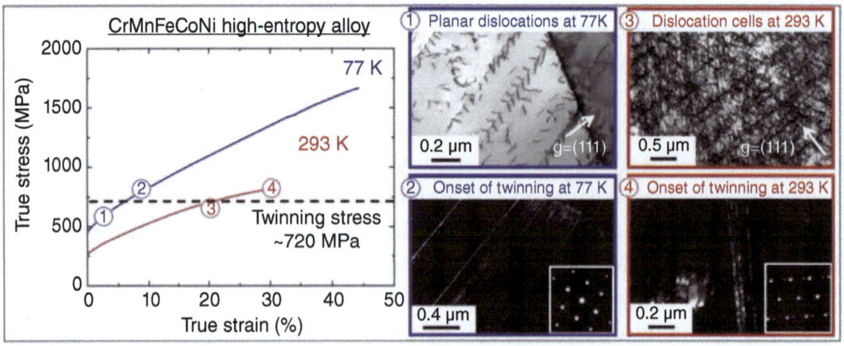

Figure 1.15 Representative true stress–strain curves of tensile tests at 77 K and 293 K, and TEM micrographs showing twin evolution with true tensile strain [77]. *Source:* [77] / with permission of Elsevier.

sites for the adsorption and activation of molecules, improving the material's ability to facilitate redox reactions. The presence of vacancies and stacking faults can also enhance the material's electrochemical reactivity, leading to higher charge transfer rates and catalytic efficiency. The presence of lattice defects and grain boundaries in HEMs contributes to grain boundary strengthening, which not only improves mechanical properties but also enhances the material's thermal stability and resistance to corrosion. In HEOs used for solid oxide fuel cells, defects at grain boundaries enhance ionic conductivity and electrochemical stability, leading to better fuel cell performance. The dislocation strengthening in HEAs can improve the material's resistance to wear and oxidation in demanding energy conversion applications, such as in thermal power plants and gas turbines.

Local Ordering. Local ordering in HEAs is a dynamic and crucial aspect of their structure, influencing many of their mechanical and physical properties. While HEAs are traditionally regarded as disordered solid solutions due to their high configurational entropy, more recent studies have shown that local ordering, specifically short-range ordering (SRO) and medium-range ordering (MRO), plays an important role in shaping the alloy's behavior. SRO refers to atomic pairings that form over short distances, typically a few atomic layers, where elements preferentially cluster together. MRO, on the other hand, refers to atomic arrangements that extend over several nanometers, often forming nanoscale domains or superlattice structures. These ordering phenomena, though not long-range periodic, have a substantial impact on the material's macroscopic properties.

In HEAs, local ordering is primarily driven by atomic size mismatches, differences in atomic radii, and the chemical affinity between constituent elements. For instance, elements with similar atomic sizes tend to cluster, reducing local lattice distortions and minimizing strain energy. This clustering can create short-range ordered regions within the alloy. MRO occurs as these short-range ordered regions coalesce into larger, more structured domains that extend over multiple atomic layers, often leading to the formation of nanoscale heterostructures. These features can significantly influence the material's behavior during deformation. For example, in FCC-based HEAs, local ordering can alter the SFE, making it easier or harder for dislocations to glide. This, in turn, influences the activation of mechanisms like twinning, phase transformations, and work hardening [78].

The presence of both SRO and MRO has significant consequences for the mechanical properties of HEAs. The enhanced dislocation resistance due to local ordering strengthens the alloy by increasing the energy barrier for dislocation motion, which improves the material's yield strength. Furthermore, the coexistence of SRO and MRO promotes more complex dislocation behaviors, such as wavy slip and cross-slip, which are beneficial for achieving high strain hardening and ductility. This interplay between local ordering and dislocation dynamics is particularly important for improving the alloy's performance in demanding conditions, such as at cryogenic temperatures. For example, in FeMnCoCr-based HEAs, MRO domains help the alloy maintain its mechanical integrity under low-temperature deformation, where traditional alloys would typically become brittle. MRO also facilitates the formation of stable nanostructures, such as nanoscale precipitates, that enhance the alloy's strength without sacrificing ductility. This is particularly useful for tailoring the performance of HEAs for applications requiring high strength and toughness at both ambient and extreme temperatures. The structural stability imparted by MRO allows these materials to withstand severe environmental conditions without undergoing detrimental phase transitions or losing their mechanical properties. As a result, HEAs with well-controlled local ordering can offer a superior combination of strength, ductility, and stability, making them ideal candidates for applications in aerospace, energy storage, and cryogenic technologies [72].

Moreover, the ability to manipulate the degree of SRO and MRO in HEAs opens up new avenues for designing materials with specific properties. This can be achieved through careful control of the alloy's composition, processing conditions, and thermal treatments. For instance, elemental substitutions, such as replacing Rh with Ru in noble-metal-based HEAs, can dramatically change the local atomic arrangement, shifting the material from a random solid solution to one with localized clustering (Figure 1.16a). Similarly, thermomechanical treatments, including annealing or cold working, can modify the extent of local ordering, resulting in alloys with enhanced strength, toughness, and resistance to thermal creep. Additionally, the incorporation of small interstitial atoms like nitrogen or oxygen can further tune the local chemical order, improving both strength and ductility [79].

While significant progress has been made in understanding and manipulating local ordering in HEAs, challenges remain in fully characterizing and quantifying this phenomenon. Advanced techniques such as APT [80], high-resolution STEM (Figure 1.16b) [81], and first-principles simulations [82] are crucial to unraveling the complex relationship between local ordering and alloy properties. These tools will be essential in developing a deeper understanding of how local ordering contributes to the unique performance of HEAs and in guiding the design of next-generation materials with tailored properties for demanding applications in aerospace, energy, and cryogenic environments. Local ordering plays a critical role in determining the mechanical, thermal, and structural properties of HEAs. Understanding and controlling these ordering phenomena provide a pathway to design HEAs with tailored properties for advanced applications.

In summary, the microstructures of HEMs, influenced by grain size, phase distribution, secondary phase precipitation, and the presence of defects, are key determinants of their overall performance. By carefully controlling these microstructural features through processing techniques, it is possible to create materials with a tailored balance of strength, toughness, hardness, and thermal stability, suitable for a wide range of advanced engineering applications. The field of HEMs has expanded from alloys to include a wide range of compounds such as oxides, nitrides, carbides, and intermetallics. The common principles of high configurational entropy, sluggish diffusion, lattice distortion, and the cocktail effect apply to both HEAs and HECs, influencing their crystal structures, phase transformations, and microstructures. These factors contribute to the exceptional mechanical, thermal, and chemical properties of HEMs, making them highly suitable for demanding applications in aerospace, energy, and advanced.

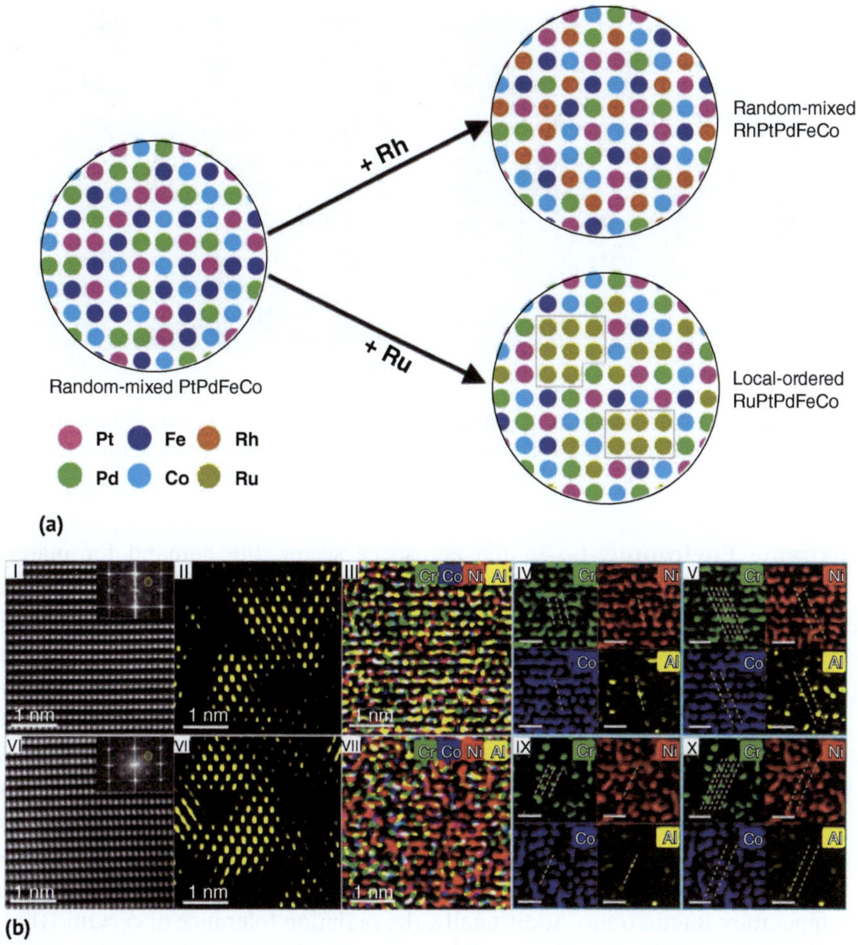

Figure 1.16 (a) Composition-structure relationship in HEAs with one constituent difference [79]. *Source*: [79] / with permission of American Chemical Society. (b) I. HAADF image under [112] z.a. (zone axis) before tensile deformation. Inset: Fast Fourier Transform (FFT) pattern. II. Corresponding Inverse Fast Fourier Transform (IFFT) image showing CMRO regions. III. EDS map. IV and V. Close-up views of Cr, Ni, Co, and Al maps in individual CSRO and CMRO regions, respectively. All dashed white/yellow lines: Cr-enriched and Cr-depleted $\{\bar{3}11\}$ planes. VI–X. Corresponding results after tensile deformation. Scale bars in IV, V, IX, and X: 0.5 nm [83]. *Source*: Reproduced with permission from [83] / Springer Nature / CC BY 4.0.

1.4 Diverse Applications of Functional HEAs and Compounds

HEMs, with their multicomponent structure, have opened new frontiers in materials science, offering a plethora of properties that can be fine-tuned for various applications. This section briefly introduces the functional applications of HEAs and compounds, exploring how their unique characteristics can be harnessed across different industries. More details will be presented in subsequent chapters.

(a) (b)

Figure 1.17 (a) Schematic picture of a space engine. (b) Schematic picture of bulk HEAs in a high-pressure and high-temperature environment. Generated with AI using Doubao.

Extreme Environments. In the aerospace sector, the demand for materials with high strength-to-weight ratios, excellent corrosion resistance, and the ability to withstand extreme temperatures is ever-increasing. HEAs, with their remarkable combination of strength, ductility, and corrosion resistance, are being considered for critical structural components (Figure 1.17). Their potential use in aircraft engines, where materials must endure high temperatures and mechanical stress, is currently a topic of intensive research. Besides, the thermal management capabilities of HEMs are particularly intriguing for industries that require efficient heat dissipation or insulation. In aerospace and automotive applications, HEMs with low thermal expansion coefficients can be used to create thermal barriers that protect sensitive components from extreme temperature fluctuations. Additionally, the radiation tolerance of certain HEMs makes them suitable for use in nuclear reactors, where materials are exposed to high levels of radiation. Their ability to maintain structural integrity and resist radiation-induced degradation is crucial for the safe and efficient operation of nuclear facilities [84–86].

Corrosion Resistance. HEMs offer exceptional corrosion resistance, making them ideal for use in harsh environments such as marine structures, chemical processing equipment, and offshore platforms. Their corrosion resistance arises from mechanisms like the formation of stable oxide layers, solid-solution strengthening, and resistance to galvanic corrosion. These features prevent degradation and reduce maintenance costs, enhancing the material's durability and lifespan. In applications like offshore oil rigs, chemical reactors, and nuclear industries, HEMs' ability to resist acidic, alkaline, and oxidative conditions ensures long-term performance. The materials' high-temperature stability further enhances their suitability for extreme environments, where traditional materials would fail more rapidly. The economic benefits of using HEMs include cost-effective solutions through reduced maintenance, extended service life, and fewer material replacements, offering sustainability and reliability in critical [87, 88].

Functional Materials. The magnetocaloric effect (MCE) refers to the reversible change in a material's temperature (T) and entropy (S) induced by the application or removal of an external magnetic field. When magnetization or demagnetization occurs adiabatically, it results in an adiabatic temperature change, ΔT_{ad}, with the total entropy remaining constant. Under isothermal conditions, the magnetization or demagnetization leads to a heat transfer, where the heat transferred is $Q = T \cdot \Delta S_m$, with ΔS_m representing the isothermal magnetic entropy change. The MCE is typically maximized near the phase transition temperature, where an ideal transition should be sharp and nonhysteretic, resembling both first-order and second-order transitions. In an ideal case, the magnetization is initially high and drops at the transition temperature [89].

For energy-efficient and sustainable magnetic refrigeration, the ideal magnetocaloric material should combine a large, reversible MCE in fields around 1 T, along with properties such as nontoxicity, noncriticality, high corrosion resistance, mechanical stability, and machinability. Currently, most magnetic refrigerators use elemental heavy rare-earth (RE) materials like Gd, which are critical and expensive. Gd-based MCEs are governed by a second-order phase transition, limiting the reversible entropy change and heat transfer. Materials such as $La(Fe,Si)_{13}$ alloys, which rely on a first-order phase transition, can enhance heat transfer but often suffer from functional fatigue and poor machinability.

HEMs could address these challenges. Recent studies have explored equiatomic compositions using RE elements or mixtures of transition metals, with some thermoelectric HEMs surpassing conventional materials in properties. However, most HEM-based magnetocaloric materials show gradual changes in magnetization, typical of second-order phase transitions. Engineering HEMs toward first-order transitions remains a major challenge. Promising systems like all-d-Heusler-type, MM′X-type, and Fe_2P-type compounds are already multielement materials that could be further optimized [90, 91].

To enhance the MCE, a deeper understanding of the underlying physical mechanisms—including crystal structure, electronic configurations, microstructure, and configurational entropy—is essential. Corrosion resistance is also crucial, as magnetocaloric materials undergo millions of cycles in contact with heat exchange fluids, typically water. HEMs offer the potential to reconcile these multifunctional requirements, integrating enhanced MCE with durability. Given the complexity of synthesis methods (e.g., conventional melting, nonequilibrium processes, and AM), guidance from theoretical models and machine learning will be invaluable for optimizing these materials.

In thermoelectrics, materials convert thermal gradients (e.g., from waste heat) into electricity through the Seebeck effect. Their efficiency is characterized by the figure of merit (ZT), given by $ZT = S^2\sigma T/\kappa$, where S is the Seebeck coefficient, σ is electrical conductivity, T is the absolute temperature, and κ is thermal conductivity. To enhance efficiency, two key properties must be optimized: reducing thermal conductivity through phonon scattering to maintain the temperature gradient and increasing charge carrier mobility and density to improve electrical conductivity. Achieving this balance is challenging due to electron–phonon interactions. Additionally, thermoelectric materials must withstand harsh environments, requiring high damage tolerance, oxidation resistance, good mechanical properties, low cost, and thermal stability [92, 93].

HEMs can be used to optimize thermoelectric materials for higher efficiency and safer operation, with potential benefits from entropy engineering and composition modification. Some thermoelectric HEMs have already demonstrated better performance than conventional materials, indicating the promise of this design pathway. For example, increasing configurational entropy can stabilize solid-solution ranges beyond traditional limits, promoting lattice distortion and enhancing phonon scattering to lower thermal conductivity. In Pb—Sb—Se—Te—S thermoelectric HEMs, increasing S and Te content triggers a multiphase state due to phase separation, which is stabilized upon adding Sn. This entropy-driven phase stabilization results in significant lattice strain (~0.45%) and a maximum ZT of 1.8 at 900 K [94]. The high configurational entropy and sluggish diffusion stabilize materials at high temperatures, although further research is needed to assess their long-term performance under thermal cycling.

HEMs can also enhance the Seebeck coefficient by tuning electronic band features close to the Fermi level, such as band degeneracy or convergence, at optimal carrier concentrations. The increased configurational entropy can lead to higher-symmetry crystal structures, which enhance the electronic density of states and effective mass, improving thermoelectric properties.

However, a challenge in using HEMs for thermoelectrics is the potential decrease in electrical conductivity due to Anderson localization in highly disordered systems, leading to a trade-off between thermal and electrical conductivity. Recent research has demonstrated that entropy engineering can enhance thermal conductivity without sacrificing electrical properties, as seen in GeTe-based HEMs with a ZT value of 2.7 at 750 K. This was explained by entropy-induced disorder suppressing transverse phonons, while increased crystal symmetry delocalized the electron distribution, improving electrical properties (Figure 1.18).

Applications in Biomedical Fields. HEMs are emerging as highly promising materials in the biomedical field due to their remarkable combination of mechanical strength, corrosion resistance, and biocompatibility. These alloys, characterized

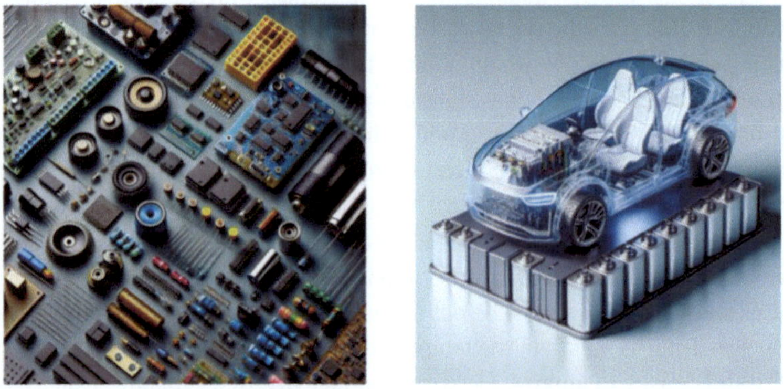

(a) (b)

Figure 1.18 (a) Schematic picture of HEMs used in electronic components. (b) Schematic picture of HEMs for automotive applications. Generated with AI using Doubao.

by the inclusion of multiple elements in near-equal proportions, exhibit unique properties that make them particularly suited for a range of biomedical applications. For instance, HEAs made from biocompatible elements such as titanium (Ti), nickel (Ni), chromium (Cr), and molybdenum (Mo) demonstrate excellent resistance to corrosion in physiological environments. This resistance is crucial for biomedical implants and prosthetics that are exposed to harsh conditions within the human body. By maintaining their integrity in these corrosive environments, HEMs ensure the long-term durability and reliability of medical devices.

The inherent mechanical properties of HEMs, such as high strength, toughness, and wear resistance, make them ideal candidates for load-bearing applications, particularly in bone and dental implants. These implants are subject to significant stress and strain during daily activities, requiring materials that not only offer superior mechanical performance but also maintain functionality over extended periods. HEAs meet these demands by providing a high-performance solution that resists wear and fatigue, which are common challenges in traditional biomaterials. This makes them particularly suitable for high-stress applications, where material degradation could compromise the structural integrity of implants. Moreover, HEAs offer considerable advantages in promoting osseointegration, the process by which implants bond with bone tissue, an essential factor for the success of orthopedic and dental implants. The flexibility in the composition of HEMs allows for the development of tailored surface properties, enabling better biomolecule adsorption and enhancing tissue integration. This promotes a stronger and more stable bond between the implant and surrounding tissues, which is crucial for the long-term success of implants. In addition, HEAs can be engineered to possess antimicrobial properties, a critical feature for preventing infections that can occur postsurgery. By incorporating specific elements or compounds into the alloy, antimicrobial coatings can be applied to medical devices, reducing the risk of infection and promoting quicker recovery for patients.

Another significant advantage of HEMs in the biomedical field is their compositional flexibility and ability to be precisely engineered for individual needs. With the rise of 3D printing technologies in medicine, HEMs can be tailored to create patient-specific implants, allowing for a perfect match to an individual's anatomical structure. This customization not only improves the fit and function of implants but also reduces the risk of complications that may arise from ill-fitting devices. By incorporating 3D printing, it becomes possible to design complex geometries that would be difficult or impossible to achieve with traditional manufacturing methods, thereby enhancing the surgical outcome and improving the overall quality of life for patients. Furthermore, the unique properties of HEMs also offer potential for the development of advanced drug delivery systems. These materials can be designed to carry and gradually release therapeutic agents, providing targeted treatments for a range of conditions. By engineering the surface properties of the alloys, drug release profiles can be carefully controlled, optimizing the therapeutic effects while minimizing side effects. This combination of mechanical performance, biocompatibility, and versatility makes HEMs an exciting avenue for the development of next-generation biomedical devices that can significantly improve patient care and treatment outcomes.

High-Performance Catalysts. The polyelemental nature of HEMs offers a unique platform for designing catalytically active materials or surfaces. Key reactions in electrolyzers and fuel cells—such as the HER, OER, and oxygen reduction reaction—require multifunctional electrocatalysts. These catalysts must be not only active and stable but, ideally, made from sustainable elements and their oxides. HEM electrocatalysts can meet these conflicting demands by combining multiple functionalities from their diverse constituents, allowing for tunability in properties through composition. Additionally, the complex chemical features of HEM surfaces make them ideal candidates for designing electrocatalysts for other critical reactions like CO_2 reduction and N_2 fixation, as well as more intricate multistep and cascade reactions. While the promising properties of HEM electrocatalysts were only recently recognized, numerous studies have demonstrated their potential across a variety of catalytic processes. The development of HEM surfaces also holds promise for multifunctional materials in broader electrochemical applications (Figure 1.19) [45, 95, 96].

HEMs, formed from compositionally complex solid solutions, offer unprecedented access to a diverse range of surface atom arrangements (SAAs) in the top few atomic layers. These SAAs refer to the spatial arrangement of surface and subsurface atoms, as well as their chemical identities and the binding configurations (on-top, bridge, hollow) for reaction intermediates. The abundance of SAAs on HEM surfaces provides an expansive design space to optimize multifunctionality for a wide array of electrochemical reactions. Through careful tuning of the elemental distribution within these SAAs, the binding energies for reaction intermediates can be adjusted, offering a means to tailor catalytic properties. This dynamic tuning can influence the shape of volcano plots, which illustrate the relationship between catalyst activity and the binding energy of reactants or intermediates. Volcano plots serve as a tool for identifying catalysts with optimal binding energy, which, according to the Sabatier principle, leads to the highest activity [97, 98].

The key difference between conventional single-element or binary catalysts (often composed of scarce and expensive noble metals such as Ir, Pt, and Rh) and HEMs lies in the statistical diversity of SAAs on the surface. In HEMs, the vast array of SAAs increases the likelihood of having active sites on the surface, which are essential for high electrochemical activity. These active sites can be engineered through

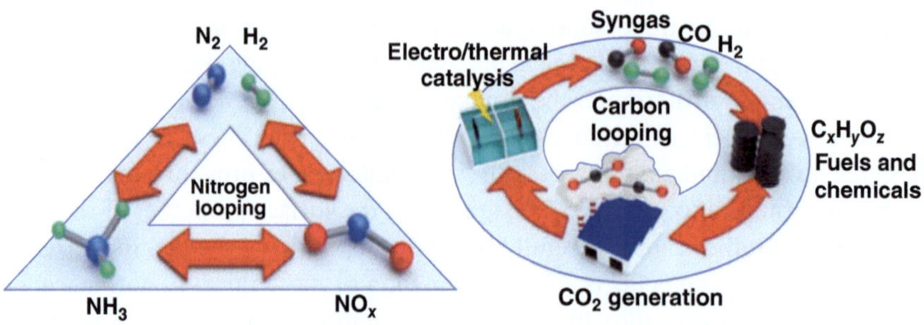

Figure 1.19 The applications of HEMs in nitrogen and carbon looping reactions [102]. *Source:* [102] / John Wiley & Sons / CC BY 4.0.

the manipulation of SAAs, overcoming the limitations of traditional catalysts and paving the way for sustainable materials with excellent multifunctionality—combining high activity, stability, and selectivity. Importantly, certain electrochemical functionalities typically dependent on rare noble metals can now be achieved with compositionally complex solid solutions containing minimal or no noble elements. This reduction or replacement of noble metals allows for finer control over the active site properties by adjusting the binding energy distribution. Moreover, replacing unstable yet highly active elements with more stable, less active ones can enhance overall catalyst performance. Recent research has used both experimental and theoretical methods to identify optimal HEM systems for the OER, demonstrating the potential for significant activity improvements when adjusting the composition of Ru, Pd, and Ir.

The vast abundance of SAAs on the surface of HEMs increases the chances of identifying active sites that remain stable during operation. Rather than relying on a single "perfect" active site, HEMs offer a multitude of potential sites, each close to the desired optimal catalytic activity. This flexibility suggests that such solid solutions might offer robust surface synthesis options and long-term operational stability.

A challenge in utilizing HEMs lies in efficiently exploring the immense surface composition space and identifying the most effective SAAs. High-throughput calculations and experiments, particularly those combining density functional theory and machine learning-enhanced computations, are key strategies for addressing this challenge. These methods allow for the correlation of surface compositions with electrochemical activity, using descriptors like adsorption energy to predict the most promising candidates [11, 99–101]. Another critical factor in the performance of electrocatalysts made from complex solid solutions is the control of surface composition. Surface segregation effects must be carefully considered to achieve the desired surface properties. Stability, which is as crucial as activity in electrocatalysis, is a key strength of HEMs. The entropy stabilization inherent in highly multinary HEMs may help overcome the common trade-off between activity and stability. Additionally, metastable phases in HEMs, achieved through nonequilibrium methods, offer rich possibilities for designing new catalysts. However, metastable materials may eventually demix into competing phases during operation, posing challenges related to long-term stability, such as oxidation state changes, contamination, and material loss during extended use. Addressing these challenges requires in situ monitoring to assess the stability and performance of HEM catalysts over time.

HEMs have also shown promise in photocatalysis, where catalysts must absorb photons to generate electron–hole pairs and drive redox reactions. The diversity of surface elements in HEMs is expected to enhance photocatalytic yields, create new catalytic sites, modify electronic structures, and improve surface properties. These improvements can increase light absorption, promote surface reactions, and ultimately enhance photocatalytic efficiency. Mechanisms that may contribute to these improvements include the generation of new surface reconstruction features, a broader spectrum of active sites, tunable band gaps, and improved charge separation.

In low-dimensional forms, HEMs are being explored for electrocatalytic applications, particularly in reducing barriers for redox reactions crucial for energy conversion. The advantages of low-dimensional materials—such as high electrical conductivity and multiple active sites—are amplified in chemically complex systems. These materials offer increased charge carrier densities, a wider selection of active sites, and enhanced chemical and thermal stability. However, challenges such as material robustness under cycling, exposure to complex reactants, and stability in real-world environments must be addressed. Furthermore, good electron mobility alone is insufficient for a successful electrocatalyst—other factors, including long-term stability and environmental resilience, must be considered for practical applications.

Energy Storage Applications. HEMs have garnered significant attention for their potential in energy storage applications, particularly in lithium-ion batteries (LIBs) and other advanced energy storage systems. One of the key advantages of HEMs is their high electrochemical stability, which improves performance during repeated charge and discharge cycles. This is particularly critical in applications where materials must endure cyclic stress without significant degradation. The inherent stability of HEMs, combined with their unique structural properties, makes them highly resilient to the degradation mechanisms that often limit the lifespan of conventional battery materials, offering a longer operational lifetime and reduced maintenance needs.

The multiphase microstructures that are often present in HEMs offer several advantages in ion storage and ion diffusion. These materials can accommodate a wide range of ion sizes, facilitating faster ion diffusion and enhanced ionic conductivity. In LIBs, the rapid transport of lithium ions is critical for fast charge–discharge cycles and high energy density. By tailoring the alloy composition, HEMs can optimize ion movement paths, allowing for higher throughput of ions and faster charge rates. This reduces internal resistance and improves the overall efficiency of the energy storage system, enabling faster charging and higher power output without compromising the material's longevity.

Another significant benefit of HEMs in energy storage systems is their mechanical durability, which is particularly valuable for high-performance applications like electric vehicles (EVs) and grid energy storage. The combination of multiple elements in HEAs or HECs can be tuned to resist mechanical degradation under extreme operating conditions, such as high temperatures, large current loads, and stress cycling. These conditions can cause material failure in traditional battery electrodes, but HEMs maintain their structural integrity during repeated charging cycles, significantly reducing the risk of failure due to mechanical stress, cracking, or material fatigue. This durability ensures longer battery life, as the materials retain their structural and electrochemical properties over extended periods of use.

The synergistic effects between various elements in HEMs contribute to increased energy capacity and improved efficiency. Unlike traditional alloys or materials, which are limited by the properties of a few elements, HEMs exploit the diverse combination of metals to achieve enhanced electrochemical performance.

For example, in battery anodes or cathodes, the presence of refractory elements can increase thermal stability, while more active elements can enhance capacity and voltage output. Additionally, the multiphase structures in HEMs are less prone to phase transitions that can adversely affect performance, such as dendrite formation or volume expansion that occurs in conventional materials like graphite or silicon. By minimizing these issues, HEMs offer a more stable and reliable option for high-capacity energy storage.

HEMs are poised to revolutionize next-generation energy storage systems due to their high capacity, long cycle life, and excellent thermal and chemical stability. These materials can be tailored for a range of applications, from EVs requiring fast charging and long-range capacity to grid storage systems that demand high efficiency and long-term performance. Moreover, HEMs' resistance to chemical degradation and their ability to maintain high electrochemical performance under demanding conditions make them an ideal choice for portable electronics, which require both lightweight and high-performance energy storage solutions. The development of HEM-based materials could help overcome some of the major limitations of current battery technologies, leading to more sustainable, efficient, and durable energy storage solutions for a wide range of industries and applications.

In conclusion, the diverse applications of functional HEAs and compounds are a testament to their versatility and potential impact on various industries (Figure 1.20). As research continues to uncover new properties and improve the understanding of HEMs, their applications are expected to expand, leading to innovative solutions in engineering, technology, and science.

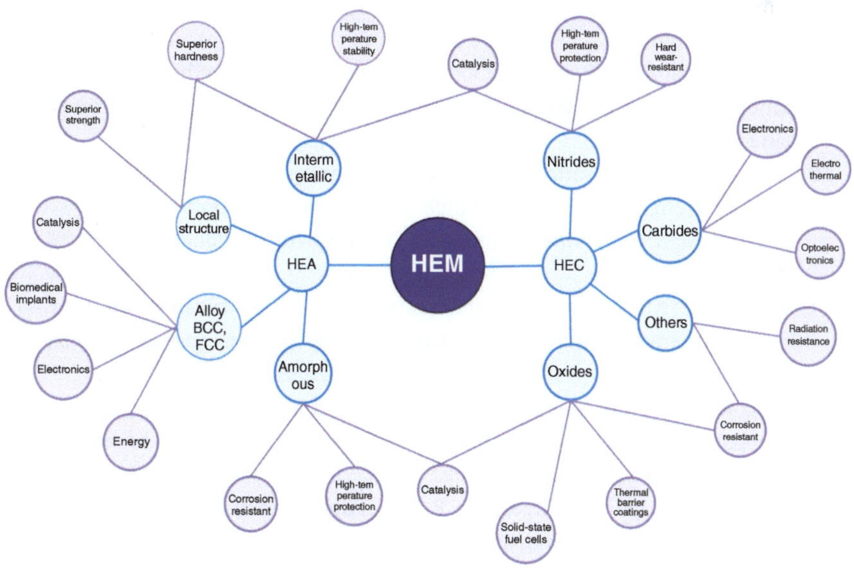

Figure 1.20 Schematic diagram of different classifications and applications of high-entropy functional materials.

References

1 Yeh J W, Chen S K, Lin S J, et al. Nanostructured high-entropy alloys with multiple principal elements: novel alloy design concepts and outcomes. *Advanced Engineering Materials*, 2004, 6: 299–303. https://doi.org/10.1002/adem.200300567

2 He Q F, Ding Z Y, Ye Y F, et al. Design of high-entropy alloy: a perspective from nonideal mixing. *JOM*, 2017, 69: 2092–2098. https://doi.org/10.1007/s11837-017-2452-1

3 Tong C J, Chen Y L, Yeh J W, et al. Microstructure characterization of $Al_xCoCrCuFeNi$ high-entropy alloy system with multiprincipal elements. *Metallurgical and Materials Transactions A*, 2005, 36, 881–893. https://doi.org/10.1007/s11661-005-0283-0

4 Wu J M, Lin S J, Yeh J W, et al. Adhesive wear behavior of $Al_xCoCrCuFeNi$ high-entropy alloys as a function of aluminum content. *Wear*, 2006, 261: 513–519. https://doi.org/10.1016/j.wear.2005.12.008

5 Wang K, Liu H, Zhang L, et al. Unveiling the nucleation and evolution of twinned intermetallic nanocrystals for CO-tolerant selective hydrogenation. *Inorganic Chemistry Frontiers*, 2023, 10: 1027–1036. https://doi.org/10.1039/d2qi02262a

6 Hsu W L, Tsai C W, Yeh A C. Clarifying the four core effects of high-entropy materials. *Nature Reviews Chemistry*, 2024, 8: 471–485. https://doi.org/10.1038/s41570-024-00602-5

7 Tung C C, Yeh J W, Shun T T, et al. On the elemental effect of AlCoCrCuFeNi high-entropy alloy system. *Materials Letters*, 2007, 61: 1–5. https://doi.org/10.1016/j.matlet.2006.03.140

8 Zhang R, Wang C, Zou P, et al. Compositionally complex doping for zero-strain zero-cobalt layered cathodes. *Nature*, 2022, 610: 67–73. https://doi.org/10.1038/s41586-022-05115-z

9 Yao Y, Huang Z, Xie P, et al. Carbothermal shock synthesis of high-entropy-alloy nanoparticles. *Science*, 2018, 359: 1489–1494. https://doi.org/10.1126/science.aan5412

10 Chen S, Aitken Z H, Pattamatta S, et al. Simultaneously enhancing the ultimate strength and ductility of high-entropy alloys via short-range ordering. *Nature Communications*, 2021, 12: 4953. https://doi.org/10.1038/s41467-021-25264-5

11 Yao Y, Liu Z, Xie P, et al. Computationally aided, entropy-driven synthesis of highly efficient and durable multi-elemental alloy catalysts. *Science Advances*, 2020, 6: eaaz0510. https://doi.org/10.1126/sciadv.aaz0510

12 Wang B, Wang C, Yu X, et al. General synthesis of high-entropy alloy and ceramic nanoparticles in nanoseconds. *Nature Synthesis*, 2022, 1: 138–146. https://doi.org/10.1038/s44160-021-00004-1

13 Nellaiappan S, Katiyar N K, Kumar R, et al. High-entropy alloys as catalysts for the CO_2 and CO reduction reactions: experimental realization. *ACS Catalysis*, 2020, 10: 3658–3663. https://doi.org/10.1021/acscatal.9b04302

14 Zhan C, Xu Y, Bu L, et al. Subnanometer high-entropy alloy nanowires enable remarkable hydrogen oxidation catalysis. *Nature Communications*, 2021, 12: 6261. https://doi.org/10.1038/s41467-021-26425-2

15 Moniri S, Yang Y, Ding J, et al. Three-dimensional atomic structure and local chemical order of medium- and high-entropy nanoalloys. *Nature*, 2023, 624: 564–569. https://doi.org/10.1038/s41586-023-06785-z

16 Liu H, Zhang Y, Zhang L, et al. Unveiling atomic-scaled local chemical order of high-entropy intermetallic catalyst for alkyl-substitution-dependent alkyne semihydrogenation. *Journal of the American Chemical Society*, 2024, 146: 20193–20204. https://doi.org/10.1021/jacs.4c05295

17 Yao Y, Dong Q, Brozena A, et al. High-entropy nanoparticles: synthesis-structure-property relationships and data-driven discovery. *Science*, 2022, 376: eabn3103. https://doi.org/10.1126/science.abn3103

18 Zhang W, Liaw P K, Zhang, Y. Science and technology in high-entropy alloys. *Science China Materials*, 2018, 61: 2–22. https://doi.org/10.1007/s40843-017-9195-8

19 Pickering E J, Muñoz-Moreno R, Stone H J. Precipitation in the equiatomic high-entropy alloy CrMnFeCoNi. *Scripta Materialia*, 2016, 113: 106–109. https://doi.org/10.1016/j.scriptamat.2015.10.025

20 Otto F, Dlouhý A, Pradeep K G, et al. Decomposition of the single-phase high-entropy alloy CrMnFeCoNi after prolonged anneals at intermediate temperatures. *Acta Materialia*, 2016, 112: 40–52. https://doi.org/10.1016/j.actamat.2016.04.005

21 Rost C M, Sachet E, Borman T, et al. Entropy-stabilized oxides. *Nature Communications*, 2015, 6: 8485. https://doi.org/10.1038/ncomms9485

22 Huang H, Wu Y, He J, et al. Phase-transformation ductilization of brittle high-entropy alloys via metastability engineering. *Advanced Materials*, 2017, 29: 1701678. https://doi.org/10.1002/adma.201701678

23 Allmann R, Hinek R. The introduction of structure types into the inorganic crystal structure database ICSD. *Acta Crystallographica Section A*, 2007, 63: 412–417. https://doi.org/10.1107/S0108767307038081

24 Mackay, A L. On complexity. *Crystallography Reports*, 2001, 46: 524–526. https://doi.org/10.1134/1.1387117

25 Yeh J W, Chang S Y, Hong Y D. Anomalous decrease in X-ray diffraction intensities of Cu–Ni–Al–Co–Cr–Fe–Si alloy systems with multi-principal elements. *Materials Chemistry and Physics*, 2007, 103: 41–46. https://doi.org/10.1016/j.matchemphys.2007.01.003

26 Wang H, He Q, Gao X, et al. Multifunctional high entropy alloys enabled by severe lattice distortion. *Advanced Materials*, 2024, 36: 2305453. https://doi.org/10.1002/adma.202305453

27 Lin K H, Tseng C M, Chueh C C, et al. Different lattice distortion effects on the tensile properties of Ni-W dilute solutions and CrFeNi and CoCrFeMnNi concentrated solutions. *Acta Materialia*, 2021, 221: 117399. https://doi.org/10.1016/j.actamat.2021.117399

28 Cao B X, Wang C, Yang T. Cocktail effects in understanding the stability and properties of face-centered-cubic high-entropy alloys at ambient and cryogenic temperatures. *Scripta Materialia*, 2020, 187: 250–255. https://doi.org/10.1016/j.scriptamat.2020.06.008

29 Jien-Wei Y. Recent progress in high entropy alloys. *Annales De Chimie – Science des Materiaux*, 2006, 31: 633–648.

30 Tsai K Y, Tsai M H, Yeh J W. Sluggish diffusion in Co–Cr–Fe–Mn–Ni high-entropy alloys. *Acta Materialia*, 2013, 61: 4887–4897. https://doi.org/10.1016/j.actamat.2013.04.058

31 Zhang C, Zhang F, Jin K, et al. Understanding of the elemental diffusion behavior in concentrated solid solution alloys. *Journal of Phase Equilibria and Diffusion*, 2017, 38: 434–444. https://doi.org/10.1007/s11669-017-0580-5

32 Hsu W.-L, Murakami H, Araki H, et al. A study of $NiCo_{0.6}Fe_{0.2}Cr_xSiAlTi_y$ high-entropy alloys for applications as a high-temperature protective coating and a bond coat in thermal barrier coating systems. *Journal of the Electrochemical Society*, 2018, 165: C524. https://doi.org/10.1149/2.1191809jes

33 Yeh J-W. Recent progress in high-entropy alloys. *European Journal of Control*, 2006, 31: 633–648. https://doi.org/10.3166/acsm.31.633-648

34 Wu D, Kusada K, Yamamoto T, et al. Platinum-group-metal high-entropy-alloy nanoparticles. *Journal of the American Chemical Society*, 2020, 142(32): 13833–13838. https://doi.org/10.1021/jacs.0c04807

35 Wu D, et al. Noble-metal high-entropy-alloy nanoparticles: atomic-level insight into the electronic structure. *Journal of the American Chemical Society*, 2022, 144: 3365–3369. https://doi.org/10.1021/jacs.1c13616

36 Hao J, Zhuang Z, Cao K, et al. Unraveling the electronegativity-dominated intermediate adsorption on high-entropy alloy electrocatalysts. *Nature Communications*, 2022, 13: 2662. https://doi.org/10.1038/s41467-022-30379-4

37 Wang L, Zhang F, Nie Z, et al. Abundant polymorphic transitions in the $Al_{0.6}CoCrFeNi$ high-entropy alloy. *Materials Today Physics*, 2019, 8: 1–9. https://doi.org/10.1016/j.mtphys.2018.12.002

38 Wang Y, Chen S, Zhang Y, et al. Effect of multi-element synergistic addition on the microstructure evolution and performance enhancement of laser cladded Al–Mg–Zn-(Mn–Cu–Er–Zr) alloy. *Materials Science and Engineering: A*, 2024, 896: 146256. https://doi.org/10.1016/j.msea.2024.146256

39 Löffler T, Ludwig A, Rossmeisl J, et al. What makes high-entropy alloys exceptional electrocatalysts? *Angewandte Chemie International Edition*, 2021, 60: 26894–26903. https://doi.org/10.1002/anie.202109212

40 Hilhorst A, Jacques P J, Pardoen T. Towards the best strength, ductility, and toughness combination: High entropy alloys are excellent, stainless steels are exceptional. *Acta Materialia*, 2023, 260: 119280. https://doi.org/10.1016/j.actamat.2023.119280

41 Couzinié J P, Dirras, G. Body-centered cubic high-entropy alloys: From processing to underlying deformation mechanisms. *Materials Characterization*, 2019, 147: 533–544. https://doi.org/10.1016/j.matchar.2018.07.015

42 Uporov S, Estemirova S K, Bykov V A, et al. A single-phase ScTiZrHf high-entropy alloy with thermally stable hexagonal close-packed structure. *Intermetallics*, 2020, 122: 106802. https://doi.org/10.1016/j.intermet.2020.106802

43 Wang H, Yang P Y, Zhao W J, et al. Lattice distortion enabling enhanced strength and plasticity in high entropy intermetallic alloy. *Nature Communications*, 2024, 15: 6782. https://doi.org/10.1038/s41467-024-51204-0

44 Zerdoumi R, Ludwig A, Schuhmann W. High entropy intermetallic compounds: a discovery platform for structure-property correlations and materials design principles in electrocatalysis. *Current Opinion in Electrochemistry*, 2024, 48: 101590. https://doi.org/10.1016/j.coelec.2024.101590

45 Aamlid SS, Oudah M, Rottler J, et al. Understanding the role of entropy in high entropy oxides. *Journal of the American Chemical Society*, 2023, 145: 5991–6006. https://doi.org/10.1021/jacs.2c11608

46 Tomboc G M, Zhang X, Choi S, et al. Stabilization, characterization, and electrochemical applications of high-entropy oxides: critical assessment of crystal phase–properties relationship. *Advanced Functional Materials*, 2022, 32: 2205142. https://doi.org/10.1002/adfm.202205142

47 Sarkar A, Velasco L, Wang D I, et al. High entropy oxides for reversible energy storage. *Nature Communications*, 2018, 9, 3400. https://doi.org/10.1038/s41467-018-05774-5

48 Lun Z, Ouyang B, Kwon D H, et al. Cation-disordered rocksalt-type high-entropy cathodes for Li-ion batteries. *Nature Materials*, 2021, 20: 214–221. https://doi.org/10.1038/s41563-020-00816-0

49 Nguyen T X, Liao Y C, Lin C C et al. Advanced high entropy perovskite oxide electrocatalyst for oxygen evolution reaction. *Advanced Functional Materials*, 2021, 31: 2101632. https://doi.org/10.1002/adfm.202101632

50 Cai T, Cai M, Mu J, et al. High-entropy layered oxide cathode enabling high-rate for solid-state sodium-ion batteries. *Nano-Micro Letters*, 2023, 16: 10. https://doi.org/10.1007/s40820-023-01232-0

51 Gild J, Zhang Y, Harrington T, et al. High-entropy metal diborides: a new class of high-entropy materials and a new type of ultrahigh temperature ceramics. *Scientific Reports*, 2016, 6, 37946. https://doi.org/10.1038/srep37946

52 Sarker P, Harrington T, Toher C, et al. High-entropy high-hardness metal carbides discovered by entropy descriptors. *Nature Communications*, 2018, 9: 4980. https://doi.org/10.1038/s41467-018-07160-7

53 Jin T, Sang X, Unocic R R, et al. Mechanochemical-assisted synthesis of high-entropy metal nitride via a soft urea strategy. *Advanced Materials*, 2018, 30: 1707512. https://doi.org/10.1002/adma.201707512

54 Peng C, Gao X, Wang M, et al. Diffusion-controlled alloying of single-phase multi-principal transition metal carbides with high toughness and low thermal diffusivity. *Applied Physics Letters*, 2019, 114: 011905. https://doi.org/10.1063/1.5054954

55 Zhang H, Akhtar F. Processing and characterization of refractory quaternary and quinary high-entropy carbide composite. *Entropy*, 2019, 21: 474.

56 Gild J, Braun J, Kaufmann K, et al. A high-entropy silicide: $(Mo_{0.2}Nb_{0.2}Ta_{0.2}Ti_{0.2}W_{0.2})Si_2$. *Journal of Materiomics*, 2019, 5: 337–343. https://doi.org/10.1016/j.jmat.2019.03.002

57 Qin Y, Liu J X, Li F, et al. A high entropy silicide by reactive spark plasma sintering. *Journal of Advanced Ceramics*, 2019, 8: 148–152. https://doi.org/10.1007/s40145-019-0319-3

58 Ma Y, Ma Y, Wang Q, et al. High-entropy energy materials: challenges and new opportunities. *Energy & Environmental Science*, 2021, 14: 2883–2905. https://doi.org/10.1039/D1EE00505G

59 Chen R, Qin G, Zheng H, et al. Composition design of high entropy alloys using the valence electron concentration to balance strength and ductility. *Acta Materialia*, 2018, 144: 129–137. https://doi.org/10.1016/j.actamat.2017.10.058

60 Guo S, Ng C, Lu J, et al. Effect of valence electron concentration on stability of fcc or bcc phase in high entropy alloys. *Journal of Applied Physics*, 109: 103505. https://doi.org/10.1063/1.3587228

61 Liu W H, Lu Z P, He J Y, et al. Ductile $CoCrFeNiMo_x$ high entropy alloys strengthened by hard intermetallic phases. *Acta Materialia*, 2016, 116: 332–342. https://doi.org/10.1016/j.actamat.2016.06.063

62 Guo S, Hu Q, Ng C. More than entropy in high-entropy alloys: forming solid solutions or amorphous phase. *Intermetallics*, 2013, 41: 96–103. https://doi.org/10.1016/j.intermet.2013.05.002

63 Santodonato L J, Zhang Y, Feygenson M, et al. Deviation from high-entropy configurations in the atomic distributions of a multi-principal-element alloy. *Nature Communications*, 2015, 6: 5964. https://doi.org/10.1038/ncomms6964

64 Niu C, LaRosa C R, Miao J. Magnetically-driven phase transformation strengthening in high entropy alloys. *Nature Communications*, 2018, 9: 1363. https://doi.org/10.1038/s41467-018-03846-0

65 Zhang F, Lou H, Cheng B, et al. High-pressure induced phase transitions in high-entropy alloys: a review. *Entropy*, 2019, 21: 239.

66 Cui M, Yang C, Hwang S, et al. Multi-principal elemental intermetallic nanoparticles synthesized via a disorder-to-order transition. *Science Advances*, 2022, 8: eabm4322. https://doi.org/10.1126/sciadv.abm4322

67 Yang T, Zhao Y L, Tong Y, et al. Multicomponent intermetallic nanoparticles and superb mechanical behaviors of complex alloys. *Science*, 2018, 362: 933–937. https://doi.org/10.1126/science.aas8815

68 Xing F, Ma J, Shimizu K.-I, et al. High-entropy intermetallics on ceria as efficient catalysts for the oxidative dehydrogenation of propane using CO_2. *Nature Communications*, 2022, 13: 5065. https://doi.org/10.1038/s41467-022-32842-8

69 He X, Zhang Y, Gu X, et al. Pt-induced atomic-level tailoring towards paracrystalline high-entropy alloy. *Nature Communications*, 2023, 14: 775. https://doi.org/10.1038/s41467-023-36423-1

70 Wu G, Liu C, Yan Y Q, et al. Elemental partitioning-mediated crystalline-to-amorphous phase transformation under quasi-static deformation. *Nature Communications*, 2024, 15: 1223. https://doi.org/10.1038/s41467-024-45513-7

71 Wang H, He Q-F, Yang Y. High-entropy intermetallics: from alloy design to structural and functional properties. *Rare Metals*, 2022, 41: 1989–2001. https://doi.org/10.1007/s12598-021-01926-7

72 Wu Y, Zhang F, Yuan X, et al. Short-range ordering and its effects on mechanical properties of high-entropy alloys. *Journal of Materials Science & Technology*, 2021 62: 214–220. https://doi.org/10.1016/j.jmst.2020.06.018

73 George E P, Curtin W A, Tasan C C. High entropy alloys: a focused review of mechanical properties and deformation mechanisms. *Acta Materialia*, 2020, 188: 435–474. https://doi.org/10.1016/j.actamat.2019.12.015

74 An Z, Mao S, Liu Y, et al. Hierarchical grain size and nanotwin gradient microstructure for improved mechanical properties of a non-equiatomic CoCrFeMnNi high-entropy alloy. *Journal of Materials Science & Technology*, 2021, 92: 195–207. https://doi.org/10.1016/j.jmst.2021.02.059

75 Seol J B, Bae J W, Li Z, et al. Boron doped ultrastrong and ductile high-entropy alloys. *Acta Materialia*, 2018, 151: 366–376. https://doi.org/10.1016/j.actamat.2018.04.004

76 Luan H, Huang L, Kang J, et al. Spinodal decomposition and the pseudo-binary decomposition in high-entropy alloys. *Acta Materialia*, 2023, 248: 118775. https://doi.org/10.1016/j.actamat.2023.118775

77 Laplanche G, Kostka A, Horst O M. Microstructure evolution and critical stress for twinning in the CrMnFeCoNi high-entropy alloy. *Acta Materialia*, 2016, 118: 152–163. https://doi.org/10.1016/j.actamat.2016.07.038

78 Sun L, He Z, Jia N, et al. Local chemical order enables an ultrastrong and ductile high-entropy alloy in a cryogenic environment. *Science Advances*, 2024, 10: eadq6398. https://doi.org/10.1126/sciadv.adq6398

79 Huang Z, Li T, Li B, et al. Tailoring local chemical ordering via elemental tuning in high-entropy alloys. *Journal of the American Chemical Society*, 2024, 146: 2167–2173. https://doi.org/10.1021/jacs.3c12048

80 Yang Y, Zhou J, Zhu F, et al. Determining the three-dimensional atomic structure of an amorphous solid. *Nature*, 2021, 592: 60–64. https://doi.org/10.1038/s41586-021-03354-0

81 Chen X, Wang Q, Cheng Z, et al. Direct observation of chemical short-range order in a medium-entropy alloy. *Nature*, 2021, 592: 712–716. https://doi.org/10.1038/s41586-021-03428-z

82 Zhao S. Effects of local elemental ordering on defect-grain boundary interactions in high-entropy alloys. *Journal of Alloys and Compounds*, 2021, 887: 161314. https://doi.org/10.1016/j.jallcom.2021.161314

83 Wang J, Jiang P, Yuan F. Chemical medium-range order in a medium-entropy alloy. *Nature Communications*, 2022, 13: 1021. https://doi.org/10.1038/s41467-022-28687-w

84 Orhan O K, Hendy M, Ponga M. Electronic effects on the radiation damage in high-entropy alloys. *Acta Materialia*, 2023, 244: 118511. https://doi.org/10.1016/j.actamat.2022.118511

85 Zhang H, Li T, Wang Q, et al. Effects of nitrogen doping on microstructures and irradiation resistance of Ti–Zr–Nb–V–Mo refractory high-entropy alloy. *Acta Metallurgica Sinica (English Letters)*, 2024, 37: 1007–1018. https://doi.org/10.1007/s40195-024-01686-0

86 Kombaiah B, Zhou Y, Jin K, et al. Nanoprecipitates to enhance radiation tolerance in high-entropy alloys. *ACS Applied Materials & Interfaces*, 2023, 15: 3912–3924. https://doi.org/10.1021/acsami.2c17540

87 Duan J M, Jiang Z N, Zeng X Q, et al. Unveiling the underlying corrosion/corrosion-resistance mechanism of a novel corrosion-resistant high entropy alloy by experimental means and AIMD simulations. *Corrosion Science*, 2024, 227: 111782. https://doi.org/10.1016/j.corsci.2023.111782

88 Shi Y, Yang B, Liaw P K. Corrosion-resistant high-entropy alloys: a review. *Metals*, 2017, 7: 43.

89 Gottschall T, Skokov KP, Fries M, et al. Making a cool choice: the materials library of magnetic refrigeration. *Advanced Energy Materials*, 2019, 9: 1901322. https://doi.org/10.1002/aenm.201901322

90 Tegus O, Brück E, Buschow KH, et al. Transition-metal-based magnetic refrigerants for room-temperature applications. *Nature*, 2002, 415: 150–152. https://doi.org/10.1038/415150a

91 Fries M, Pfeuffer L, Bruder E, et al. Microstructural and magnetic properties of Mn-Fe-P-Si (Fe_2 P-type) magnetocaloric compounds. *Acta Materialia*, 2017, 132: 222–229. https://doi.org/10.1016/j.actamat.2017.04.040

92 Snyder G J, Toberer E S. Complex thermoelectric materials. *Nature Materials*, 2008, 7: 105–114. https://doi.org/10.1038/nmat2090

93 Perumal S, Roychowdhury S, Negi D S, et al. High thermoelectric performance and enhanced mechanical stability of p-type $Ge_{1-x}Sb_xTe$. *Chemistry of Materials*, 2015, 27: 7171–7178. https://doi.org/10.1021/acs.chemmater.5b03434

94 Jiang B, Yu Y, Cui J, et al. High-entropy-stabilized chalcogenides with high thermoelectric performance. *Science*, 2021, 371, 830–834. https://doi.org/10.1126/science.abe1292

95 Xin Y, Li S, Qian Y, et al. High-entropy alloys as a platform for catalysis: progress, challenges, and opportunities. *ACS Catalysis*, 2020, 10: 11280–11306 https://doi.org/10.1021/acscatal.0c03617

96 Han X, Wu G, Zhao S, et al. Nanoscale high-entropy alloy for electrocatalysis. *Matter*, 2023, 6: 1717–1751. https://doi.org/10.1016/j.matt.2023.03.034

97 Lu Z, Chen Z W, Singh C V. Neural network-assisted development of high-entropy alloy catalysts: decoupling ligand and coordination effects. *Matter*, 2020, 3: 1318–1333. https://doi.org/10.1016/j.matt.2020.07.029

98 Batchelor T A, Pedersen J K, Winther S H, et al. High-entropy alloys as a discovery platform for electrocatalysis. *Joule*, 2019, 3: 834–845. https://doi.org/10.1016/j.joule.2018.12.015

99 Bell T E, Torrente-Murciano L. H_2 production via ammonia decomposition using non-noble metal catalysts: a review. *Topics in Catalysis*, 2016, 59: 1438–1457. https://doi.org/10.1007/s11244-016-0653-4

100 Xin L, Yongqiang, H, Husheng J. Pt-Rh-Pd alloy group gauze catalysts used for ammonia oxidation. *Rare Metal Materials and Engineering*, 2017, 46: 339–343. https://doi.org/10.1016/S1875-5372(17)30091-7

101 Xie P, Yao Y, Huang Z, et al. Highly efficient decomposition of ammonia using high-entropy alloy catalysts. *Nature Communications*, 2019, 10: 4011. https://doi.org/10.1038/s41467-019-11848-9

102 Yu L, Zeng K, Li C, et al. High-entropy alloy catalysts: from bulk to nano toward highly efficient carbon and nitrogen catalysis. *Carbon Energy*, 2022, 4: 731–761. https://doi.org/10.1002/cey2.228

2

Extreme Conditions and Novel Structure Design

Rongzhi Li[2], *Yong Zhang*[1,2,*]

[1] *Key Laboratory of Silicon-based Materials, The Ministry of Education, and School of Materials Science and Engineering, Fuyao University of Science and Technology, Fuzhou, China*
[2] *University of Science and Technology Beijing, State Key Laboratory for Advanced Metal Materials, Beijing, China*

*Corresponding author: drzhangy@ustb.edu.cn/yongzhang@fyust.edu.cn

2.1 Introduction to Extreme Conditions and Harsh Requirements

In a variety of industrial, scientific, and environmental applications, materials and structures are subjected to extreme conditions and demanding requirements. These conditions may include high or low temperatures, intense radiation, high pressure, corrosive environments, or a combination of these factors. Designing materials and structures to withstand these conditions is critical to ensuring the reliability, durability, and performance of various systems and components. For example, in harsh operating environments where extremely high temperatures and heavy loads may be present, such as in aerospace, aircraft engines, and high-temperature furnaces, special materials are needed to create the desirable materials and components. Conventional materials, although good in performance, often fail to have sufficient durability or multifunctionalities to endure these extreme conditions.

High-entropy materials (HEMs) have emerged as a promising class of materials due to their unique composition and properties, making them particularly advantageous for extreme conditions such as high and low temperatures, radiation exposure, and harsh environmental environments. Their favorable behavior can be attributed to a combination of structural, chemical, and thermodynamic factors. Notably, the combination of multiple elements in a high-entropy system results in a large mixing entropy, which contributes to the stabilization of a single-phase solid solution. This high configurational entropy further stabilizes the material at high temperatures and other extreme conditions. In addition, HEMs tend to have lattice distortions because of the large differences in atomic sizes of the constituent elements. These distortions and lattice mismatch contribute to solid solution strengthening, which enhances

Functional High Entropy Alloys and Compounds, First Edition. Edited by Yong Zhang, Yonggang Yao and Haojie Liu.

the material's strength and structural stability. This is particularly beneficial in extreme conditions where materials are subjected to high thermal or mechanical loads. Furthermore, the complexity and randomness in the atomic arrangement of HEMs contribute to improved radiation resistance. Under radiation exposure, especially in nuclear reactors, materials are subjected to severe damage due to the displacement of atoms caused by high-energy particles. The highly disordered atomic structure in high-entropy alloys (HEAs) increases the probability of self-healing of radiation-induced defects. Lastly, one of the most significant advantages of HEMs is their design flexibility. By varying the relative proportions of the constituent elements, researchers can optimize the material properties for specific or multifunctional applications under extreme conditions. For example, increasing the amount of refractory elements (such as W or Mo) can improve high-temperature strength; adding elements with high thermal stability or high melting points can enhance their performance in heat-intensive applications; modifying the composition to include more transition metals can increase the resilience to irradiation and corrosion. This flexibility enables the development of bespoke materials for a wide range of extreme conditions. As a result, HEMs are being explored for aerospace (high-performance components like turbines and heat shields), nuclear reactors (to withstand radiation and extreme heat), automotive systems (turbochargers, brake systems), space exploration (high-temperature alloys for rocket engines), cryogenics (low-temperature applications), and marine and coastal environments (corrosion-resistant alloys).

2.2 HEMs for Extreme Temperatures

2.2.1 The Application of HEM at High Temperatures

HEMs, especially those refractory ones, typically have excellent thermal stability and oxidation resistance and can maintain structural integrity and mechanical properties at high temperatures. This makes them ideal for parts inside rocket combustion chambers, industrial gas turbines, spacecraft, and other components (Figure 2.1) to cope with extreme operating environments. In extreme conditions, such as high temperatures, HEAs can maintain a stable solid solution at elevated temperatures, whereas phase segregation occurs in traditional alloys. This makes them resistant to

(a) (b) (c)

Figure 2.1 Applications of HEAs in high-temperature environments. (a) Rocket combustion chamber. (b) Industrial gas turbine. (c) Spacecraft. Generated with AI using Doubao.

thermal degradation and improves their performance in high-temperature applications (e.g., jet engines and nuclear reactors).

Traditional alloys, after quenching and subsequent tempering, typically exhibit significant softening. Even high-temperature-resistant steels, such as high-speed steels, have a maximum service temperature of around 550 °C, beyond which they experience considerable softening. In contrast, HEAs maintain relatively high strength and hardness even at higher temperatures. Russian scholars, Kuznetsov et al., studied the tensile properties of as-cast samples and multistage hot-forged samples of HEAs with atomic compositions such as AlCrCuNiFeCo at various temperatures, as shown in Figure 2.2 [1]. The results indicate that, with increasing temperature, the strength of the alloy samples decreases, but their tensile plasticity significantly increases. Analysis reveals that the ductile-to-brittle transition temperature (DBTT) of the as-cast samples is in the range of 700–800 °C, while the DBTT of the hot-forged samples falls between 600 and 700 °C. Furthermore, after hot forging, the alloy exhibits a greater amount of deformation at higher temperatures. In the temperature range of 800–1000 °C, the alloy demonstrates superplastic behavior, as shown in Figure 2.4c.

Moreover, a study by Senkov et al. in 2011 at the U.S. Air Force Research Laboratory investigated the microstructures and mechanical properties of two high-melting-point HEAs, $Nb_{25}Mo_{25}Ta_{25}W_{25}$ and $V_{20}Nb_{20}Mo_{20}Ta_{20}W_{20}$ [2]. The

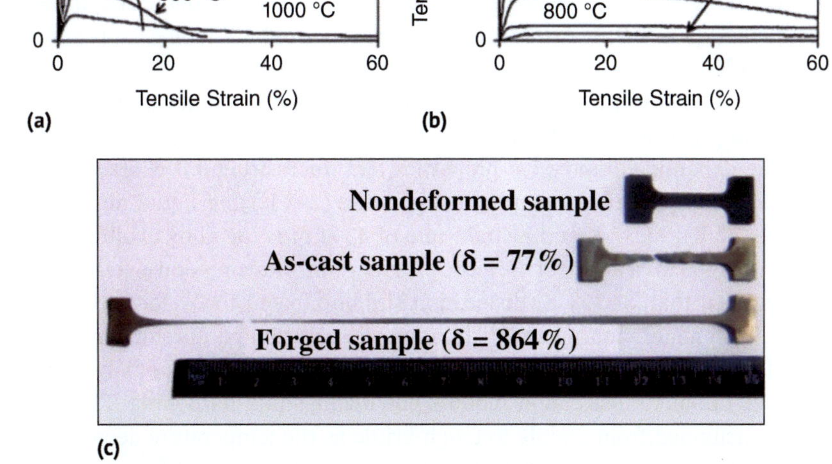

Figure 2.2 The tensile stress–strain curves of AlCrCuNiFeCo alloy at different temperatures. (a) As-cast samples. (b) Thermally forged samples and (c) morphology of the alloy samples after tensile deformation at 1000 °C [1]. *Source*: Reproduced with permission from [1]/Elsevier.

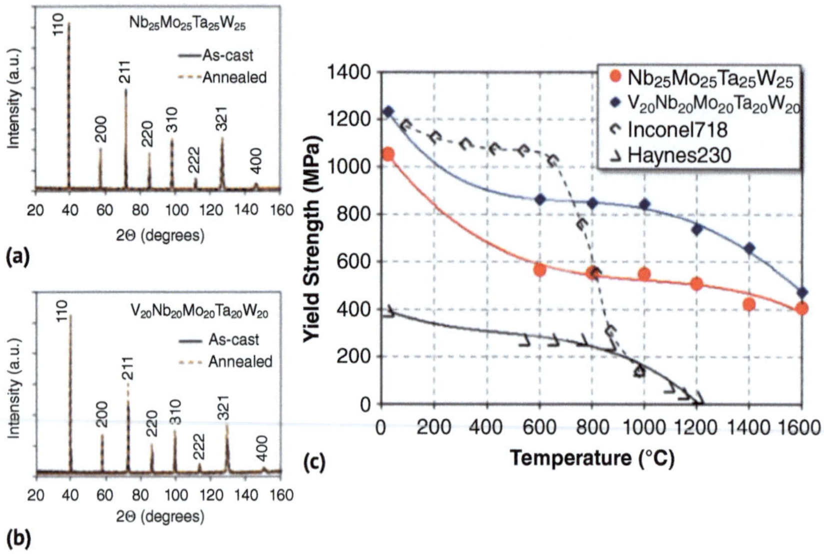

Figure 2.3 Neutron diffraction patterns of two high-melting-point HEAs in the as-cast and annealed states: (a) $Nb_{25}Mo_{25}Ta_{25}W_{25}$. (b) $V_{20}Nb_{20}Mo_{20}Ta_{20}W_{20}$ and (c) Compressive stress–strain curves of the alloys at different temperatures [2]. *Source*: [2] / with permission of Elsevier.

research demonstrated that both alloys possess a simple body-centered cubic (BCC) structure, which remains stable even after 19 hours of annealing at 1400 °C, as shown in Figure 2.3. In terms of mechanical properties, although both alloys exhibit limited room temperature plasticity, their plastic flow increases with temperature. After exceeding 600 °C, the yield strength of both alloys stabilizes, indicating excellent thermal stability. Compared to traditional Ni-based superalloys, these two high-melting-point HEAs exhibit superior resistance to high-temperature softening above 800 °C.

Eutectic high-entropy alloys (EHEAs) with excellent mechanical properties and good casting performance have promising applications. Zhong et al. [3] investigated the dynamic compressive properties and microstructural characteristics of $Al_{1.19}Co_2CrFeNi_{1.81}$ EHEAs at room temperature (298 K) and liquid nitrogen temperature (77 K). At 77 K and a strain rate of 4300 m/s, the alloy exhibited a yield strength of 1365 MPa and over 40% true plastic strain. Transmission electron microscopy revealed that, at 298 K, dislocation slip dominated the deformation, while, at 77 K, both nanotwinning and dislocation slip contributed. Interactions between stacked layer dislocations, dislocation–twin boundaries, and nanotwin–nanotwin structures inhibited dislocation motion, enhancing strain hardening. The fracture mode transitioned from ductile to tough-brittle as the temperature decreased, and adiabatic shear zones formed at the late stage of deformation, as shown in Figure 2.4.

Otto and others from Oak Ridge National Laboratory in the United States also studied the tensile mechanical properties of CoCrFeNiMn multicomponent HEAs

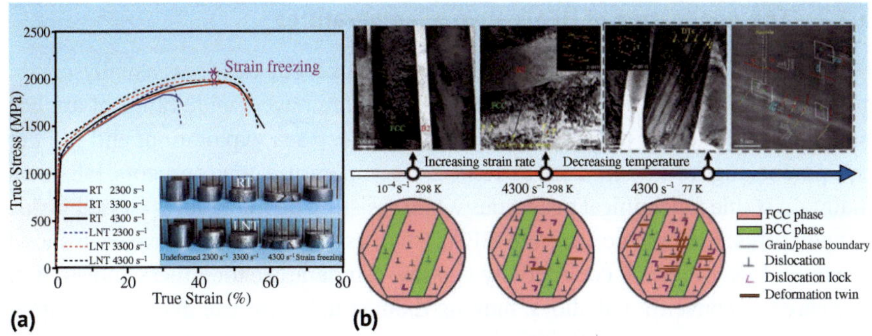

Figure 2.4 Mechanical properties and microscopic deformation mechanisms of $Al_{1.19}Co_2CrFeNi_{1.81}$ EHEA. (a) Curves of EHEA under compressive true stress–strain dynamic loading. The inset shows the deformation of the specimen under the corresponding loading conditions. (b) Microstructural evolution and associated deformation mechanisms of EHEA under different loading conditions [3]. *Source*: Reproduced with permission from [3]/Elsevier.

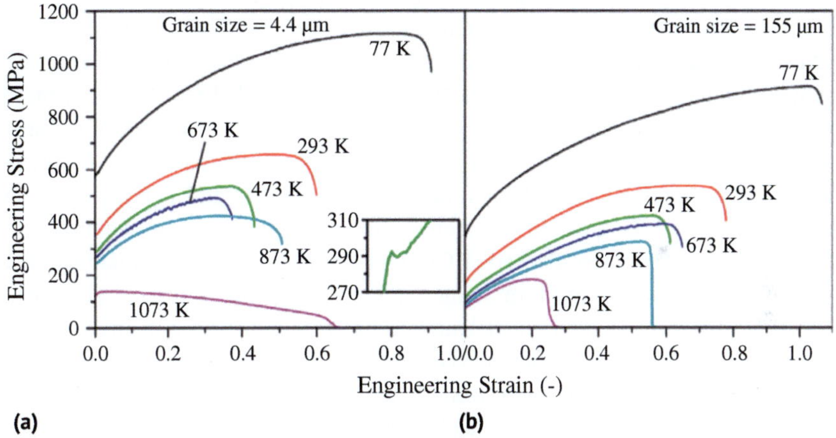

Figure 2.5 Tensile stress–strain curves of CoCrFeNiMn multicomponent HEAs with different grain sizes at different temperatures: (a) grain size of 4.4 μm; (b) grain size of 155 μm [4]. *Source*: [4] / with permission of Elsevier.

with different grain sizes at various temperatures, as shown in Figure 2.5. They concluded that the good plastic deformation behavior of CoCrFeNiMn multicomponent HEAs at low temperatures is attributed to the formation of deformation twins (DTs) during the deformation process [4].

Overall, the exceptional mechanical properties of HEAs at elevated temperatures make them ideal candidates for high-temperature applications, especially in aerospace, energy, and other high-temperature engineering fields. These alloys not only maintain high strength and hardness but also exhibit good plastic deformability, allowing them to withstand more demanding operating conditions.

2.2.2 The Application of HEMs at Low Temperatures

HEMs also have excellent properties at low temperatures. They usually exhibit good toughness and impact resistance and therefore have a wide range of applications in fields such as polar scientific research, deep sea exploration, and military equipment (Figure 2.6). In these extreme low-temperature environments, HEAs can maintain stable mechanical properties while resisting brittle failures caused by low temperatures. For instance, certain HEAs (such as CoCrFeMnNi) have been shown to exhibit lower thermal conductivity and low-temperature toughness and strength compared to conventional alloys, making them suitable for use in cryogenic applications such as aerospace, liquefied gas storage, and low-temperature mechanical systems. Due to their strong solid solution phase, they can maintain mechanical properties at low temperatures by avoiding phase transitions that might embrittle other alloys.

Metal foams are highly demanded for extreme environments with low temperatures and high strain rates due to their unique properties. However, reduced ductility and impact toughness at these conditions limit their application. Meng et al. [5] tested the quasi-static and dynamic compressive properties of CoCrFeMnNi HEA composite foams at liquid nitrogen temperature (−196 °C). The alloy showed ultra-high strength and energy absorption, particularly excellent resistance to embrittlement at low temperatures. Microstructural analysis revealed that the foam matrix exhibited significant twinning activity and shear localization. Dense DTs and shear bands formed a braided microstructure, dispersing deformation and enhancing energy absorption. The DTs also strengthened the matrix and delayed crack initiation and growth (Figure 2.7). Under dynamic loading, the face-centered cubic (FCC) to Hexagonal Close-Packed (HCP) phase transition formed a nanolamellar FCC/HCP structure, further improving toughness. Together with DTs, HCP phase transition, dislocations, and shear bands, these mechanisms compensate for the limitations of dislocation motion at low temperatures and high strain rates, resulting in composite foams with high strength and energy absorption capacity.

Ma et al. [6] studied the mechanical properties of $Ti_2(CrFeNi)_{98}$ multiprincipal element alloy (MPEA) at 293 and 77 K, including yield strength, ultimate tensile strength, and ductility, as shown in Figure 2.5. Results showed that, as the temperature decreased, yield strength increased from 800 to 980 MPa, ultimate

Figure 2.6 Applications of HEAs in low-temperature environments. (a) Polar research icebreaker. (b) Deep sea exploration. (c) Extremely cold military equipment. Generated with AI using Doubao.

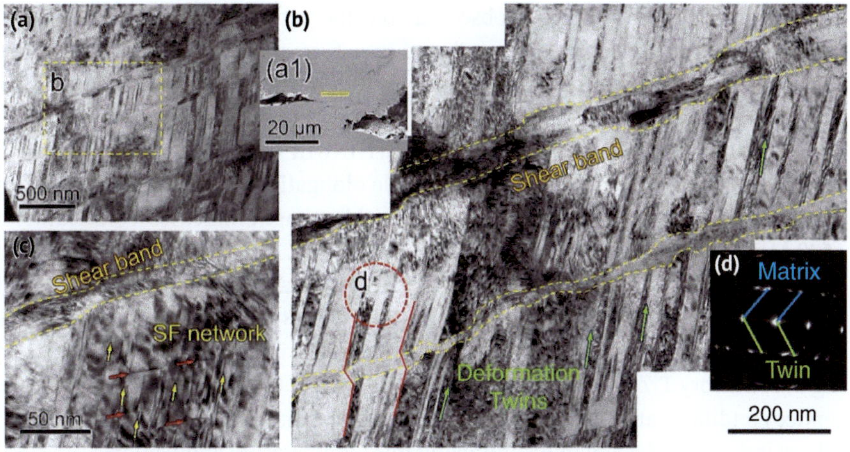

Figure 2.7 TEM photographs of crack-adjacent regions under quasi-static compression at −196 °C. (a) Low-magnification bright-field image showing DTs and shear bands in the sample; (a1) An SEM image showing the sampling location. (b) A montage showing the zoomed-in morphology of the shear bands and the DTs in the yellow-framed region in (a). (c) The zoomed-in BF image showing the layer misalignment network in the vicinity of the shear bands. (d) The SAED (Selected-Area Electron Angular Diffraction) pattern of the red dashed-circle region in (b) indicating the DTs in the matrix [5]. *Source*: Reproduced with permission from [5]/Elsevier.

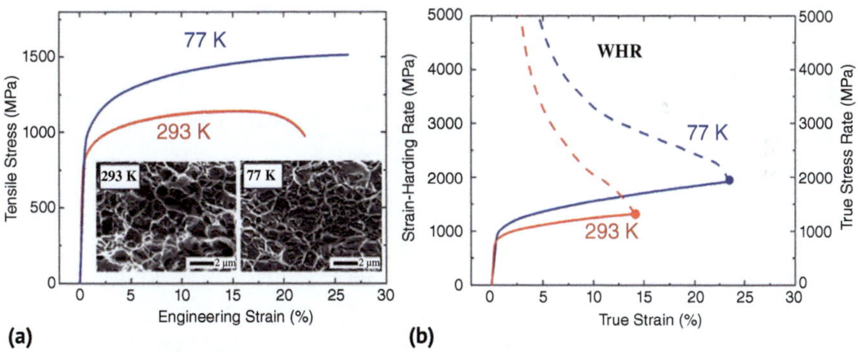

Figure 2.8 (a) Tensile engineering stress–strain curves and fracture morphology of specimens of $Ti_2(CrFeNi)_{98}$ alloy at 293 and 77 K. (b) True stress–strain curves and corresponding work-hardening rates (dσ/dε) at 293 and 77 K [6]. *Source*: Reproduced with permission from [6]/Elsevier.

tensile strength rose from 1.1 to 1.5 GPa, and ductility increased from 20% to 25%. Microstructural analysis revealed that the enhanced ductility at 77 K was due to DTs in the FCC matrix near BCC precipitates, as shown in Figure 2.8. Finite element analysis confirmed that DTs formed due to localized stresses surpassing the critical twinning stress at low temperatures. This study offers a design strategy for cobalt-free intermetallics with good strength and ductility at extremely low temperatures.

Isotropic TiZrHfNbTa alloys are among the few refractory HEAs that exhibit tensile plasticity at room temperature, and their deformation is dominated only by dislocation slip. Wang et al. [7] observed that the activation of [8] nanotwinning was accompanied by deformation-induced transformation of the BCC structure to the non-closed stacked hexagonal ω-phase as well as the dislocation slip during the low-temperature tensile deformation, which suggests that the intrinsic mechanical instability of the single-phase BCC TiZrHfNbTa solid solution. As shown in Figure 2.9, the alloy maintains a high tensile elongation of 20.8% as the temperature decreases from 277 to 77 K, while the yield strength increases significantly to 1549 MPa without significant tough-brittle transition. This particular combination of high strength and ductility at low temperatures can be explained by considering the synergistic effects of spiro dislocation slip [8] mechanical twinning, and BCC → ω phase transition. These results provide new insights into the understanding of refractory TiZrHfNbTa-based alloys and extend their applications to extreme service conditions at low temperatures in aerospace, marine shipbuilding, and natural gas industries, despite their promising high-temperature applications.

Dual-phase microstructures produced by eutectic reactions offer several advantages: ease of casting, composite material properties, and tunable phase characteristic length scales. EHEAs with multicomponent solution phases and hard intermetallic phases are candidates for identifying alternative high-temperature materials. Alloying elements can be selected to improve resistance to laminated microstructural degradation and coarsening. In this work, Shah et al. [9] presented a Calculation of PHAse Diagrams (CALPHAD)-guided design of a sub-EHEA, $CoCrFeNiTa_{0.395}$, which has an incipient dendritic FCC phase and a fine eutectic

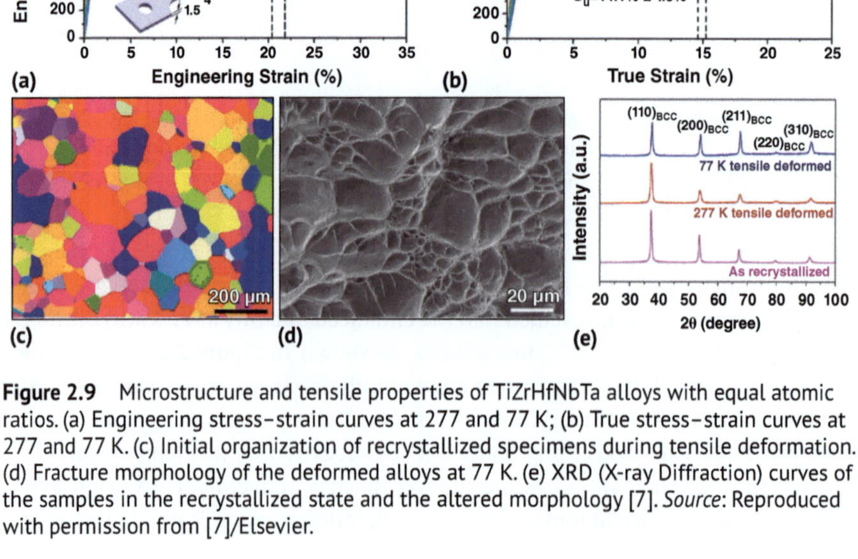

Figure 2.9 Microstructure and tensile properties of TiZrHfNbTa alloys with equal atomic ratios. (a) Engineering stress–strain curves at 277 and 77 K; (b) True stress–strain curves at 277 and 77 K. (c) Initial organization of recrystallized specimens during tensile deformation. (d) Fracture morphology of the deformed alloys at 77 K. (e) XRD (X-ray Diffraction) curves of the samples in the recrystallized state and the altered morphology [7]. *Source*: Reproduced with permission from [7]/Elsevier.

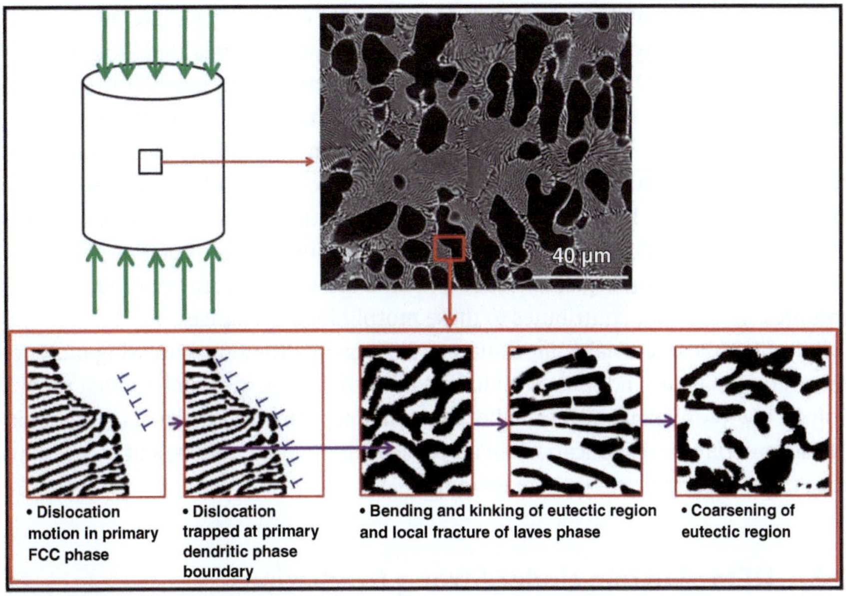

Figure 2.10 A possible deformation mechanism, where the darker lining is the Laves phase and the brighter lining is the FCC phase [9]. *Source*: Reproduced with permission from [9]/Elsevier.

Figure 2.11 Applications of HEMs in irradiated environments. (a) Nuclear power plant. (b) Space radiation environment. Generated with AI using Doubao.

(FCC solution + Laves phase) organization with good high-temperature mechanical properties. The primary FCC phase fraction was 0.42 ± 0.02. The eutectic layer spacing was 0.69 ± 0.12 μm. The alloy exhibited an equilibrium yield strength of 1303 ± 18 MPa, a fracture strength of 2237 ± 23 MPa, and a fracture strain of 0.3 under room temperature compression tests (Figure 2.10).

The mechanism of lamellar morphology changes during deformation above 0.8TE is explained through finite element simulations, showing a good correlation with observed microstructural gradients as shown in Figure 2.11. Nanoscale precipitates form at low strain rates due to thermal effects, and high-temperature precipitation of L12- and Ni_3Ta-type precipitates in the FCC phase extends the alloy temperature range. The mechanisms behind degraded structures and lamellar morphology changes include the following: initial dislocation motion in the FCC phase allows plastic deformation, while dislocation accumulation at dendritic phase boundaries causes strain hardening. In later stages, the eutectic zone undergoes plastic strain, and the Laves phase layer fractures. Localized eutectic colonies bend/twist, and dislocations in the FCC lamellae near the kinking region indicate bending. The Laves phase in the kinking region fractures, with high-temperature conditions during compression, further contributes to these morphological changes.

By tuning the chemical composition and choosing different processing methods, researchers can design HEMs with tailored thermal properties such as high melting points, low coefficients of thermal expansion, and enhanced thermal conductivity. These materials find applications in aerospace, energy, electronics, and other industries that encounter temperature extremes.

2.3 HEMs Sustain Under Strong Irradiations

In addition to extreme temperatures, materials used in nuclear energy, space, and high-energy physics applications must be able to withstand intense radiation from irradiation sources such as gamma rays, neutrons, and charged particles. HEMs offer potential solutions for mitigating radiation-induced damage and maintaining structural integrity under radiation, as shown in Figure 2.11. The radiation tolerance of HEMs stems from their unique atomic arrangements, which impede irradiation-induced defect formation and migration. In addition, the high conformational entropy in these materials promotes the retention of defects in an energetically favorable conformation, thereby reducing their detrimental effects on mechanical properties.

Energy is a vital material foundation for the survival and development of human society. Controlled nuclear fusion reactors, using hydrogen isotopes like deuterium and tritium as raw materials, can provide vast amounts of energy. These raw materials are abundant and easily obtained from the oceans, with the main product being nonradioactive helium, making it an ideal new energy source. China has already proposed the China fusion engineering test reactor (CFETR) plan to develop controlled nuclear fusion reactors. In a nuclear reactor, the blanket structure is one of the most crucial components. It can shield radiation and prevent the leakage of radioactive materials from the reactor. However, the operational environment of fusion reactor blanket materials is harsh, as they must withstand high radiation doses and high thermal loads, which can significantly degrade the mechanical properties of the material, such as strength and toughness, reduce service life, and affect operational safety. The recent Fukushima nuclear leak incident in Japan once again reminds humanity of the importance of nuclear safety. Therefore, the performance requirements for blanket materials are very high, to say the least. They must not only resist radiation, have low activation, and maintain stable mechanical properties but also possess resistance to high-temperature creep and corrosion. According to the CFETR phase II plan, the radiation damage resistance of blanket materials will need to be further enhanced, with higher operating temperatures and stronger corrosive heat transfer media.

Traditional blanket materials include low-activation ferritic-martensitic (RAFM) steels, oxide dispersion strengthened (ODS) steels with dispersed oxides in RAFM steels, vanadium alloys, and silicon carbide composites. Among these, ODS steels produced by mechanical alloying show better performance, but, due to limitations in manufacturing methods, they are costly and inefficient. Vanadium alloys have low-temperature limitations due to thermal creep and helium embrittlement and are incompatible with hydrogen isotopes. The connection technology and mass production of silicon carbide composites are limited and unable to meet current application needs. Therefore, there is an urgent need to develop new materials to solve these issues.

HEAs possess unique properties due to their composition and structure, such as high-entropy effects, thermodynamically, severe lattice distortion structurally, slow diffusion kinetics, and cocktail effects in terms of performance. These properties give HEAs excellent stability and mechanical properties at both room and high temperatures, such as high hardness, high strength, resistance to high-temperature softening, and corrosion resistance. They are promising new materials that can meet the above requirements. As a result, HEAs have attracted significant attention in the field of nuclear materials research, from institutions such as Beijing University of Science and Technology, Oak Ridge National Laboratory in the United States, the University of Michigan, and Osaka University in Japan. This chapter will introduce the current research status on the irradiation behavior of HEAs.

2.3.1 Anti-Irradiation Mechanism of HEAs

Under irradiation conditions, metal materials are affected by irradiated particles (such as ions, neutrons, and electrons), causing changes in the lattice atoms, creating defects like interstitial atoms and vacancies. These point defects can further migrate, recombine, and cluster, leading to noticeable irradiation defects that impact the service performance. Due to the inability to conduct irradiation experiments within a nuclear reactor, research on the irradiation damage of HEAs typically uses ion irradiation and electron irradiation, which have higher irradiation doses and efficiency. Irradiation leads to many defects, the most significant being volume swelling, followed by irradiation hardening, embrittlement, phase transitions, irradiation creep, and high-temperature helium embrittlement. These are known as the five major challenges faced by materials in real fission and fusion reactor environments.

The key to structural materials in irradiation is excellent anti-irradiation performance, where the material does not undergo significant microstructural changes that drastically reduce mechanical properties, thus ensuring service safety. The main reason for volume swelling in metal materials under irradiation is that high-energy particles displace atoms from their original lattice positions, and these displaced atoms diffuse quickly in the metal matrix, while the vacancies formed at the lattice nodes move slower. As irradiation continues, vacancies gradually form clusters and aggregate into large voids, leading to volume swelling. To inhibit the formation and growth of these voids, traditional materials often introduce high-density defect traps to absorb irradiation-induced point defects, preventing their aggregation. For instance, small precipitates, nanocrystalline grain boundaries (GBs), and high-density dislocations are incorporated into radiation-resistant materials. Currently, research on irradiation defect formation mechanisms of HEAs involves both experimental and simulation studies.

Japanese scholars Nagase et al. and Egami et al. [10–13] were among the first to research the irradiation behavior of HEAs and discovered that HEAs have a "self-healing" function in terms of anti-irradiation mechanisms. Egami, through molecular dynamics simulations, found that multielement alloys have atomic-level stresses due to differences in atomic sizes. These stresses lead to partial amorphization, and irradiation accumulates heat, causing local melting and recrystallization in the alloy, reducing the dislocation defects in the process, this is the "self-healing" function. Nagase further explained this "self-healing" mechanism experimentally. Using electron irradiation on CoCrCuFeNi HEA, he observed that irradiation generated point defects (vacancies and interstitial atoms) in the alloy. In traditional single-element metals, these vacancies and interstitial atoms would aggregate into rings or voids, causing significant irradiation damage, and making pure metals prone to failure under irradiation. However, in HEAs, due to the high degree of lattice distortion, it is difficult for vacancies and interstitial atoms to form, and they cannot easily migrate to form rings or voids. Under irradiation, some regions of the HEA undergo partial ordering, similar to intermetallic compounds. Upon further irradiation, these ordered regions transition to amorphous states. Since the amorphous state is unstable, it spontaneously crystallizes, completing the entire "self-healing" process, which gives HEAs higher resistance to irradiation [12]. A comparison of the damage mechanisms of traditional alloys and the "self-healing" mechanism of HEAs under irradiation conditions is shown in Figure 2.12. Generally, the "self-healing" mechanism in HEAs under irradiation conditions is just one explanation for their radiation resistance. To further reveal the irradiation mechanisms, more detailed and systematic approaches are needed. Furthermore, for the nanocrystalline HEAs studied by the aforementioned authors, high defect concentration and defect traps (a mechanism commonly used to enhance traditional materials' irradiation performance) contribute to improved irradiation resistance. Therefore, further exploration of the irradiation resistance mechanisms of HEAs is necessary.

Jin et al. [14] studied the variation of volume swelling rate with the increase in the number of elements in multicomponent HEAs under ion irradiation conditions. They found that the more components the alloy contains, the lower its volume swelling rate under ion irradiation, as shown in Figure 2.13. This further reveals the role of entropy in the resistance of HEAs to ion irradiation. However, the exact role of entropy under ion irradiation conditions still requires further in-depth research.

In recent years, HEAs have received much attention, and their excellent properties under extreme conditions have been explored, leading to their consideration as

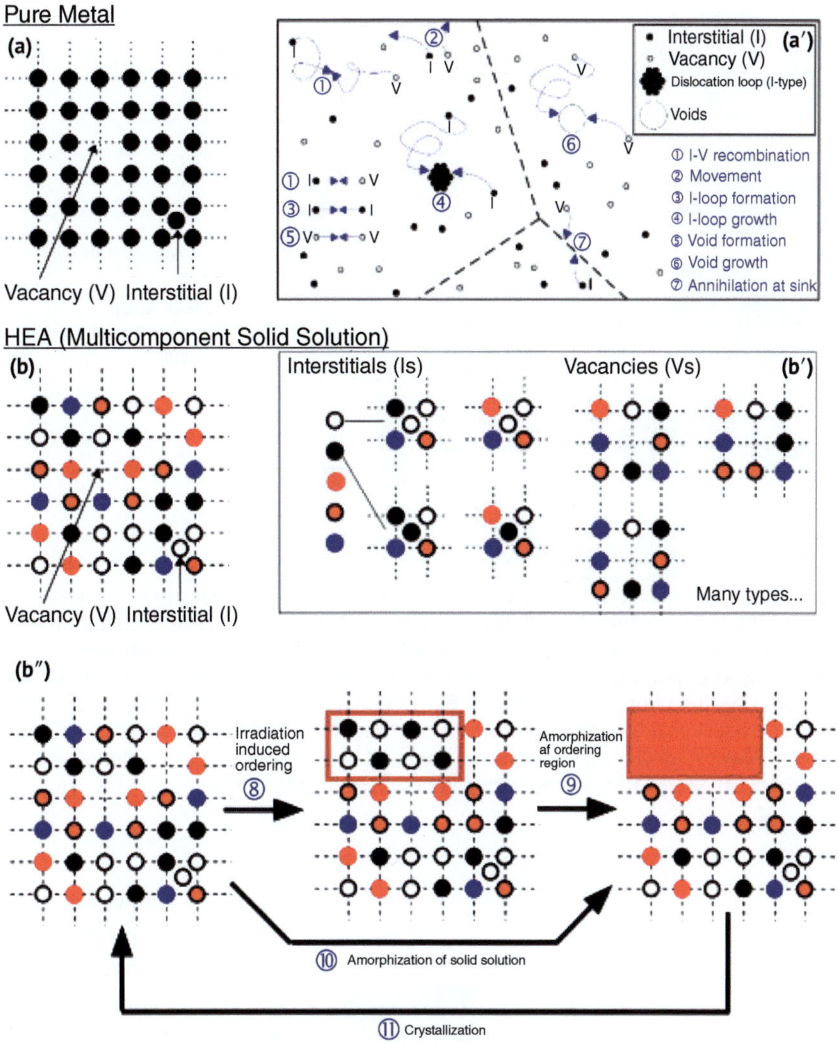

Figure 2.12 Schematic illustration of the simplest type of defects introduced by the electron knock-on effect under MeV electron irradiation and defect recovery processes. (a) (a′) Single-component crystal, namely, the pure metal. (b) (b′) Multicomponent solid solution in HEAs. (b″) (b′) The schematic illustration of the defect recovery via the amorphization–crystallization process [12]. *Source*: [12]/with permission of Elsevier.

candidate materials for fusion and fission conditions. However, modeling of irradiation hardening behavior in HEAs is still lacking. Fang et al. [15] investigated the hardening behavior of irradiated FeNiMnCr HEAs using the crystal plasticity theory, considering the effect of severe lattice distortion. The results show that the yield strengths of irradiated FeNiMnCr HEAs obtained by the hardening model at irradiation doses of 0, 0.03, and 0.3 dpa are in good agreement with the experimental data.

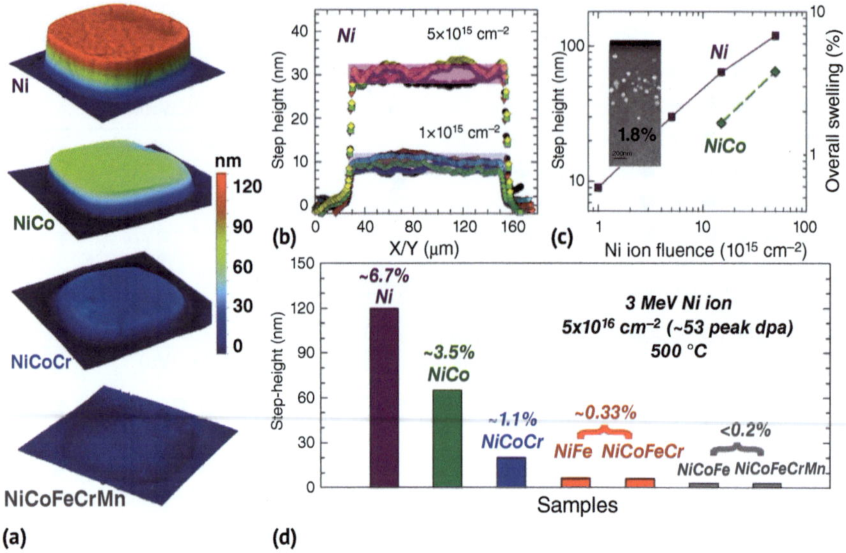

Figure 2.13 Schematic comparison of volume swelling rate of HEAs with an increasing number of principal elements. (a) Surface step measurements of Ni, NiCo, NiCoCr, and NiCoFeCrMn after ion irradiations. (b) Step-height profiles for Ni under low-fluence irradiations. The spots are the data, and the bands indicate the averages and uncertainties. (c) Overall swelling from step height as a function of ion fluence in Ni and NiCo. The inset is the cross-sectional transmission electron microscopy image for Ni irradiated at a fluence of 5×10^{15} cm^{-2}. (d) Comparison of step height and overall swelling among the seven materials [14]. *Source*: Reproduced with permission from [14]/Elsevier.

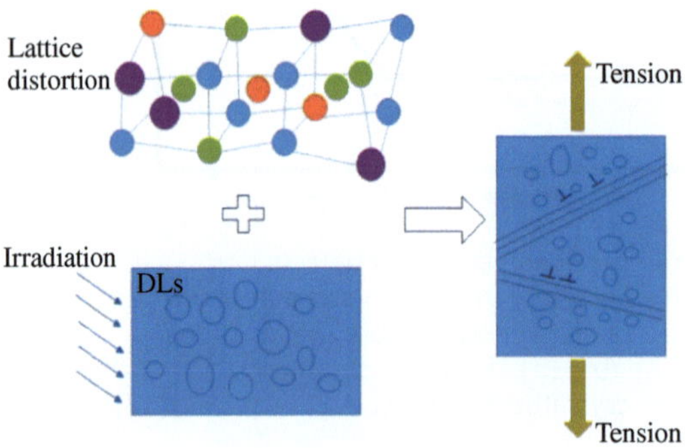

Figure 2.14 Hardening mechanism model [15]. *Source*: [15] / Elsevier / CC BY 4.0.

Figure 2.14 shows a model of the hardening mechanism of HEAs considering severe lattice distortions, dislocations, and irradiation-induced dislocation rings. The model and results can provide an effective theoretical way for yield strength modulation and alloying design of advanced HEAs to meet irradiation properties.

The enhanced helium ion (He^+) irradiation resistance and mechanical behavior of HEAs have attracted great research interest. Pu et al. [16] prepared ultrafine nanocrystalline $Al_{1.5}CoCrFeNi$ HEA thin films with single-phase structure. The effects of ultrafine nanocrystals on He evolution behavior, irradiation resistance, and mechanical behavior in $Al_{1.5}CoCrFeNi$ HEA films were investigated. The results show that (i) He clusters are preferentially distributed along GBs rather than within the grains due to the defect sink effect at the nanocrystalline boundaries. When the injection amount was 1×10^{17} cm^{-2}, no obvious He bubbles were observed in the peak damage region of the films even when the He concentration was as high as 8.50 at.%; (ii) some grains with limited irradiation damage retained their stability and integrity by eliminating the irradiation defects at the GBs and exhibited enhanced irradiation resistance in the ultrafine nanocrystalline HEA films; (3) by reducing the He cluster size, ultrafine nanocrystalline HEA films exhibit higher resistance to swelling than bulk coarse-crystalline HEA materials. The irradiated ultrafine crystalline HEA films have higher relative hardness (H/H0) compared to the coarse-crystalline HEA.

2.3.2 Evolution of Microstructure and Defects

Since alloys in the Ni, NiFe, NiCoFe, NiCoCr, and NiCoCrFeMn systems all have an FCC structure, these alloys are commonly used as research objects when exploring the impact of principal component and concentration on the irradiation performance of alloys. Zhang et al. used first-principles methods to compare the electronic structures of alloys in different systems, such as Ni, NiCo, NiFe, and CoCrFeNi, as shown in Figure 2.15. The study found that, as the number of principal elements

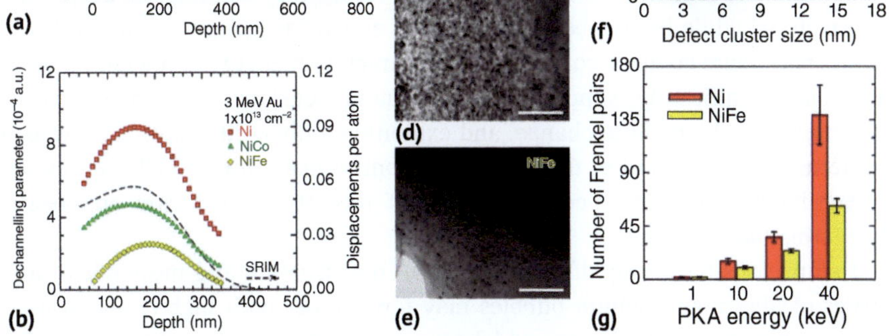

Figure 2.15 Comparison of irradiation performance of Ni, NiCo, and NiFe. (a) Rutherford backscattering spectra along (001) direction, showing random and virgin spectra for damage-free and fully amorphous states. (b) Ion irradiation damage distribution from channeling spectra and Kinchon–Pease model in Stopping and Range of Ions in Matter (SRIM). Binary alloys, especially NiFe, show improved irradiation resistance under 3 MeV Au ions at 1×10^{13} cm^{-2} fluence. Bright-field images of (c) nickel, (d) NiCo, and (e) NiFe irradiated with 3 MeV Au ions at 2×10^{13} cm^{-2} fluence; scale bars, 40 nm. (f) Defect cluster size distribution. (g) Comparison of Frenkel pairs in nickel and NiFe. Ion–solid interactions are stochastic, with error bars showing uncertainty from simulations [17]. *Source*: Reproduced with permission from [17]/Springer Nature / CC BY 4.0.

and the complexity of the composition increase, the mean free path of electrons becomes shorter, and both electrical conductivity and thermal conductivity decrease [17]. This reduces energy dissipation, extends the thermal peak time, and provides a longer time for point defect recombination, thus inhibiting the aggregation and growth of defects. As a result, HEAs exhibit excellent resistance to irradiation damage. Molecular dynamics calculations and corresponding experimental results also show that, as the compositional complexity of the alloy increases, the size of internal defects (such as stacking fault tetrahedra and dislocation rings) gradually decreases. Both experimental and simulation perspectives reveal that, in these alloy systems, as the compositional complexity increases, the alloy's resistance to irradiation becomes more pronounced.

Zhang [17] and Jin et al. [18] have conducted systematic work in this field and found that under irradiation, alloys with relatively complex compositions experience a gradual reduction in lattice damage, with a shallower depth of irradiation damage. From an experimental perspective, it has been shown that, in the above alloy systems, as the complexity of the principal components increases, the evolution rate of irradiation-induced defects slows down, leading to better irradiation resistance. Lu et al. [19] found that in Ni and NiCo alloys, vacancies accumulate in the shallow irradiation cascade collision zones, forming larger voids, while interstitial atoms migrate to deeper regions, forming dislocations. In contrast, in more complex principal element alloys such as NiFe, NiCoFe, NiCoFeCr, and NiCoFeCrMn, the shallow irradiation cascade collision zones form high-density dislocations, and a small number of vacancies accumulate deeper, forming small-sized voids, as shown in Figure 2.16. This is due to the different migration behaviors of point defects in different materials. In alloys with simpler compositions, interstitial atoms have lower migration energy, allowing them to move a greater distance in a single direction, exhibiting one-dimensional movement. As a result, they can migrate deeper into the material while leaving behind numerous excess vacancies that form voids. In alloys with more complex compositions, the migration energy of interstitial atoms increases significantly, suppressing their one-dimensional movement, causing the migration direction to change, and exhibiting a three-dimensional movement pattern. Therefore, most of the interstitial atoms remain in the shallower regions. When interstitial atoms encounter vacancies, recombination occurs, suppressing the formation of voids.

Helium ions have a significant impact on the structure and mechanical properties of alloys, and helium bubbles may form under both high-temperature or room-temperature irradiation, which can significantly reduce the mechanical properties of the alloys. Wang et al. investigated the formation of defects in Ni, NiCo, NiFe, NiCoCr, and NiCoFe alloys under He ion irradiation [20]. The study found that helium bubbles were produced in all of the aforementioned alloys, with bubble sizes in the following order from largest to smallest: Ni > NiCo > NiCoCr > NiFe > NiCoFe. This generally indicates that, as the complexity of the alloy composition increases, the helium bubble defects induced by irradiation become smaller. However, helium bubbles in NiCoCr alloy were relatively large, and the authors explained this from the perspective of vacancy exchange mechanisms:

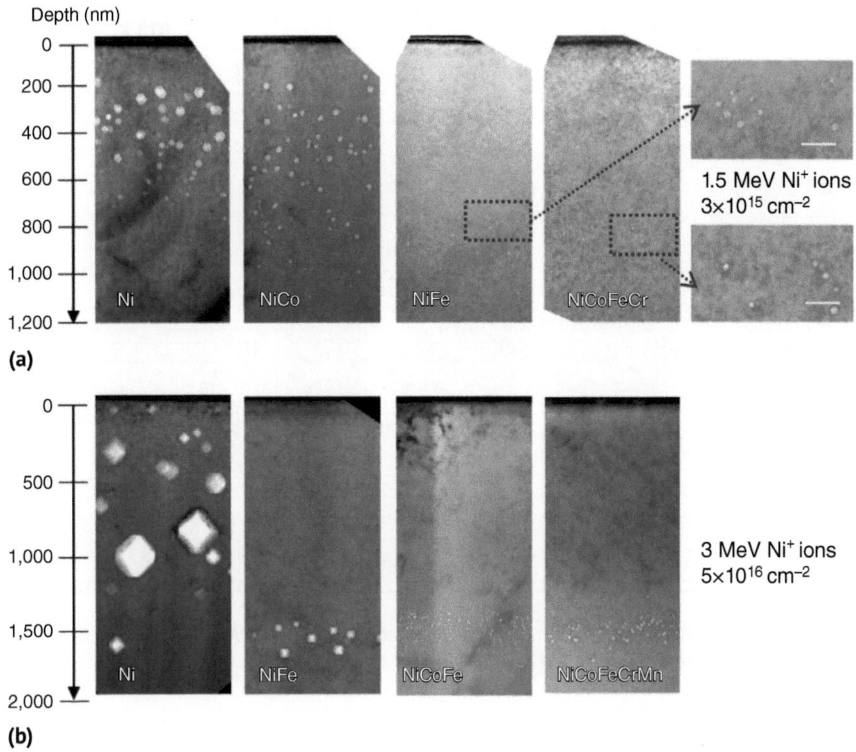

Figure 2.16 The pore distribution map caused by irradiation [19]. (a) Cross-sectional TEM images of nickel, NiCo, NiFe, and NiCoFeCr irradiated with 1.5 MeV Ni^+ ions to 3×10^{15} cm^{-2} at 773 K; scale bars in the zoomed images are 50 nm. (b) Cross-sectional TEM images of nickel, NiFe, NiCoFe, and NiCoFeCrMn irradiated with 3 MeV Ni^+ ions to 5×10^{16} cm^{-2} at 773 K. The ions enter the specimen from the top of the images. *Source*: Reproduced with permission from [19]/Springer Nature / CC BY 4.0.

vacancies migrate too quickly by exchanging with Cr, thus promoting the growth of helium bubbles. Chen et al. [21, 22] investigated the irradiation behavior of CoCrFeNi HEAs under He ions at temperatures ranging from 250 to 700 °C. The study found that, compared to pure Ni, the helium bubble size and volume fraction in CoCrFeNi alloys were both smaller, as shown in Figure 2.17. This indicates that CoCrFeNi alloys suppressed the formation of helium bubbles, exhibiting better irradiation resistance.

In the search for new materials that can withstand severe irradiation and mechanical extremes for advanced applications (e.g., fission and fusion reactors, space applications), the design, prediction, and control of advanced materials that go beyond the current material designs become critical. Atwan et al. [23] designed a nanocrystalline refractory high-entropy alloy system through a combination of experimental and simulation methods. As shown in Figure 2.18, the compositions evaluated in extreme environments and in situ electron microscopy showed high thermal stability and radiation resistance, grain refinement under heavy-ion irradiation was

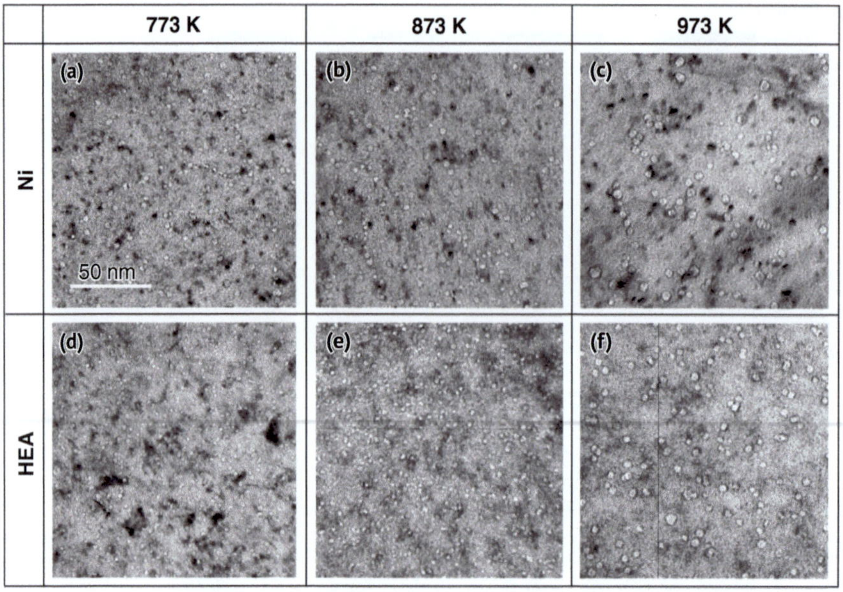

Figure 2.17 Comparison of helium bubbles in Ni and CoCrFeNi HEAs [22]. *Source*: Reproduced with permission from [22]/Elsevier.

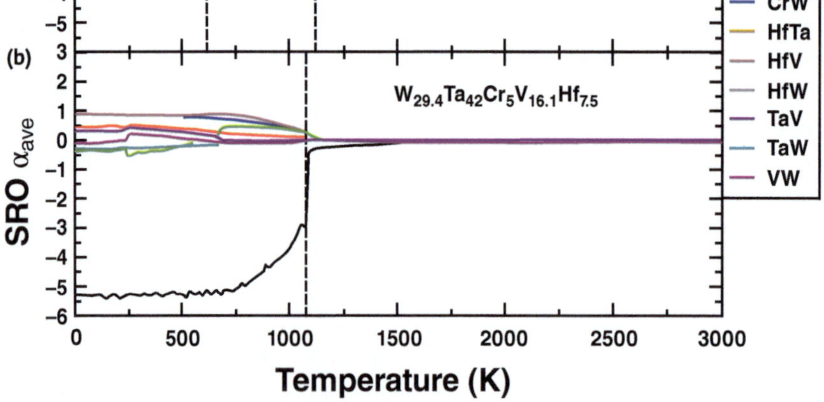

Figure 2.18 Chemical short-range ordering and order to disorder transition temperatures [23]. *Source*: [23] / Springer Nature / CC BY 4.0.

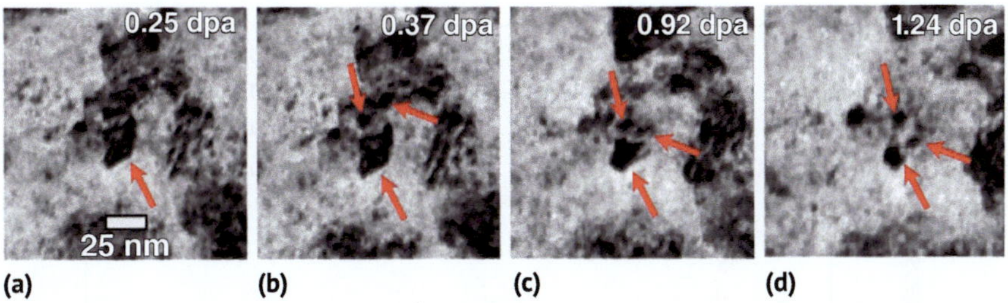

Figure 2.19 BF-TEM snapshots from in situ irradiated TEM videos tracking cascade-induced grain fragmentation during irradiation. (a) 0.25 dpa, (b) 0.37 dpa, (c) 0.92 dpa, and (d) 1.24 dpa [23]. *Source*: Reproduced with permission from [23]/Springer Nature/ CC BY 4.0.

observed, as well as resistance to dual-beam irradiation and helium ion implantation in the form of low-defect generation and evolution and undetectable grain growth, shown in Figure 2.19.

2.3.3 Mechanical Changes Caused by Irradiation

The above text primarily discusses the effects of irradiation on the microstructure and defects of structural materials, and how these microstructures and defects influence the material properties ultimately reflected in the mechanical performance. Current research on the mechanical properties of HEAs under irradiation mainly includes tests such as nanoindentation and nanopillar compression. Kumar et al. analyzed the post-irradiation hardness of CrFeMnNi HEAs under different irradiation conditions, as shown in Figure 2.20. The study found that at the same irradiation temperature, as the irradiation dose increased, the hardness of the alloy also increased [24]. However, at the same irradiation dose, an increase in irradiation temperature weakened the irradiation hardening of the alloy.

Sadeghilaridjani et al. [25] conducted a high-dose irradiation (40 dpa) study on the BCC-structured HfTaTiVZr refractory HEA at room temperature. They found that the irradiation hardening effect, measured by nanoindentation, was not significant, about 13%, which is much lower than the 50% observed in austenitic stainless steel. Moschetti et al. [26] studied the irradiation hardening effect of the HfTaTiNbZr refractory HEA and found about 10% irradiation hardening from nanoindentation tests, with little change in hardness, indicating good resistance to irradiation hardening.

Sadeghilaridjani et al. [25] also found that, after high-dose irradiation at room temperature, the nanopillars of the HfTaTiVZr HEA showed a more pronounced irradiation hardening effect (28%) in compression tests compared to nanoindentation, where the hardening effect was about 13%. Li et al. performed neutron irradiation experiments on the $Fe_{27}Mn_{27}Ni_{28}Cr_{18}$ HEA at room temperature [27]. The results indicated that the mechanical properties of $Fe_{27}Mn_{27}Ni_{28}Cr_{18}$ were similar to those of conventional austenitic stainless steel, and, with increasing irradiation

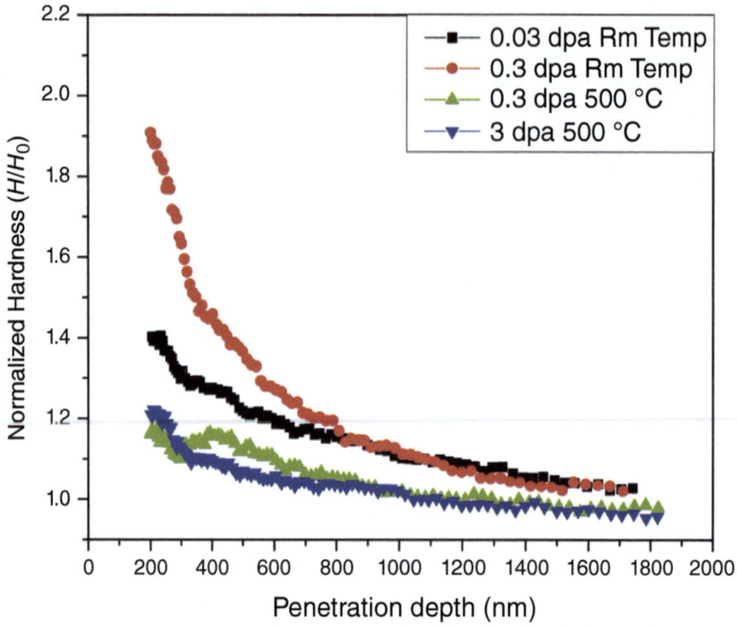

Figure 2.20 Comparison of hardness values of CrFeMnNi HEA under different irradiation conditions [24]. *Source*: [24] / with permission of Elsevier.

dose, phenomena such as irradiation hardening, embrittlement, and yield strength reduction occurred.

Moschetti et al. also investigated the irradiation hardening effect of the HfTaTiNbZr refractory HEA and found that the yield strength measured by microtensile tests changed by about 14%, demonstrating good resistance to irradiation [26]. It is worth noting that for the nanocrystalline samples obtained by high-pressure torsion, although significant irradiation hardening occurred, the fracture elongation did not show significant changes.

In addition to the focus on the hardness, tensile, and compression properties of HEAs after irradiation, there is the consideration of creep behavior. A study employed in situ TEM creep testing to investigate the creep behavior of CoCrFeMnNi HEA under irradiation conditions [28]. The study found that, at temperatures ranging from 23 to 500 °C, irradiation creep was inversely proportional to grain size. A comparison with traditional materials revealed that its behavior was closer to that of copper alloys.

2.3.4 High-Entropy Ceramics

Materials with complex compositions show promise for extraordinary structural robustness in extreme environments. The most common of these are HEAs, whose chemical complexity confers a harmonious combination of hardness, ductility, and thermo-elasticity. In contrast to these metal–metal bonding systems, the incorporation of ionic and covalent bonds has led to the discovery of high-entropy ceramics or compounds (HECs). These materials also offer excellent structural, thermal, and chemical robustness but with a wider variety of functional properties, enabling access to continuously controllable magnetic, electronic, and optical phenomena for functional applications in extreme environments.

Tunes et al. [29] proposed an extension of the HEA concept for the synthesis and characterization of novel HECs as emergency materials for the application of energy-containing particle irradiation in the environment. A novel carbide was synthesized in the five-member refractory system CrNbTaTiW by magnetron sputtering technique. The material exhibits nanocrystalline grains, a single-phase crystal structure, and a C content of 50 at.% or so. The irradiation response of the novel HEC at 573 K was evaluated using heavy-ion irradiation combined with in situ transmission electron microscopy and compared with the HEA within the system. As shown in Figure 2.21, the BF-TEM photographs from A to D and from E to H show the microstructural evolution of HEC and HEA, respectively, under irradiation at doses up to 10 dpa, with no displacement damage effects.

Surprisingly, HEC did not amorphize under the studied conditions. Xenon was injected into the material and nucleated in bubbles, but they are potential candidates for use in nuclear energy due to their smaller size compared to conventional nuclear materials. The above results suggest that HECs with suitable components may have excellent irradiation resistance. If a certain HEA can combine excellent high-temperature strength, toughness, good oxidation resistance, and low neutron absorption, it will have a promising application in nuclear materials. Researchers are exploring various strategies to enhance the radiation resistance of HEMs,

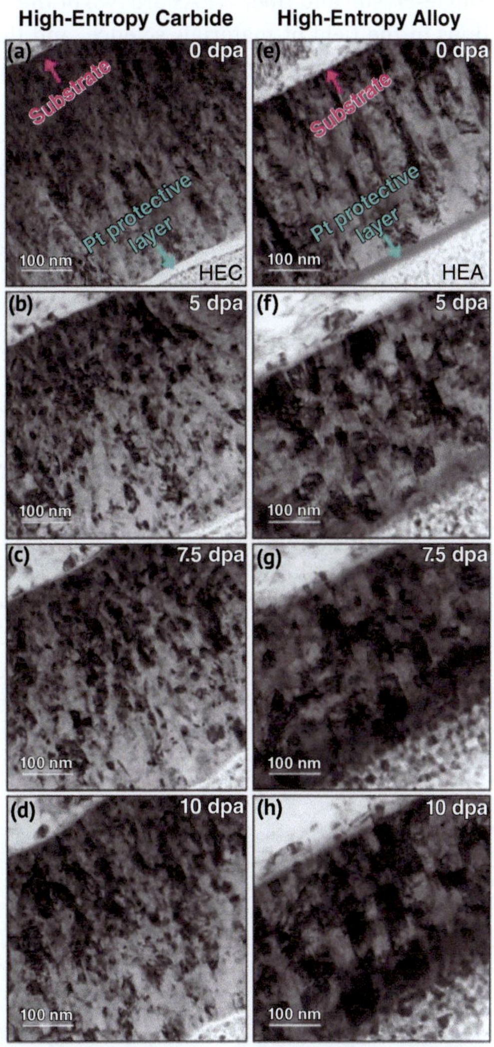

Figure 2.21 In situ TEM study of microstructural evolution under heavy-ion irradiation [29]. *Source:* Fang et al. *Source*: Reproduced with permission from [29]/Elsevier.

including alloying with radiation-tolerant elements, nano-structuring, and introducing defects or traps. By understanding the underlying mechanisms that control radiation damage in HEMs, scientists aim to develop materials that can withstand extreme radiation environments for extended periods.

2.4 Fiber and Single-Crystal HEMs

In addition to traditional polycrystalline HEMs, both fiber and single-crystal HEAs offer alternatives to meet the requirements of extreme conditions. Single-crystal and fiber HEAs are two of the current high-profile research directions in the field of materials science, and they show great potential in materials design and applications. While traditional alloy materials tend to be based on fixed atomic composition and crystalline structure, single-crystal and fiber HEAs could break this tradition by introducing multiple atomic compositions and complex crystalline structures, which provide new possibilities for material properties.

2.4.1 HEA Fibers

For HEAs, three-dimensional bulk alloys are typically considered rigid materials, while high-entropy fibers or thin ribbons with certain flexibility are considered flexible materials. Currently, most research on HEAs, especially in terms of mechanical properties, has focused on bulk materials, with relatively few studies on fibers and related forms. With the development and growing need for advanced materials, materials like high-strength fibers have become increasingly important. Traditional high-strength metal wires, such as pearlite steel wire, have been extensively studied due to their ultrahigh tensile strength and are widely used in critical infrastructure, such as large cable-stayed bridges, suspension bridges, and special rescue equipment. Figure 2.22 shows the application of HEA fibers. The ultrahigh tensile strength of pearlite steel wire is attributed to strengthening mechanisms such as GB strengthening, dislocation strengthening, solid solution hardening, and amorphous phase transformation.

(a) (b) (c)

Figure 2.22 Application scenarios of HEA fibers. (a) Crane sling. (b) Megasuspension bridge. (c) Aircraft carrier arresting cable. Generated with AI using Doubao.

However, due to the inherent design limitations, ultrahigh strength often leads to a significant reduction in plasticity, which in turn reduces its safety and reliability in service. HEAs or MPEAs, with their novel design concepts, have become strong candidates to overcome the strength-ductility trade-off. For wire materials, an increase in deformation leads to a significant reduction in diameter. When the characteristic size and working space of plastic deformation carriers, such as dislocations and twinning defects, are of similar magnitude to the external geometric dimensions or microstructural dimensions, the material may exhibit plastic deformation behavior that differs from that of macroscopic materials, such as scale effects. These unusual plastic behaviors are crucial for the development and application of microscale materials and thus have attracted widespread attention. For example, in tensile tests on polycrystalline silver wires, it was found that the tensile strength showed a clear dependence on wire diameter and grain size. In tensile tests on polycrystalline copper wires, the yield strength was found to be influenced by both grain size and the ratio of wire diameter to grain size, highlighting the increasing importance yet limited research on HE fibers or similar materials.

Figure 2.23 shows the main preparation methods for HEA wires, which include the methods of spinning, stretching, low-temperature rolling, and glass coating. The cast HEA samples for appropriate processing deformation and heat treatment are indispensable. The heat processing process not only reduces casting defects, such as shrinkage and segregation, but also appropriately regulates the alloy's microstructure and various properties of its microstructure. The plastic processing of alloys mainly includes rolling, forging, extrusion, and high-pressure twisting. At present, there are some HEA preparation processes applied to the rolling and forging process; forging can make the alloy organization more uniform, and grain refinement and cold rolling deformation can be very effective in improving the hardness of the alloy and even changing the phase composition of the alloy.

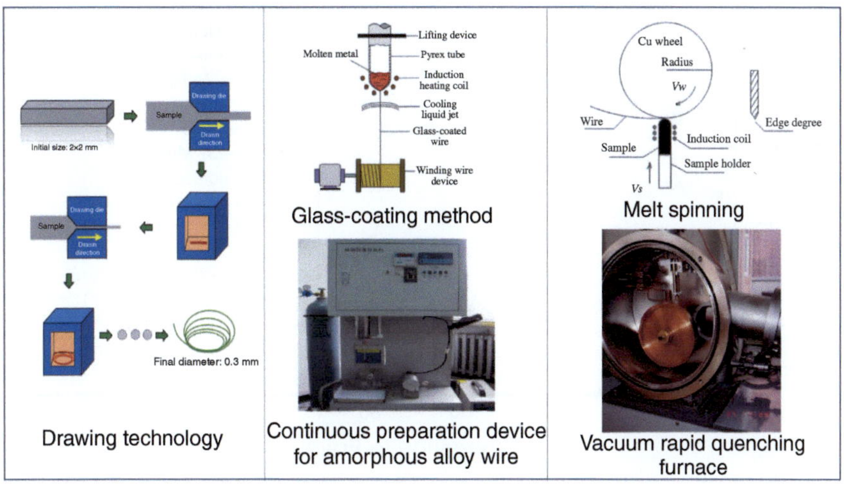

Figure 2.23 Common HEA wire preparation methods and their corresponding equipment.

Li et al. [30] improved the ductility of $Al_{0.3}CoCrFeNi$ high-entropy fibers by annealing the microstructure. A previous study on $Al_{0.3}CoCrFeNi$ high-entropy fibers reported a tensile breaking strength of 1200 MPa and an elongation of 8% at room temperature. $Al_{0.3}CoCrFeNi$ fibers with diameters of Φ1.0 mm and Φ1.6 mm were annealed at 900 °C at different times. The resulting microstructure is a fine FCC structure with a grain size of less than 3 μm, strengthened by dense NiAl-type ordered BCC (B2) precipitation. The fracture strength of these fibers at room temperature is greater than 900 MPa and the elongation exceeds 25%. The comparison of the mechanical properties between conventional material wires and bulk materials is shown in Figure 2.24.

In further studies, HEA fibers with $Al_{0.3}CoCrFeNi$ (atomic percent, at.%) composition in the diameter range of 1–3.15 mm were successfully prepared by hot drawing method to determine the tensile strength and plasticity of the fibers at 298 K (1207 MPa/7.8%) and 77 K (1600 MPa/17.5%) [31]. As shown in Figure 2.25, detailed TEM analysis showed that the increase in mechanical properties at 77 K (i.e., improved strength and ductility) was due to the change in the deformation mechanism from planar slip of dislocations to nanotwinning. These properties are favorable for low-temperature applications.

Zhou et al. [32] investigated the successful construction of a bionic bamboo fiber heterogeneous microstructure in the ultrafine crystalline EHEA $AlCoCrFeNi_{2.1}$ by multipass cold drawing and subsequent annealing treatment. This bionic

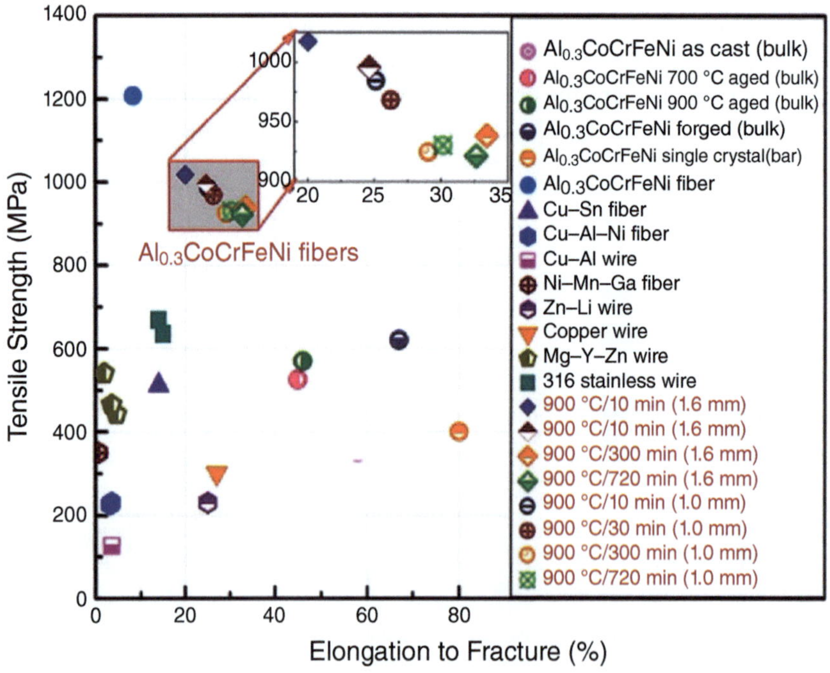

Figure 2.24 Comparison of mechanical properties of $Al_{0.3}CoCrFeNi$ HEA blocks, fibers, and other conventional alloy fibers [30]. *Source*: [30] / with permission of Elsevier.

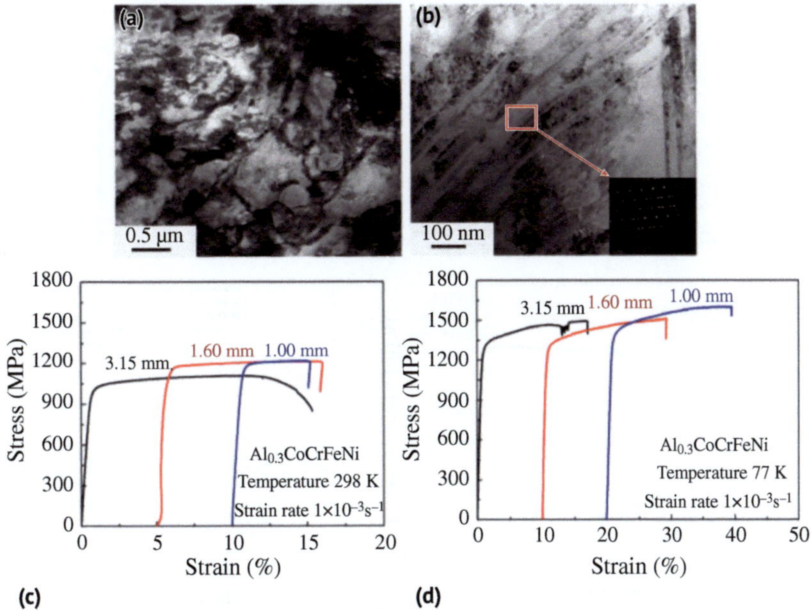

Figure 2.25 $Al_{0.3}$CoCrFeNi fibers with a diameter of 3.15 mm tested at (a) 298 K and (b) 77 K. TEM micrographs. Mechanical property curves of $Al_{0.3}$CoCrFeNi fibers at (c) 298 K and (d) 77 K [31]. *Source*: Reproduced with permission from [31]/Elsevier.

microstructure has hard B2 fibers embedded in a soft FCC matrix along the axial direction, achieving a synergistic effect of excellent strength and ductility. Stemming from this structural inhomogeneity, deformation incompatibility, and strain gradients between the B2 fiber and FCC matrix phases arise, leading to the formation of significant heterogeneous deformation-induced hardening. This hardening effect not only strengthens the FCC matrix, resulting in excellent yield strength, but also enhances high-strain hardening and delays strain localization of brittle B2 fibers, which in turn improves tensile ductility. The comparison of the microstructure and properties of biomimetic bamboo fiber EHEAs is shown in Figure 2.26.

Metallic wires with high strength and ductility at room and low temperatures have been sought after for engineering applications [33]. AlCoCrFe$Ni_{2.1}$ EHEA wires are carefully designed to achieve excellent synergistic effects in the multistage deep-drawing heat treatment process. Its organizational characteristics are similar to Figure 2.26a, which is a hard gradient-distributed B2 lamellae layer embedded in a soft FCC matrix. After medium-temperature annealing, the EHEA filaments successfully exhibited an excellent strong plastic accumulation balance at 293 K, with tensile strength and uniform elongation of about 1.85 GPa and 12.1%, respectively. Interestingly, when the temperature is lowered from 293 to 77 K, the EHEA filaments can even obtain very high strength, with an increase in tensile strength to 2.52 GPa, while the uniform elongation can be slightly increased to 14.3%, demonstrating excellent mechanical properties at low temperatures (Figure 2.27).

Reinforcing HEMs by fibrillation combines the high strength and toughness of the fibers with the unique properties of the HEM matrix, providing enhanced

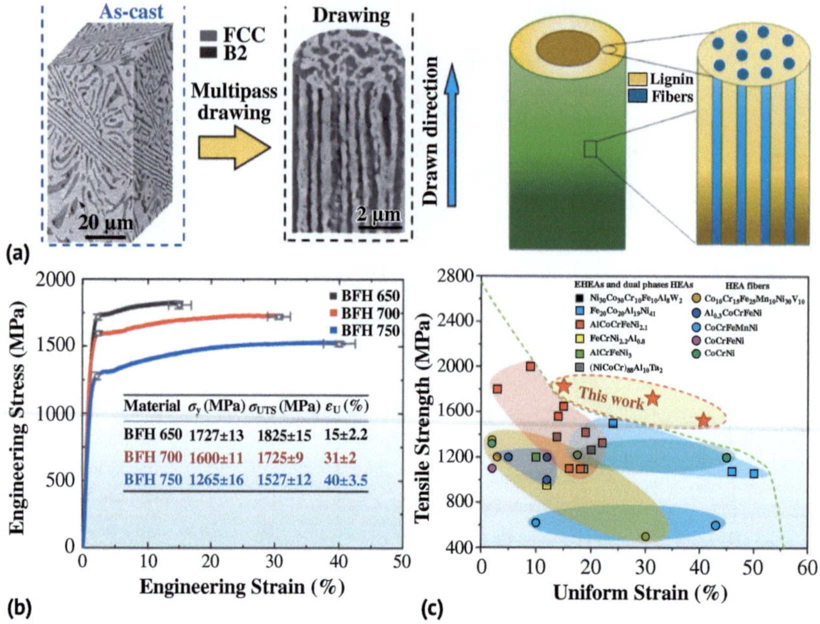

Figure 2.26 (a) Bionic bamboo fiber-like weave. (b) Mechanical properties of $AlCoCrFeNi_{2.1}$ EHEA. (c) Comparison of mechanical properties of common HEA fibers [32]. *Source*: Reproduced with permission from [32]/Elsevier.

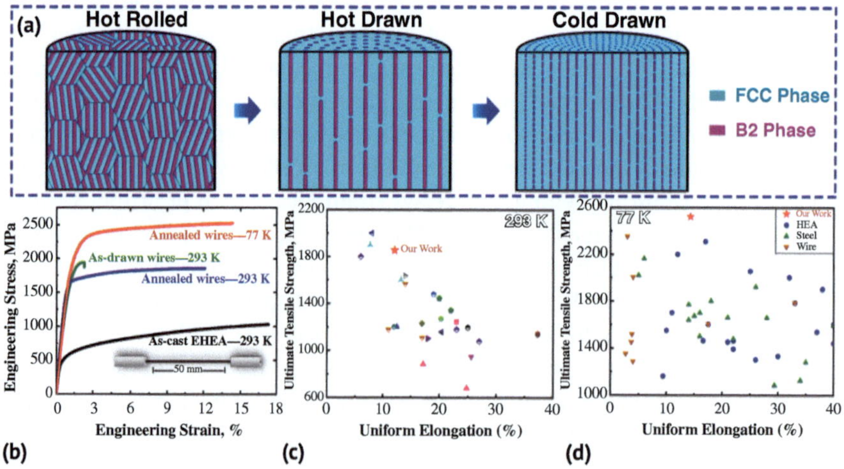

Figure 2.27 (a) Organizational structure. (b) Room and low-temperature mechanical properties of $AlCoCrFeNi_{2.1}$ EHEA [33]. *Source*: [33] / with permission of Elsevier.

mechanical properties and damage tolerance. These fiber filaments find applications in aerospace, automotive, and structural engineering where they can provide high-strength and tough lightweight materials with superior mechanical properties.

2.4.2 Single-Crystal HEA

Single crystals of HEAs exhibit a uniform atomic arrangement without GBs, a structure that is susceptible to degradation under extreme conditions. By growing single crystals of HEAs, researchers can eliminate the problems associated with GBs and utilize the full potential of these materials in high-temperature and radiation environments. However, there are still challenges in synthesizing large, defect-free single crystals of HEAs and optimizing their properties for practical applications. Feuerbacher et al. [34] used Bridgman's technique to grow single crystals of an isotropic FeCoCrMnAl HEA with a diameter of about 1 cm and a length of about 6.6 cm. The volume reaches more than 1 cm^3, which is more than that of the same system of Pradeep et al. [35] of 0.35 cm^3, which is a great improvement.

Severe mid-temperature embrittlement, as well as rapid softening and oxidation at high temperatures, are important problems faced by HEAs with polycrystalline structures, which greatly limits their application in modern industry. In this study, a novel Co-rich single-crystal HEA $Co_{41}Ni_{35}Al_{11.5}Ta_{2.5}Cr_4Ti_6$ (at.%). After isothermal aging at 1250 °C, the alloy exhibits an ultrahigh cubic L12-type nanoprecipitated phase volume fraction (~80%), which is higher than that of previous Co-based high-temperature alloys (Figure 2.28). More importantly, this HEA exhibits excellent mechanical properties at both room and elevated temperatures, and no mid-temperature embrittlement or rapid softening was observed. Notably, the yield strengths can still be maintained at ~800 and ~500 MPa when tested at 900 and 1000 °C, which is superior to other conventional single-crystal alloys. In addition, due to the formation of a dense sandwich-structured oxide layer, extraordinary antioxidant properties can be obtained at 1000 °C.

Comparison of single crystals and nanocrystals revealed that the plastic deformation along the [001] direction in single-crystal samples was dominated by nanotwinning and dislocation slip [36]. At an impact velocity of 1.0 km/s, Figure 2.29 shows the atomic-scale FCC → HCP structural transition and the formation of twin crystals. In Figure 2.29a, the dislocation slip produces many stacked laminar faults (SFs). These SFs for the HCP structure are represented by red atoms. In addition, blue, green, and white atoms represent BCC, FCC, and disordered structures, respectively. The FCC → BCC structural transition occurs almost simultaneously with the FCC → HCP structural transition, and the BCC structure is located at the front end of the SFs. With further deformation, twin crystals are formed in the region where SFs are aggregated, and the subsequent growth of the twin crystals produces twin bands, as shown in Figure 2.29c. Figure 2.29d is an enlarged picture of the atoms in the black dashed box in Figure 2.29c.

In summary, the development of HEMs and innovative structural designs holds great promise for addressing the extreme conditions and demanding requirements faced in a variety of fields. By utilizing the unique properties of HEMs and exploring novel preparation techniques, researchers aim to create materials and structures that can withstand the most demanding environments and facilitate new technologies and advances.

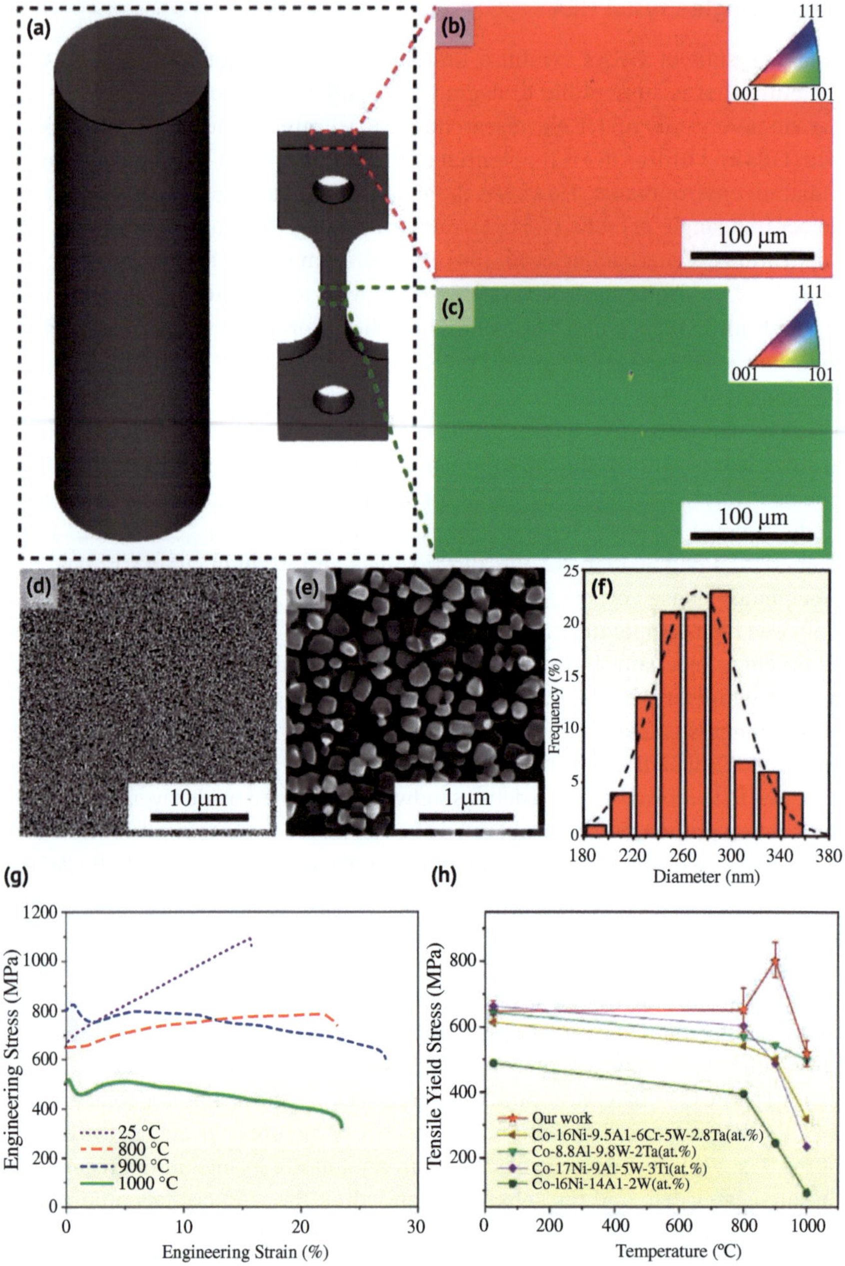

Figure 2.28 $Co_{41}Ni_{35}Al_{11.5}Ta_{2.5}Cr_4Ti_6$ HEA single-crystal samples and their mechanical properties: (a–c) Schematic diagrams of the tensile specimens. (d–e) Typical SEM images of the high-density nanoprecipitates. (f) Histograms of the statistical nanoprecipitates diameters of the L12 precipitates. (g) Engineered stress–strain curves of the test specimens. (h) Comparison of the yield strengths of single-crystal alloys with other single-crystal cobalt-based high-temperature alloys for yield strength comparison [35]. *Source*: Reproduced with permission from [35]/Elsevier.

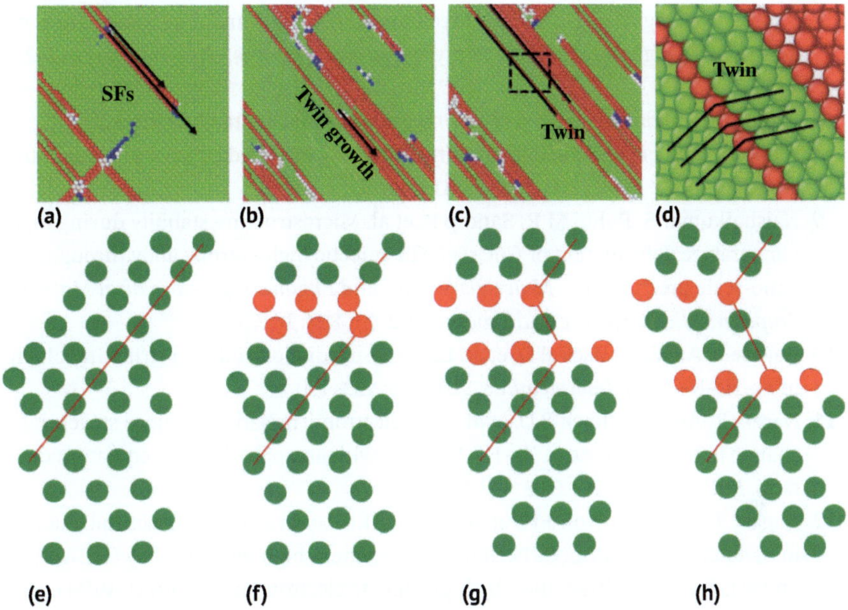

Figure 2.29 Snapshots of the structural transformation of single-crystal HEA along the [001] direction under impact loading: (a) stacked laminar dislocation, (b) twin growth, (c) twin bands, (d) increase of atoms in the black dashed rectangle in (c). (e)–(h) represent the twin formation process. The FCC, BCC, HCP, and twisted structures are green, blue, red, and white, respectively [36]. *Source*: Reproduced with permission from [36]/Elsevier.

References

1 Kuznetsov A V, Shaysultanov D G, Nikita S, et al. Tensile properties of an AlCrCuNiFeCo high-entropy alloy in as-cast and wrought conditions. *Materials Science and Engineering: A*, 2012, 533: 107.

2 Miracle D B, Senkov O N, Jon S, et al. Mechanical properties of $Nb_{25}Mo_{25}Ta_{25}W_{25}$ and $V_{20}Nb_{20}Mo_{20}Ta_{20}W_{20}$ refractory high entropy alloys. *Intermetallics*, 2011, 19: 698.

3 Xianzhe Z, Qingming Z, Mingzhen M, et al. Dynamic compressive properties and microstructural evolution of $Al_{1.19}Co_2CrFeNi_{1.81}$ eutectic high entropy alloy at room and cryogenic temperatures. *Materials & Design*, 2022, 219: 110724.

4 Otto F, Dlouhý A, Christoph S, et al. The influences of temperature and microstructure on the tensile properties of a CoCrFeMnNi high-entropy alloy. *Acta Materialia*, 2013, 61: 5743.

5 Jin M, Yu Q, Yan C, et al. A high-entropy alloy syntactic foam with exceptional cryogenic and dynamic properties. *Materials Science and Engineering A-structural Materials Properties Microstructure and Processing*, 2023, 876: 145146.

6 Ding M, Xu H, Jili D, et al. Improved strength and ductility of $Ti_2(CrFeNi)_{98}$ multi-principal element alloy at cryogenic temperature with deformation twins induced by BCC precipitates. *Materials Science and Engineering A-structural Materials Properties Microstructure and Processing*, 2023, 862: 144489.

7 Shubin W, Mingxu W, Da S, et al. Mechanical instability and tensile properties of TiZrHfNbTa high entropy alloy at cryogenic temperatures. *Acta Materialia*, 2020, 201: 517.

8 Guijun B, Youxiang C, Fei W, et al. Process study and characterization of properties of FerCrNiMnCo high-entropy alloys fabricated by laser-aided additive manufacturing. *Proc.SPIE*, 2018, 108130V.

9 Naishalkumar S, Rahul M R, Sandip B, et al. Microstructure stability during high temperature deformation of CoCrFeNiTa eutectic high entropy alloy through nano-scale precipitation. *Materials Science and Engineering A-structural Materials Properties Microstructure and Processing*, 2021, 824: 141793.

10 Nagase T, Anada S, Rack P D, et al. Electron-irradiation-induced structural change in Zr–Hf–Nb alloy. *Intermetallics*, 2012, 26: 122–130.

11 Nagase T, Anada S, Rack P D, et al. MeV electron-irradiation-induced structural change in the bcc phase of Zr–Hf–Nb alloy with an approximately equiatomic ratio. *Intermetallics*, 2013, 38: 70–79.

12 Nagase T, Rack P D, Noh J H, et al. In-situ TEM observation of structural changes in nano-crystalline CoCrCuFeNi multicomponent high-entropy alloy (HEA) under fast electron irradiation by high voltage electron microscopy (HVEM). *Intermetallics*, 2015, 59: 32–42.

13 Egami T, Guo W, Rack P D, et al. Irradiation resistance of multicomponent alloys. *Metallurgical and Materials Transactions A*, 2014, 45(1): 180–183.

14 Ke J, Chenyang L, Wang L M, et al. Effects of compositional complexity on the ion-irradiation induced swelling and hardening in Ni-containing equiatomic alloys. *Scripta Materialia*, 2016, 119: 65.

15 Qihong F, Jing P, Yang C, et al. Hardening behaviour in the irradiated high entropy alloy. *Mechanics of Materials*, 2021, 115: 103744.

16 Pu G, Liwei L, Ran A, et al. Outstanding radiation tolerance and mechanical behavior in ultra-fine nanocrystalline $Al_{1.5}$CoCrFeNi high entropy alloy films under He ion irradiation. *Applied Surface Science*, 2020, 516: 146129.

17 Yongfeng Z, Stocks G M, Ke J, et al. Influence of chemical disorder on energy dissipation and defect evolution in concentrated solid solution alloys. *Nature Communications*, 2015, 6: 8736.

18 Ke J, Hongbin B, Yongfeng Z. Ion irradiation induced defect evolution in Ni and Ni-based FCC equiatomic binary alloys. *Journal of Nuclear Materials*, 2016, 471: 193.

19 Chenyang L, Lei N, Nanjun C, et al. Enhancing radiation tolerance by controlling defect mobility and migration pathways in multicomponent single-phase alloys. *Nature Communications*, 2016, 7: 13564.

20 Xing W, Ke J, Di C, et al. Investigating effects of alloy chemical complexity on helium bubble formation by accurate segregation measurements using atom probe tomography. *Microscopy and Microanalysis*, 2019, 25: 1558.

21 Chen D, Tong Y, Li H, et al. Helium accumulation and bubble formation in FeCoNiCr alloy under high fluence He^+ implantation. *Journal of Nuclear Materials*, 2018, 501: 208–216.

22 Chen D, Zhao S, Sun J, et al. Diffusion controlled helium bubble formation resistance of FeCoNiCr high-entropy alloy in the half-melting temperature regime. *Journal of Nuclear Materials*, 2019, 526: 151747.

23 Osman E-A, Vo H T, Matheus A T, et al. A quinary WTaCrVHf nanocrystalline refractory high-entropy alloy withholding extreme irradiation environments. *Nature Communications*, 2023, 14: 2516.

24 Kumar N A P K, Li C, Leonard K J, et al. Microstructural stability and mechanical behavior of FeNiMnCr high entropy alloy under ion irradiation. *Acta Materialia*, 2016, 113: 230–244.

25 Sadeghilaridjani M, Ayyagari A, Muskeri S, et al. Ion irradiation response and mechanical behavior of reduced activity high entropy alloy. *Journal of Nuclear Materials*, 2020, 529: 151955.

26 Moschetti M, Xu A, Schuh B, et al. On the room-temperature mechanical properties of an ion-irradiated TiZrNbHf Ta refractory high entropy alloy. *JOM*, 2020, 72(1): 130–138.

27 Li C, Hu X, Yang T, et al. Neutron irradiation response of a Co-free high entropy alloy. *Journal of Nuclear Materials*, 2019, 527: 151838.

28 Jawaharram G S, Barr C M, Monterrosa A M, et al. Irradiation induced creep in nanocrystalline high entropy alloys. *Acta Materialia*, 2020, 182: 68–76.

29 Tunes M A, Fritze S, Osinger B, et al. From high-entropy alloys to high-entropy ceramics: the radiation-resistant highly concentrated refractory carbide (CrNbTaTiW)C. *Acta Materialia*, 2023, 250: 118856.

30 Li D, Gao M C, Hawk J A, et al. Annealing effect for the $Al_{0.3}$CoCrFeNi high-entropy alloy fibers. *Journal of Alloys and Compounds*, 2019, 778: 23–29.

31 Dongyue L, Chang-jiu L, Tao F, et al. High-entropy $Al_{0.3}$CoCrFeNi alloy fibers with high tensile strength and ductility at ambient and cryogenic temperatures. *Acta Materialia*, 2017, 123: 285.

32 Shiwei Z, Cuixia D, Huaxing H, et al. A remarkable toughening high-entropy-alloy wire with a bionic bamboo fiber heterogeneous structure. *Scripta Materialia*, 2023, 226: 115234.

33 Jinxi C, Tong L, Yan C, et al. Ultra-strong heavy-drawn eutectic high entropy alloy wire. *Acta Materialia*, 2023, 243: 118515.

34 Feuerbacher M, Étienne W, András K, et al. Single-crystal growth of a FeCoCrMnAl high-entropy alloy. *Materials Research Letters*, 2016, 5: 128.

35 Marshal A, Pradeep K G, Music D, et al. Combinatorial synthesis of high entropy alloys: introduction of a novel, single phase, body-centered-cubic FeMnCoCrAl solid solution. *Journal of Alloys and Compounds*, 2017, 691: 683–9.

36 Shanshan L, Genzhu F, Lijun X, et al. Shock-induced dynamic response in single and nanocrystalline high-entropy alloy FeNiCrCoCu. *International Journal of Mechanical Sciences*, 2023, 239: 107859.

3

Corrosion and Antioxidation

Di Ouyang[2], *Cheng Zhang*[1,*]

[1] *School of Materials Science and Engineering, State Key Laboratory of Materials Processing and Die & Mould Technology, Huazhong University of Science and Technology, Wuhan, P.R. China*
[2] *Department of Metallic Engineering, School of Materials Science and Engineering, Wuhan University of Technology, Wuhan, P.R. China*

*Corresponding author: czhang@hust.edu.cn

3.1 Introduction to Corrosion and Antioxidation Scenario

Corrosion and oxidation are pervasive and complex phenomena that significantly impact the performance, reliability, and longevity of materials used across various industries. These processes result from the interaction of materials with environmental elements, such as water, oxygen, acids, salts, and high temperatures, leading to the degradation of material properties. The consequences of corrosion and oxidation are particularly pronounced in industries like aerospace, automotive, energy, and marine, where components are exposed to extreme conditions. Understanding the mechanisms behind these phenomena and developing strategies to mitigate their effects is crucial for the sustainability and efficiency of modern technology.

Corrosion is a material degradation process that occurs when a material, usually a metal, reacts chemically with its environment, often leading to the formation of corrosion products. This reaction can take many forms, including uniform corrosion, pitting, crevice corrosion, galvanic corrosion, and stress corrosion cracking. Uniform corrosion is the most common and occurs evenly across the surface while pitting corrosion results in a localized attack, forming holes or pits. Crevice corrosion takes place in confined spaces where the material is exposed to stagnant solutions, whereas galvanic corrosion occurs when two different metals are in electrical contact in a conductive environment, accelerating the corrosion of the less noble material. The impact of corrosion extends far beyond material loss; it affects infrastructure, transportation systems, and

Functional High Entropy Alloys and Compounds, First Edition. Edited by Yong Zhang, Yonggang Yao and Haojie Liu.

manufacturing processes, leading to substantial maintenance costs, operational downtime, and safety risks. The economic burden of corrosion is staggering, with studies estimating that corrosion-related costs represent a significant percentage of global GDP. For example, in the United States alone, the annual cost of corrosion is estimated to exceed $400 billion. These figures highlight the urgent need for effective corrosion prevention and mitigation strategies, as the continued degradation of materials underpins a wide range of technological and industrial limitations.

Oxidation, which specifically refers to the reaction between a material and oxygen, is a form of corrosion that leads to the formation of oxide films on the surface of metals. In many cases, the oxide layer can serve as a protective barrier, slowing further degradation. However, in many metals, particularly at elevated temperatures, this oxide layer may not be protective and can even accelerate material failure. This is particularly problematic in environments such as gas turbines, engines, or reactors, where materials experience both high temperatures and reactive atmospheres. A particularly relevant example is the behavior of stainless steel in high-temperature, high-pressure steam environments. Stainless steel forms a chromium oxide layer, which provides resistance to corrosion; however, under extreme conditions, such as in power plants or nuclear reactors, the oxide layer can degrade, leading to accelerated corrosion. This has led to numerous studies focused on enhancing the protective nature of oxide layers through material composition modifications. One such study demonstrated that incorporating high levels of nitrogen into the composition of stainless steel improves the formation of a more stable and adherent oxide layer, significantly increasing its corrosion resistance.

Antioxidation, or corrosion inhibition, involves a variety of approaches designed to protect materials from degradation by preventing or slowing down the oxidation process. One of the most widely used methods is the application of protective coatings, which can act as physical barriers to oxygen, moisture, and corrosive chemicals. These coatings can range from simple organic paints to complex composite materials, with research continuously improving their durability and effectiveness. Another strategy involves alloying, where elements such as chromium, nickel, or zinc are added to metals to enhance their oxidation resistance. For instance, stainless steel contains chromium, which forms a passive oxide layer that protects the underlying metal from further corrosion, a property that has made it invaluable in applications ranging from architecture to medicine.

In addition to these traditional methods, advances in nanotechnology have provided novel solutions to corrosion prevention. Nanomaterials, such as nanoparticles and nanotubes, are being integrated into coatings to create ultra-durable, self-healing, and highly impermeable barriers. These materials not only enhance the mechanical properties of the coatings but also provide additional functionalities, such as the ability to repair damage automatically when exposed to corrosive environments. The incorporation of nanomaterials into corrosion-resistant coatings has shown remarkable potential in enhancing the longevity of materials under extreme conditions, opening new possibilities for their use in fields like aerospace, marine, and energy production.

Furthermore, the role of corrosion inhibitors—chemicals that, when added to an environment, reduce the rate of corrosion—has been a topic of extensive research. These inhibitors function by forming a protective layer on the metal surface or by disrupting the electrochemical processes that drive corrosion. They are commonly used in applications such as cooling systems, oil pipelines, and concrete reinforcement, where exposure to corrosive agents is inevitable. The development of green inhibitors, which are environmentally friendly and nontoxic, has become a priority as industries seek sustainable and safer alternatives to conventional chemical inhibitors.

High-entropy alloys (HEAs) typically exhibit excellent corrosion resistance, especially in extreme environments such as marine or acid-based environments. This is because the high-entropy effect can stabilize the oxide layers on the surface, reducing the likelihood of corrosion. Many high-entropy materials (HEMs), especially those with transition metals and refractory elements, exhibit a remarkable resistance to oxidation. The oxide layer that forms on the surface of these materials is usually protective and prevents further oxidation. This is particularly important in extreme environments where oxidative conditions, such as high-temperature combustion or exposure to reactive gases, could degrade materials quickly.

3.2 Anticorrosion Design in High-Entropy Materials

The corrosion behavior of HEAs is influenced by their complex compositional structure, which can lead to distinct electrochemical properties compared to conventional alloys. Studies have shown that, in some cases, HEAs exhibit superior resistance to corrosion in aggressive environments, such as acidic or saline solutions, compared to traditional alloys like stainless steel. For instance, the addition of elements like chromium, which is known for its role in forming protective oxide layers, can enhance the corrosion resistance of HEAs. However, the corrosion resistance of HEAs is not uniform across all compositions; it can vary significantly depending on the specific elemental combination and the microstructure that forms as a result of the alloying process. For example, the corrosion resistance of an HEA may be dictated by the formation of a passivating oxide layer that protects the underlying material from further degradation, similar to the mechanism observed in stainless steel. In some HEAs, however, the formation of heterogeneous phases can lead to localized galvanic corrosion, where different phases of the material exhibit varying electrochemical behaviors.

For HEAs, the random arrangement of multiple elements in solid solutions results in a particular locally disordered chemical environment, which is expected to lead to unique corrosion-resistant properties [1]. The HEAs systems, which contain passivating elements, such as Cr, Ni, and Mo, revealed equivalent or superior corrosion-resistant properties compared with conventional alloys [2], as shown in Figure 3.1. Furthermore, the exceptional properties, such as the outstanding strength-ductility synergy, improved fatigue resistance, high fracture toughness, and strong thermal stability, make HEAs attractive in many extreme service environments such as in the nuclear, turbine, and aerospace industries [3]. Based on

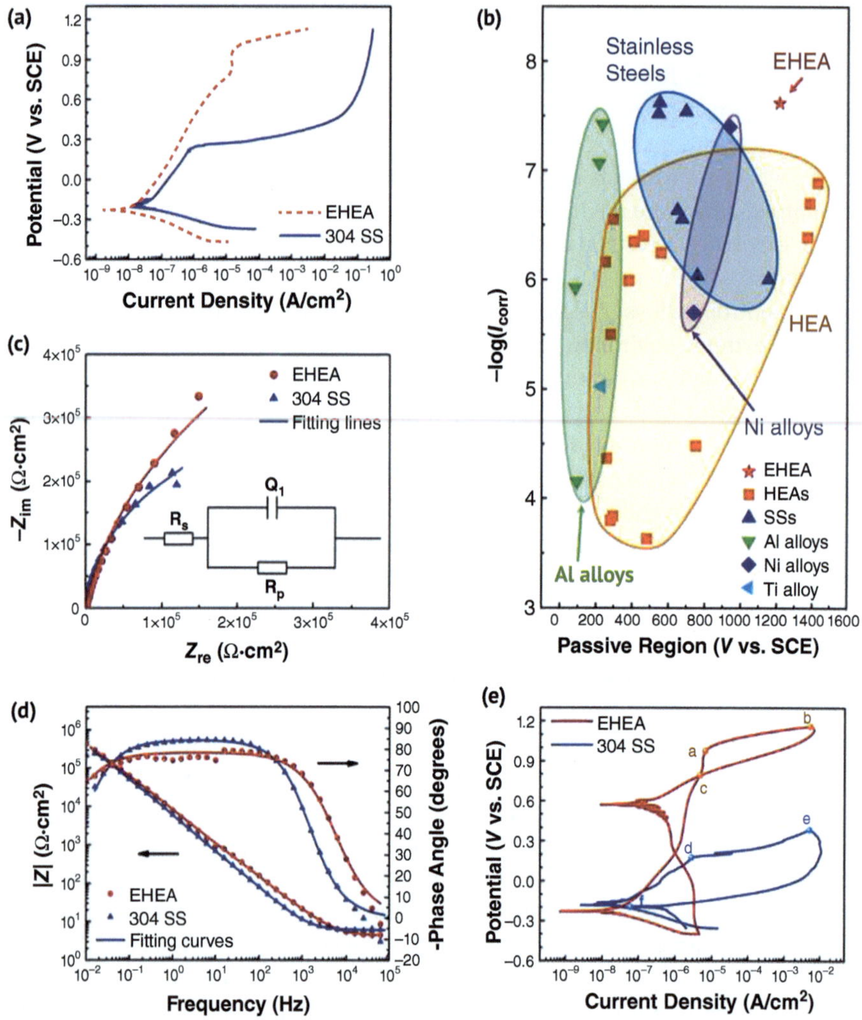

Figure 3.1 The comparison of corrosion-resistant properties of eutectic HEA with other alloys in 1M NaCl solution. *Source*: [2]/with permission of Elsevier.

the research progress of the corrosion-resistant HEAs, this section summarizes the effects of alloying elements and processing methods on the corrosion-resistant properties of HEAs.

3.2.1 Effect of Alloying Elements

The corrosion-resistant properties of HEAs usually depend on the alloying elements. Different alloying elements can affect corrosion resistance by altering the elemental distribution and phase structure, further affecting the dissolution kinetics of passive film properties [4]. The effect of typical alloying elements on the corrosion of HEAs is presented herein.

Chromium (Cr). Chromium (Cr) is a primary element in stainless steels (Fe–Cr, Ni–Cr, Fe–Cr–Ni alloys, etc.), because of its features of high potential and passivation tendency. The addition of Cr makes the alloy stainless by forming a protective Cr-based oxide passive film on the alloy surface. Commonly, the Cr concentration must be higher than 12 at.% in the case of Fe–Cr, Ni–Cr, and Fe–Cr–Ni alloys for stainless behavior, and the increase in content will improve the corrosion resistance of stainless steels [5]. The Cr concentration in Cr-containing HEAs is typically higher than 12%, and it is therefore undisputed that these HEAs have better corrosion resistance. However, in some HEAs, things are different [6]. First, other elements dispersed into the chromium oxide film may reduce the protective performance of the passive film. Thus, the surface films formed on HEAs need to be extensively characterized to estimate their properties and their correlation with the alloy composition. On the other hand, with more Cr addition, the alloy may have an interdendritic microstructure (such as in AlCoCrFeNi system), with a Cr-depleted dendritic region and Cr-rich interdendritic region. It was proposed that the Cr-depleted regions become anodic and the Cr-rich regions become cathodic during exposure to an aqueous solution, resulting in the selective dissolution of the Cr-depleted regions in the HEA.

Nickel (Ni). Nickel (Ni) is also an important element used in the manufacturing of stainless steel. An important role for Ni addition is to stabilize the austenite phase, accompanied by a high Cr content. Ni can be alloyed with a high content of metals known to improve corrosion resistance, such as Mo, Cr, and Cu, while retaining its ductile face-centered cubic (FCC) structure [7]. Qiu and Liu [8] investigated the effect of Ni addition on the corrosion resistance of the $Al_2CrFeCoCuTiNi_x$ ($x = 0$, 0.5, 1, 1.5, and 2) HEA in 3.5 wt.% NaCl and 1M NaOH and found that there is critical Ni content beyond which Ni has a detrimental impact on the corrosion of the HEA. It is verified that Ni may combine with Al to form the (Al, Ni)-rich B2 phase, which may also deteriorate the corrosion resistance of Ni-containing HEA. In addition, the inferior corrosion resistance of the Ni-containing HEAs is regarded as related to the small atomic radius of Ni, compared to the other constitutional elements. This would lead to a large lattice distortion in alloys with high Ni content and thereby could significantly influence the alloy microstructure, which will thus influence the corrosion resistance of such Ni-containing HEAs [9].

Molybdenum (Mo). In austenitic stainless steels, molybdenum (Mo) is introduced to improve the passivity and pitting resistance, due to the generation of protective MoO_4^{2-} based passive film or MoO_4^{2-} species [10]. However, in HEAs, Mo tends to form a Mo and Cr-rich sigma (σ) phase, which could be detrimental to the corrosion resistance of such Mo-containing HEAs if such a σ phase is present [11]. Clayton et al. [12] clarified that the Mo addition improves the pitting resistance as well as the self-repairing capabilities of the passive film in the $Co_{1.5}CrFeNi_{1.5}Ti_{0.5}Mo_x$ HEA. It was proposed that the MoO_4^{2-} species serves to inhibit the pitting of the Mo-containing HEA and promotes the formation of the Cr-based oxide inner barrier layer, thus resulting in the improved corrosion resistance of such Mo-containing HEA.

Titanium (Ti). Titanium (Ti) is nominally a passive metal in most aqueous environments, due to its spontaneously developing a protective TiO_2-based film on the alloy surface [13]. The addition of Ti to HEAs can also improve corrosion resistance when Ti is incorporated into the surface film. However, Ti tends to form several intermetallic compounds with the other elements in HEAs, such as the Laves phase, sigma phase, and R phase, which makes the Ti-containing HEA more heterogeneous and attendant ramifications for corrosion resistance [14]. Qiu et al. [15] investigated the effect of Ti content on the corrosion resistance of $Al_2CrFeNiCoCuTi_x$. They found that the Ti-containing HEAs had lower i_{corr} values than the Ti-free HEA in 0.5M HNO_3, and the i_{corr} values decreased with the increasing Ti content. This detrimental effect was attributed to the formation of intermetallic compounds in the HEAs. Overall, Ti mitigates the general corrosion of HEAs, but the presence of a Ti-containing phase may contribute to localized corrosion.

Aluminum (Al). Aluminum (Al) is a light metal with a lower galvanic series compared to the other elements. Thus, it is plausible that Al atoms may be preferentially released from a solid solution containing more noble elements during exposure to aqueous environments [16]. The corrosion rate of HEAs with Al may therefore be expected to increase with increasing Al content in depassivating conditions. Li et al. [17] investigated the corrosion characteristics of the alloy $FeCoNiCrCu_{0.5}Al_x$ ($x = 0.5$, 1, and 1.5) in both 0.5M NaCl and 0.5M H_2SO_4 solution. It was observed that the $Al_{0.5}$ and $Al_{1.0}$ alloys were found to have higher corrosion current density (i_{corr}) values than the $Al_{1.5}$ alloy in the 0.5M NaCl solution. However, the corrosion resistance increased in the order $Al_{1.0} > Al_{0.5} > Al_{1.5}$ in 0.5M H_2SO_4 solution, with the $Al_{1.0}$ alloy having a lower i_{corr} than the $Al_{0.5}$ alloy. They found that the phase structure of HEA transformed from a single FCC phase to FCC + body-centered cubic (BCC) duplex phase, further to a single BCC phase with the increase of Al content. The BCC phase was more corrosion-resistant than the FCC phase in the chloride and acidic solutions. The comparatively inferior corrosion resistance of the $Al_{1.5}$ alloy was attributed to the presence of galvanic corrosion among the BCC and FCC phases.

Copper (Cu). Copper (Cu) possesses the positive binary mixing enthalpy with the other constitutional elements (namely, Fe, Co, Ni, and Cr) in HEAs, which promotes elemental segregation, forming a Cu-rich interdendritic phase and a Cu-lean dendritic phase in the alloy matrix [18]. Thus, from the reports to date, it appears that Cu is detrimental to the corrosion resistance of HEAs. Ren et al. investigated the effect of Cu on the corrosion of the CuCrFeNiMn system in 1M H_2SO_4 [19]. They found that the $CuCr_2Fe_2Ni_2Mn_2$ alloy with low Cu content and minimal elemental segregation had the highest corrosion resistance, whereas the $Cu_2CrFe_2Ni_2Mn_2$ alloy with the higher Cu content and more substantial elemental segregation had the poorest corrosion resistance. It was proposed that the alloys experienced general corrosion and localized corrosion, and the addition of Cu favors elemental segregation, thus increasing the corrosion rate of Cu-rich HEAs.

Other Elements. Elements such as B [20], Sn [21], W [22], Nb [23], and Ta [24] can also influence the corrosion resistance of HEAs, by altering the phase structure or properties of the passivating film. Commonly, the element segregation will deteriorate the corrosion resistance of HEA. In addition, the role of a certain element in an HEA is usually influenced by other elements, as well as the resulting changes in the HEA's microstructure. This makes the impact on the corrosion resistance of HEA even more complex. Currently, research in this area is still relatively lacking.

Each alloying element plays a distinct role in the corrosion behavior of HEAs, and their effects are often interdependent. The elemental distribution, phase structure, and electrochemical properties of the material are determined by the complex interactions between these elements. Therefore, tailoring the composition of HEAs to achieve optimal corrosion resistance requires a careful balance of alloying elements, guided by both experimental studies and computational simulations. Understanding these effects is critical for designing HEAs with superior corrosion-resistant properties for specific applications.

3.2.2 Effect of Processing

The corrosion resistance of HEAs is not only determined by their composition but also significantly influenced by the processing methods used during their fabrication. Processing methods impact the microstructure, phase distribution, grain size, and surface characteristics of HEAs, all of which play a crucial role in determining their electrochemical behavior and corrosion resistance. Properly optimized processing techniques can enhance the homogeneity of elemental distribution and promote the formation of protective passive films, while suboptimal methods may introduce defects or undesirable phases that degrade corrosion performance.

Aging Treatments. Aging treatments are widely used in structural materials, by tailoring their microstructure to achieve better physical and mechanical properties. Also, this method has been used in the enhancement of corrosion resistance of HEAs. Lin and Tsai [25] investigated the effect of aging treatment on the corrosion resistance of $Al_{0.5}$CoCrFeNi alloy. They found that aging at 650 °C for 24 h can significantly decrease the corrosion rate of HEA in 3.5 wt.% NaCl solution, exhibiting the lowest corrosion current density, highest pitting potential, and noble corrosion potential. During the corrosion process, Cl^--containing electrolytes preferentially attacked the Al–Ni-rich phases, and the FCC matrix remained essentially unaffected. Aging treatment, in this case, weakened the heterogeneity of microstructure by decreasing the precipitation of Al–Ni-rich phases and consequently received better corrosion resistance. In another work, they found that, compared to aging at lower temperatures (350–1100 °C), aging at higher temperatures (1250–1350 °C) is more beneficial in improving the corrosion resistance of $Cu_{0.5}$CoCrFeNi HEA. It was proposed that aging at higher temperatures results in the dissolution of the Cu-rich FCC phase into the FCC matrix. This therefore reduces the ultimate Cu segregation in the HEA and improves corrosion resistance.

Anodization. Passive elements can produce a protective film on their surface. In the case of elements with low passivation tendencies or limited passivity, anodization is often industrially applied. The HEAs commonly comprise one or more of the

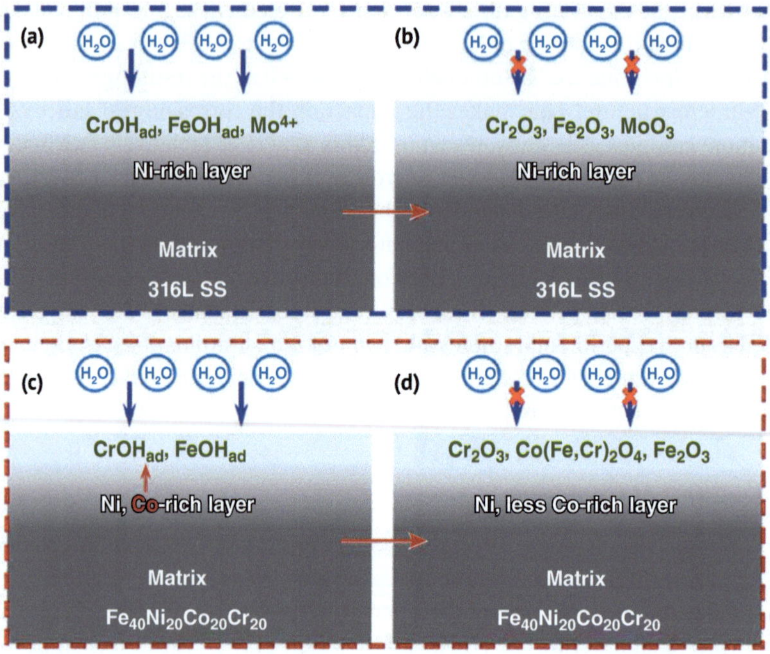

Figure 3.2 Schematic figures of the passivation process for the HEA and the 316 stainless steel. *Source*: [26] / with permission of Elsevier.

passive elements. To improve the corrosion resistance of HEAs, anodization was often performed with the view of developing a superior protective film on the alloy surface. Lee et al. [26] carried out anodization treatment to the $Al_xCrFe_{1.5}MnNi_{0.5}$ HEAs in 15% sulfuric acid solution (700 mV_{SHE} for a period of 1800 s). They found that a layer of Cr_2O_3 thin film formed on the surfaces of the anodization-treated HEAs and significantly enhanced the corrosion resistance of these alloys in 0.1M HCl solution by two orders of magnitude, compared to the untreated alloys. Thus, anodization treatment to either enrich a passive metal on the HEA surface or develop a uniform passive film on the HEA may be a useful approach to increase the corrosion resistance of HEAs (Figure 3.2) [27].

3.3 High-Temperature Oxidation Resistance

High-temperature oxidation is a critical degradation process in which metallic atoms on an alloy's surface react with oxygen to form oxides when exposed to elevated temperatures [28]. In engineering applications, especially in energy and transportation sectors, materials often operate under extreme thermal conditions where oxidation resistance is paramount. For these applications, alloys must form a continuous, dense, and adherent oxide film that acts as a barrier, preventing further oxygen penetration and protecting the underlying material from degradation. HEAs have emerged as promising candidates for such applications due to their unique compositional complexity and microstructural stability, which contribute to superior high-temperature oxidation resistance.

Oxidation of HEMs, particularly at elevated temperatures, is another critical aspect of their performance. The oxidation behavior of HEAs can be more complex than that of traditional alloys due to the interplay between the different elements in the alloy. In high-temperature environments, HEAs may form multiphase oxide layers, with some elements contributing to the formation of protective oxide scales and others potentially promoting the growth of porous or unstable oxide phases. The formation of these oxide layers is often highly dependent on the ratio of alloying elements, with elements like aluminum, chromium, and silicon contributing to the formation of dense, stable oxides that provide a protective barrier against further oxidation. For instance, the addition of aluminum in some HEAs has been shown to improve high-temperature oxidation resistance by promoting the formation of a stable, adherent Al_2O_3 oxide scale. On the other hand, elements like copper or nickel may form less protective oxide layers that can lead to material degradation at elevated temperatures.

The oxidation behavior of HEAs is governed by the interplay of their constituent elements and the resulting microstructural evolution during high-temperature exposure. Unlike conventional alloys, HEAs are composed of multiple principal elements, which can lead to the formation of complex oxides. These oxides may provide enhanced protection compared to simple binary or ternary alloys. For example, the incorporation of elements such as aluminum, chromium, and silicon is known to significantly improve the high-temperature oxidation resistance of HEAs by promoting the formation of stable, adherent oxides like Al_2O_3, Cr_2O_3, and SiO_2. These oxides are thermodynamically stable, have low oxygen permeability, and adhere strongly to the alloy surface, offering robust protection against oxidation.

Recent research into the antioxidation strategies for HEAs has focused on tailoring their composition and microstructure to enhance their resistance to both corrosion and oxidation. One promising approach involves the use of alloying elements that can promote the formation of self-healing oxide layers. These self-healing properties are particularly desirable in harsh environments, such as in gas turbines or nuclear reactors, where prolonged exposure to high temperatures and reactive gases can rapidly degrade materials. In HEAs, the inclusion of reactive elements like zirconium or hafnium can lead to the formation of oxide scales that not only protect the material but also regenerate when damaged, providing a form of intrinsic corrosion protection. Moreover, the development of coatings and surface treatments, such as ceramic or metallic coatings, has shown promise in improving the oxidation resistance of HEAs, particularly in environments where the materials are exposed to aggressive oxidizing agents.

Studies on the microstructural evolution of oxide films on HEAs have revealed several key factors influencing their high-temperature oxidation resistance [29]. One critical factor is the diffusion behavior of alloying elements during oxidation. Elements with high oxygen affinity, such as aluminum and chromium, tend to diffuse to the surface and react with oxygen to form protective oxide layers. The rate of diffusion and the ability to form a continuous oxide layer depends on the composition and phase distribution of the HEA. For example, single-phase HEAs with a homogeneous elemental distribution typically exhibit better oxidation resistance than multiphase HEAs, where interphase boundaries may act as pathways for rapid oxygen ingress.

Another important factor is the stability of the oxide film under cyclic thermal conditions. Thermal cycling can induce stresses in the oxide film due to differences in thermal expansion coefficients between the oxide and the underlying alloy. HEAs with tailored compositions can minimize these stresses by forming oxides with compatible thermal expansion properties. Additionally, the high configurational entropy of HEAs can suppress the formation of volatile oxides, enhancing their performance under high-temperature, cyclic conditions.

Recent research has also highlighted the role of minor alloying elements, such as yttrium or hafnium, in improving the high-temperature oxidation resistance of HEAs. These elements act as reactive element additions, which enhance the adhesion and growth stability of oxide films. They achieve this by segregating to grain boundaries or oxide–metal interfaces, reducing the formation of voids and cracks that could compromise the protective layer.

In summary, high-temperature oxidation resistance in HEAs is determined by their ability to form stable, adherent, and continuous oxide films. The unique compositional design of HEAs allows for the tailoring of these properties through careful selection of alloying elements. Advances in understanding the oxidation mechanisms and the factors influencing oxide film formation and stability continue to expand the potential applications of HEAs in high-temperature environments. These alloys hold promise for use in turbines, engines, and other critical components where superior oxidation resistance is required, driving innovation in the energy and transportation sectors.

3.3.1 Oxidation Mechanism

Thermodynamic Factors. The electronegativities of metallic elements as well as their Gibbs free energies of oxide are vital for the high-temperature oxidation resistance of HEAs. For that, it is related to the formation sequence of oxides. If the protective dense oxides formed first, it is possible to prevent oxygen from further reacting with other alloy elements, thus achieving an excellent oxidation resistance in HEAs. On the contrary, if the oxides formed first are porous or permeable to oxygen, the oxidation resistance of the HEAs worsens, owing to the continuous reaction of oxygen with other alloy elements [30]. Hong et al. [31] found that for the $Al_{0.3}$CoCrFeNiCu HEA, Al and Cr have fewer valence electrons and lower binding energy, which leads to the first formation of Al_2O_3 and Cr_2O_3 thin films covering the alloy surface. Thus, the high-temperature oxidation process is sluggish under these protective oxides. The formation of oxide films during high-temperature oxidation is also strongly associated with the Gibbs free energy of metallic atoms. Generally, oxides with lower Gibbs free energy are preferentially generated during oxidation. For example, the standard Gibbs free energies followed by the more negative value rank are $Al_2O_3 < SiO_2 < MnO < Cr_2O_3 < Mn_3O_4 < Mn_2O_3 < Fe_3O_4 < Fe_2O_3 < CoO < NiO < Co_3O_4$ [28], as shown in Figure 3.3. Thus, the first growth of Al_2O_3, MnO, and SiO_2 on top of the FeCoNiCrAl, FeCoNiCrMn, and FeCoNiCrSi HEAs, respectively, should be favorable under thermodynamic reasons [32]. However, all the initial-grown oxides do not completely suppress the inward

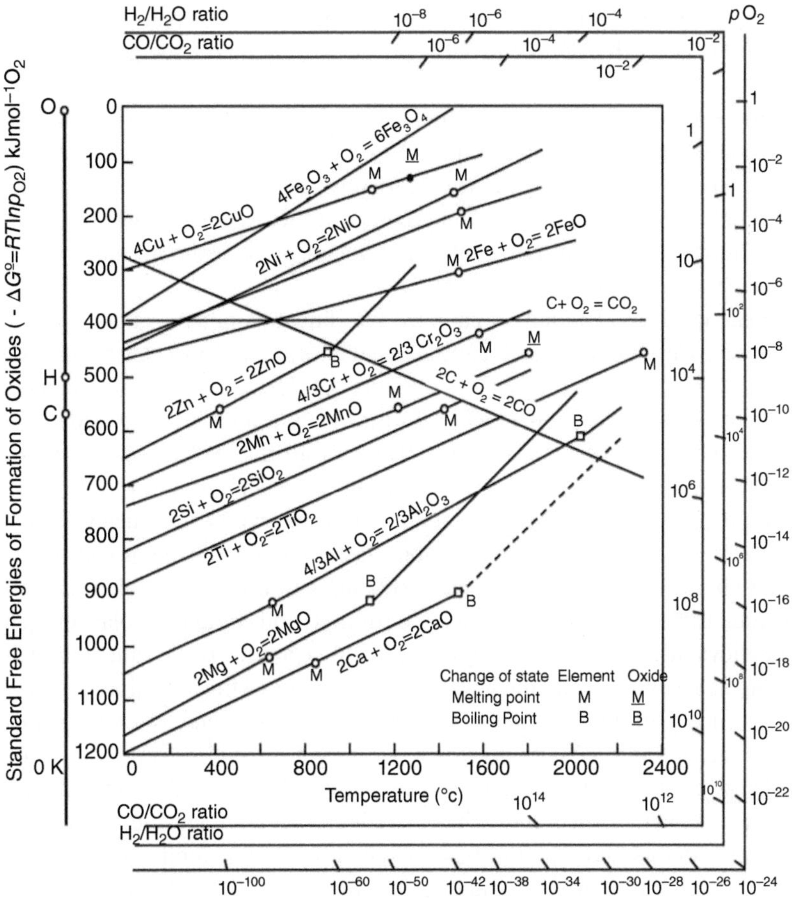

Figure 3.3 The standard free energy of formation of oxides as a function of temperature. *Source*: [27] / with permission of Cambridge University Press.

diffusion of oxygen and outward diffusion of metallic cations, which result in the formation of various later-grown oxides as the oxidation goes on. The formation of the final oxidation product is influenced not only by the initial oxidation but also by the elemental composition of the alloy. For example, Giggins et al. [33] investigated the high-temperature oxidation behavior of the Ni—Cr—Fe system within the temperature range of 1000–1200 °C. Within the ternary Ni—Cr—Al system, the oxidation products of the alloy can be categorized into three types. Type I alloys, characterized by limited Cr and Al contents, form nonprotective NiO and spinel phases on the outer surface, with discontinuous Al_2O_3 and Cr_2O_3 suboxide layers inside. Partial internal oxidation of Al further contributes to an increased oxidation rate. Type II alloys, which contain higher Cr concentrations but limited Al content, develop an external Cr_2O_3 film and an internal Al_2O_3 subscale. The oxidation rate is primarily governed by the diffusion of the aluminum-doped Cr_2O_3 film, resulting in a progressively decreasing oxidation rate; type III alloys contain a higher concentration of Al and have the most vigorous oxidation resistance. This type of alloy forms a highly protective outer Al_2O_3 film, and the oxidation rate is regulated by the diffusion of the outer Al_2O_3 film, providing the best oxidation resistance.

Kinetic Factors. The isothermal oxidation kinetics curves commonly characterize the oxidation kinetics of metals. In the oxidation process, the total mass of most alloys increases with time, and the oxidation weight gain curve represents the metal's oxidation kinetics. The oxidation kinetics of metals offer diverse insights, such as oxidation mechanisms, rate-controlling factors, protective film behavior, reaction rate constants, and activation energy.

Kinetic laws are generally categorized into linear, parabolic, cubic, logarithmic, and antilogarithmic. The parabolic and linear laws are the most commonly observed in oxidation studies [34], as shown in Figure 3.4. The linear oxidation kinetics law applies when the oxidation weight gain is directly proportional to oxidation time. Under such conditions, the oxide film is nonprotective. For instance, Shang et al. [35] investigated the high-temperature oxidation behavior of a BCC refractory HEA $(TiZrNb)_{14}AlMo$ and its carbon-containing coating, synthesized using arc melting and laser cladding technology. Oxidation tests were conducted at 800 °C. Its oxidation kinetic curve meets the linear oxidation law, indicating a high-temperature oxidation film that was porous and discontinuous and could not isolate oxygen. For most alloys, when oxidized at high temperatures, a thick oxide film is formed, following the parabolic oxidation law; that is, the square of the oxidation weight gain is proportional to the oxidation time. It shows that, with the increase of oxidation time, the oxidation rate gradually decreases, forming a protective oxide film. Diffusion mechanisms govern the oxidation process. Many HEAs exhibit high-temperature oxidation behavior consistent with the parabolic oxidation law [36].

Various factors, such as the type of metal, oxidation temperature, and exposure duration, influence the oxidation kinetics of metals. Different metals exhibit distinct oxidation behaviors. For instance, pure iron (Fe) adheres to a parabolic oxidation law at 800 °C [37]. In contrast, tantalum exhibits a linear oxidation law under the same conditions [38]. Additionally, the oxidation behavior of the same metal may conform to different kinetic laws at varying temperatures. For example,

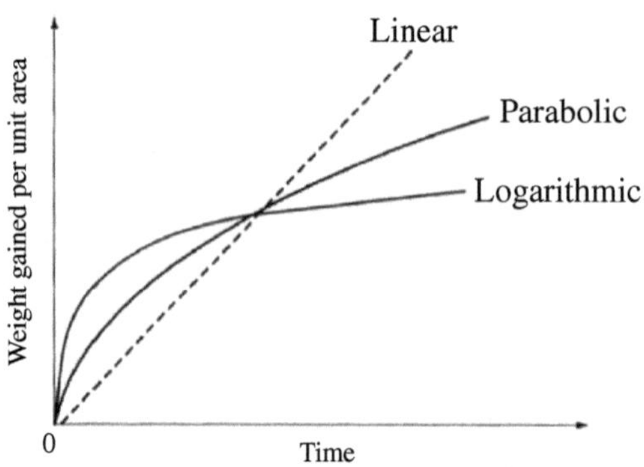

Figure 3.4 Main oxidation rate laws in oxide film growth. *Source*: [33] / with permission of Springer Nature.

Chen et al. [39] studied the high-temperature oxidation behavior of $Al_{0.6}CrFeCoNi$ and $Al_{0.6}CrFeCoNiSi_{0.3}$ HEAs at 800–1000 °C. The results showed that, at 800 °C, the oxidation weight gain of the two alloys was low and followed the linear oxidation law; at 900 °C, the two HEAs also followed the parabolic growth law; however, at 1000 °C, $Al_{0.6}CrFeCoNi$ followed parabolic oxidation, while the oxidation rate of $Al_{0.6}CrFeCoNiSi_{0.3}$ alloy almost followed the linear growth law due to the large number of oxide scale defects. Furthermore, oxidation kinetics may evolve for the same metal at the same temperature. For example, Lu et al. [40] studied the high-temperature oxidation behavior of refractory HEA $AlMo_{0.5}NbTa_{0.5}TiZr$. The oxidation kinetics results at 900 and 1000 °C showed that the initial oxidation conformed to power-law passivation oxidation but later transformed to linear oxidation. In this alloy, the oxide layer formed in the early stages served as an effective protective barrier against further oxidation. However, as stress accumulated within the oxide film, it ruptured, shifting toward linear oxidation kinetics. Even with the same kinetic law, the oxidation rate constant may change. Wang et al. [41] studied the oxidation behavior of CoCrFeMnNi HEA after plastic deformation. Their study demonstrated that deformed specimens (CR-25%, CR-50%, and CR-70%) adhered to a parabolic law but exhibited a two-stage behavior: an initial rapid oxidation phase (stage 1), followed by a slower growth phase (stage 2). Further analysis revealed that the transition time between these stages closely coincided with the completion of recrystallization, suggesting that recrystallization processes primarily drove this kinetic transition.

The diffusion kinetics of metallic cations and oxygen in oxides are other important factors in the oxidation process of HEAs. The diffusion rate is affected by whether the oxide film is continuous. If the metallic cations with high diffusion speed react with oxygen to form a dense protective oxide film, the oxidation resistance of HEAs will be improved. Alternatively, if the oxide film is discontinuous or the oxygen anions continuously pass through the oxide film and react with the alloy matrix, the oxidation resistance of the HEAs will be greatly reduced. Kai et al. [42] investigated the oxidation behavior of FeCoCrNiMn at 950 °C and found that the diffusion order of ions in this alloy is $Mn^{+2} > Fe^{+2} > Ni^{+2} > Cr^{+3}$, so the diffusion rate of Mn^{+2} is the fastest, while the Mn_3O_4 formed by the Mn element in the outer layer is discontinuous, leading to oxidation voids and accelerated oxidation. Besides, severe lattice distortion will introduce slow diffusion of HEAs, which will also have a certain effect on the high-temperature oxidation. The slow diffusion effect can reduce the combination speed between oxygen and element, increasing the activation energy of diffusion and reducing the diffusion rate, thereby improving the oxidation resistance of HEAs [43]. Grain boundaries also influence the diffusion rate. The grain boundaries of metal materials are characterized by high concentrations of defects, and the activation energy for the diffusion of ions at the grain boundaries is much lower compared to the grain interior. Grain boundaries act as fast diffusion pathways for ions compared to the grain interior [44]. The diffusion coefficient in the material is determined by the effective diffusion coefficient, which is the weighted average of the diffusion coefficients within the grain and at the grain boundaries [45]. An increased grain boundary density offers more diffusion pathways, thereby facilitating the transport of cations and anions. In addition, specific fabrication methods, such as 3D printing, may introduce defects like holes and cracks in the alloy, which serve as fast diffusion pathways for ions and alter the alloy's high-temperature oxidation properties. For example, Min et al. [46] investigated the high-temperature oxidation behavior of GH3536 alloys fabricated using the laser powder bed fusion (LPBF) method at 950 °C. The results showed that the type and density of processing defects influence the oxidation resistance of LPBF GH3536. Microcracks impair the formation of protective oxide films by increasing diffusion pathways and inducing internal oxidation, thereby acting as channels between the matrix and the interface. Minimizing manufacturing defects can effectively enhance the oxidation resistance of alloys.

3.3.2 Influencing Factors

Alloy Elements. The oxidation resistance of HEAs is determined by the features of oxide thin films on their surface, which are directly influenced by the alloy elements. Cr_2O_3, Al_2O_3, and SiO_2 are commonly accepted protective films, which are formed by the additions of Cr, Al, and Si. Continuous Cr_2O_3, Al_2O_3, and SiO_2 films can effectively prevent the internal diffusion of oxygen anions and external diffusion of metallic cations, achieving an excellent oxidation resistance of HEAs [47]. Cr_2O_3, Al_2O_3, and SiO_2 possess dense corundum-type crystal structures and high melting points, contributing significantly to their protective properties in high-temperature environments. The Al_2O_3 oxide layer is widely employed in practical applications due to its slow growth rate and long-term chemical stability. Its high thermodynamic activity allows it to preferentially react with oxygen, forming a stable, protective film. Although Al_2O_3 provides adequate oxidation protection, it is prone to spallation during the cooling process due to high thermal stress [48]. Cr is widely used in structural and tool steels to enhance oxidation resistance. Cr_2O_3 forms a protective film below 1300 °C; however, it is susceptible to rupture due to its Pilling–Bedworth ratio (PBR) exceeding 2. SiO_2 is an effective high-temperature oxidation protective layer; however, adding Si to Fe-, Co-, and Ni-based alloys adversely affects their mechanical properties. Therefore, only a small amount of Si is added to the alloy. In some HEAs containing more Al, it acts as an activator to accelerate the formation of dense oxide layers such as Al_2O_3 [49]. Similar to Al, adding a third element, such as Cr, significantly promotes the formation of a SiO_2 protective layer. Liu et al. [50] studied the corrosion behavior of Al_xCoCrCuFeNi HEA at 1000 °C and found that, without Al addition, the outer oxide layer is composed of hollow, loose Cr_2O_3 with poor adhesion, which exhibits weak protection. With increasing Al content in HEAs, the formed Al_2O_3 films become denser and more continuity, leading to better oxidation resistance. Besides the above simple protective oxide scales, complex oxide scales such as $CrTaO_4$ (rutile-type crystal structure) are also known as protective oxide by inhibiting outward diffusion of the constituent elements, which possesses a potential to protect HEAs as a relatively low mass gain even up to 1500 °C [51]. The $CrTaO_4$ protective oxide was observed in alloy systems such as MoTaTiCr [52], TaMoCrTiAl [53], and CrMoNbTaTiAlSi [32]. Despite the perspective protective properties of $CrTaO_4$, the growth constant of it is twice compared with that of Al_2O_3. In addition, relatively thick zones of internal corrosion were observed below $CrTaO_4$ layers indicating significant inward diffusion of oxygen and nitrogen [30]. Some elements will significantly deteriorate the corrosion resistance of HEAs due to the formation of liquid or volatile oxide species even at temperatures below 1000 °C. For insistence, the oxides of V_2O_5, MoO_3, WO_3, Nb_2O_5, and Ta_2O_5 are all volatile, and the evaporative temperatures are 678, 795, 1000, 1370, and 1370 °C respectively. Once the evaporation of oxides occurs, the mass of pores in the oxide film will accelerate the oxidation process due to the ampliative diffusion channel of oxygen and metallic cations [30]. Zr and Hf have high oxygen affinity and a large atomic radius, which leads to the weakening of interatomic bonds and an increase in intermetallic diffusion coefficient at high temperatures. Therefore, it is easier for oxides to form and cause the oxides to peel off from the HEA surface [54]. Nb plays a dual role in the oxidation behaviors of HEA due to the formation of the Laves phase. Huang et al. [36] found that the antioxidant capacity of $CoCrFeMnNb_{0.25}Ni$ HEA is better than that of CoCrFeMnNi HEA at low temperatures and worse at high temperatures. That is because the formed the Laves phase will hinder the outward diffusion of metal ions and become a channel for the inward diffusion of oxygen ions. The combined effect of these two factors will lead to changes in oxidation resistance in HEAs under various temperatures. As demonstrated by Li et al. [49], $AlCoCrFeNiSi_x$ HEA was synthesized to investigate its high-temperature oxidation behavior. The results showed that the appropriate Si content in the alloy facilitates the formation of a dense single-layer Al_2O_3 oxide film, thereby improving the oxidation resistance of the alloy; however, excessive Si content can result in the delamination and spallation of the oxide film, which negatively impacts the alloy's performance. Chen et al. [55] fabricated $AlCoCr_xNiTi$ alloy and studied the Cr–Al synergistic effect during the oxidation process of the alloy. It was observed that the addition of an appropriate amount of Cr acts as a structural template for the formation of α-Al_2O_3. This addition facilitates the diffusion of Al during the stable oxidation stage, contributing to the development of a continuous Al_2O_3 oxide film, which enhances the alloy's oxidation resistance. Overall, the effect of alloy elements on the oxidation resistance of HEAs is terribly complicated, which needs further in-depth research.

Growth Stress. Growth stress generated by the oxidation process in the oxide scales may lead to scale failure. Thus, the intensity of growth stress in oxides determines HEAs to form long-term protective scales. Growth stresses developed during oxidation may arise from a number of reasons. First, compositional changes resulting in growth stress by changes in the lattice parameter of the alloy occur if high amounts of oxygen are dissolved in the HEAs [56]. Second, chemical reactions between simple oxides that have been built simultaneously may lead to notable anisotropic volume expansion within the oxide scale and resultant growth stress [57]. Finally, growth stress may also arise from unfavorable values of the oxide–metal volume ratios (PBR) of corresponding metals and oxides. For example, oxides of W, Ta, Nb, and V exhibit extremely high PBR values as WO_3 (3.3), Ta_2O_5 (2.5), Nb_2O_5 (2.68), and V_2O_5 (3.29), which will give rise to high growth stress developed within the oxide scales [30].

Lattice Distortion and Slow Diffusion. Severe lattice distortion will introduce slow diffusion of HEAs, which will also have a certain effect on the high-temperature oxidation. The change of structure will affect the diffusion rate of oxygen and even transform the diffusion path of oxygen in HEAs. When the atomic radius of a metal element is large, it will lead to lattice distortion in HEAs such as Ti, Zr, and Mo [58]. Wang et al. [43] found that there are two obvious zones in the TiZrNbTa HEA during the oxidation process, namely, the TiZr-rich region and the TaNb-rich region. The region containing Zr will have serious lattice distortion, resulting in obvious lattice distortion and the dislocation tube effect. The more serious the lattice distortion in the TiZr enrichment region and the more dislocations, the higher the diffusion rate and oxidation rate of oxygen atoms. The uneven diffusion of oxygen atoms at 800 °C creates stress concentration and cracks along the TiZr zone. When the temperature is greater than 1000 °C, the oxygen atom diffuses uniformly, and the degree of lattice distortion is less than the threshold of the dislocation tube effect, which slows the diffusion rate of oxygen atoms and leads to uniform oxidation. To improve the oxidation resistance of HEA, we can seek an alloying element to produce obvious lattice distortion, leading oxygen to the vicinity of Al, Cr, and other alloy elements with good oxidation resistance similar to drainage to form Al_2O_3 and Cr_2O_3.

Influence of the Test Environment. Temperature is a critical factor influencing the high-temperature oxidation behavior of alloys, directly impacting the oxidation reaction kinetics and the chemical composition of the oxide film. With the temperature increase, the oxidation reaction accelerates significantly, resulting in a significant increase in the oxidation rate. Temperature variations influence the oxide film's composition, phase structure, and density. Laplanche et al. [59] investigated the high-temperature oxidation behavior of $Cr_{20}Mn_{20}Fe_{20}Co_{20}Ni_{20}$ alloys fabricated via vacuum induction melting in the temperature range of 600–900 °C. Their findings indicate that the oxidation kinetics adhere to a parabolic rate law within this temperature range. With increasing temperature, the parabolic oxidation rate constant rises from 0.0015 to 0.1487 $mg^2\ cm^{-4}\ h^{-1}$, and the oxidation rate of the alloy increases significantly; the oxides vary considerably with temperature. At 600 °C, the oxide scale predominantly consists of α-Mn_2O_3 (xenotime), with a thin layer of Cr_2O_3 (chromium oxide) forming near the interface. While α-Mn_2O_3 remains at 800 °C, it transforms into Mn_3O_4 (hausmannite) at 900 °C. Kim et al. [60] prepared HEAs by vacuum induction casting and investigated the high-temperature oxidation behavior of Cr—Mn—Fe—Co—Ni HEAs. The results showed that the weight gain at the three temperatures was 1.76, 4.45, and 9.08 mg/cm^2, respectively, showing a typical parabolic oxidation law. The oxide composition varies notably with increasing temperature. At 900 °C, the primary oxides are Cr_2O_3 and Mn_2O_3. However, at 1000 °C and 1100 °C, Cr_2O_3 persists while the dominant oxides shift to $(Mn,Cr)_3O_4$ and Mn_3O_4.

Oxygen partial pressure also affects the alloy's high-temperature oxidation performance. The increase in oxygen partial pressure will accelerate the oxidation reaction. Varying oxygen partial pressures influence the chemical composition and structure of the oxide film, subsequently impacting its density and protective properties. Kai et al. [42] investigated the oxidation behavior of equimolar FeCoNiCrMn

HEAs prepared via arc melting at 950 °C under varying oxygen partial pressures. The study showed that the oxidation kinetics of the alloy conforms to the parabolic rate law and the oxidation rate constant increases with the increase of oxygen partial pressure. However, the composition of the oxides remained similar across different oxygen partial pressures. Sakai et al. [61] investigated iron's oxidation products and mechanisms at 573 K under various oxygen partial pressures, identifying four distinct oxidation regions. At oxygen partial pressures between 10^{-1} and $10^{-0.5}$ Pa, magnetite is the dominant product, and the oxidation rate constant strongly depends on oxygen partial pressure. Between 1 and $10^{0.5}$ Pa, hematite forms over magnetite, with grain boundaries enhancing diffusion and maintaining a high oxidation rate constant. At pressures of 10 to $10^{0.3}$ Pa, hematite uniformly covers magnetite as a protective film, stabilizing the oxidation rate constant at a low level. However, at pressures exceeding 10^5 Pa, simultaneous growth of hematite and magnetite occurs, causing a sharp increase in the oxidation rate constant.

This chapter has provided a comprehensive exploration of the corrosion and oxidation resistance of HEAs, highlighting the significant impact of alloying elements and processing methods on their performance. The unique composition and microstructure of HEAs offer promising avenues for enhancing their resistance to both corrosion and oxidation, making them attractive candidates for applications in extreme environments such as aerospace, energy, and marine industries. The detailed analysis of the effects of various alloying elements and processing techniques has laid a solid foundation for the design and development of HEAs with superior properties. Despite the substantial progress made, several challenges remain to be addressed in the field of HEA corrosion and oxidation resistance. Future research efforts should focus on the following areas:

Optimization of Alloy Compositions. The development of high-performance HEAs requires further exploration to identify the optimal combinations of alloying elements that balance corrosion resistance, oxidation resistance, and mechanical properties. The compositional flexibility of HEAs allows for tailoring properties by incorporating elements that synergistically enhance passive film stability, improve oxide adherence, and strengthen the underlying matrix. For example, elements like chromium and aluminum contribute to protective oxide formation, while molybdenum and nickel enhance resistance to localized corrosion. However, excessive concentrations of certain elements may lead to phase segregation, brittleness, or galvanic effects, which could undermine performance. Advanced methodologies such as high-throughput experimentation and machine learning-guided design are essential to map the vast compositional space of HEAs efficiently. These approaches can identify compositions optimized for specific applications, such as aerospace components requiring both high strength and resistance to oxidative and corrosive environments.

Microstructure Performance Correlations. Understanding the microstructural evolution of HEAs during corrosion and oxidation processes is critical for improving their performance. Advanced characterization techniques such as transmission electron microscopy, atom probe tomography, and in situ synchrotron radiation studies should be employed to elucidate the mechanisms underlying oxide film formation, elemental diffusion, and phase transformations at various scales. These techniques enable researchers to capture the dynamic interactions between microstructure and environmental stressors, such as grain boundary oxidation or localized corrosion initiation. By establishing precise correlations between microstructure and performance, researchers can design alloys with tailored microstructures, such as fine-grained or single-phase HEAs, to enhance resistance to degradation. Furthermore, the development of grain boundary engineering techniques and precipitation control strategies offers additional avenues to optimize microstructural features for superior durability.

Innovative Protective Technologies. The integration of advanced protective technologies could significantly improve the performance of HEAs under corrosive and high-temperature conditions. Nanotechnology-based approaches, such as the application of nanocoatings, can enhance the barrier properties of oxide films and provide additional protection against aggressive environments. Additionally, self-healing coatings incorporating responsive materials that can repair microcracks or defects autonomously hold great promise for extending the service life of HEAs. Composite materials combining HEAs with ceramic or polymer matrices can offer hybrid solutions, combining the mechanical strength of metals with the chemical stability of ceramics or polymers. Research into multifunctional coatings that provide not only corrosion and oxidation resistance but also thermal insulation or wear resistance will further broaden the applicability of HEAs in extreme conditions.

Theoretical Models and Computational Simulations. Refining theoretical models and computational methods is essential for predicting the behavior of HEAs under diverse environmental conditions. Density functional theory calculations, molecular dynamics simulations, and phase-field modeling can provide detailed insights into oxidation kinetics, corrosion mechanisms, and phase stability. These tools can be combined with machine learning algorithms to identify patterns and predict material behavior across vast datasets. Such computational approaches reduce the reliance on costly and time-consuming experimental trials by enabling virtual screening of alloy compositions and processing parameters. Additionally, integrating multiscale modeling with experimental validation will ensure more accurate predictions of long-term performance, aiding in the design of HEAs optimized for industrial applications.

Practical Application Development. The successful deployment of HEAs in industrial applications requires addressing challenges related to scalability, manufacturability, and long-term stability. Efforts should focus on developing cost-effective production methods, such as additive manufacturing and powder metallurgy, to fabricate complex geometries and minimize material waste. Heat treatment protocols and surface finishing techniques must be optimized to ensure consistent performance across large-scale production. Long-term studies on the environmental stability of HEAs under cyclic thermal, mechanical, and chemical stresses are crucial for establishing their reliability in real-world scenarios. Collaborative research between academia and industry will play a pivotal role in overcoming these barriers, facilitating the transition of HEAs from laboratory research to practical implementation in sectors like energy, transportation, and aerospace.

By addressing these challenges through interdisciplinary research and technological advancements, HEAs can unlock unprecedented possibilities for high-performance materials capable of withstanding the most demanding conditions. Their unique combination of corrosion resistance, oxidation resistance, and mechanical robustness positions them as key enablers of innovation and sustainability in critical industrial sectors. The continued exploration of HEAs will not only expand the boundaries of materials science but also contribute to developing environmentally resilient technologies for the future.

References

1 Diao H, Santodonato L J, Tang Z. Local structures of high-entropy alloys (HEAs) on atomic scales: an overview. *JOM*, 2015, 67(10): 2321–2325.

2 Shuang S, Ding Z Y, Chung D, et al. Corrosion resistant nanostructured eutectic high entropy alloy. *Corrosion Science*, 2020, 164: 108315.

3 Zhang Y, Zuo T T, Tang Z, et al. Microstructures and properties of high-entropy alloys. *Progress in Materials Science*, 2014, 61: 1–93.

4 Qiu Y, Thomas S, Gibson MA, et al. Corrosion of high entropy alloys. *NPJ Materials Degradation*, 2017, 1(1): 11.

5 Sieradzki K, Newman R C. A percolation model for passivation in stainless steels. *Journal of the Electrochemical Society*, 1986, 133(9): 1970–1980.

6 Li Q H, Yue T M, Guo Z N. Microstructure and corrosion properties of AlCoCrFeNi high entropy alloy coatings deposited on AISI 1045 Steel by the electrospark process. *Metallurgical and Materials Transactions A*, 2012, 44(4): 1767–1778.

7 Crook P. Corrosion resistance nickel alloys part 1. *Advanced Materials & Processes*, 2007, 165: 37–39.

8 Qiu X, Liu C. Microstructure and properties of $Al_2CrFeCoCuTiNi$ high-entropy alloys prepared by laser cladding. *Journal of Alloys and Compounds*, 2013, 553: 216–220.

9 Kao Y F, Lee T D, Chen S K, et al. Electrochemical passive properties of $Al_xCoCrFeNi$ (x = 0, 0.25, 0.50, 1.00) alloys in sulfuric acids. *Corrosion Science*, 2010, 52(3): 1026–1034.

10 Ilevbare G O, Burstein, G T. The inhibition of pitting corrosion of stainless steels by chromate and molybdate ions. *Corrosion Science*, 2003, 45(7): 1545–1569.

11 Chou Y L, Yeh J W, Shih H C. The effect of molybdenum on the corrosion behaviour of the high-entropy alloys $Co_{1.5}CrFeNi_{1.5}Ti_{0.5}Mo_x$ in aqueous environments. *Corrosion Science*, 2010, 52(8): 2571–2581.

12 Clayton C R, Lu Y C. A bipolar model of the passivity of stainless steel: the role of Mo addition. *Journal of the Electrochemical Society*, 1986, 133(12): 2465.

13 Mazzarolo A, Curioni M, Thompson Vicenzo A. Anodic growth of titanium oxide: Electrochemical behaviour and morphological evolution. *Electrochimica Acta*, 2012, 75: 288–295.

14 Shun T T, Chang L Y, Shiu M H. Microstructures and mechanical properties of multiprincipal component $CoCrFeNiTi_x$ alloys. *Materials Characterization*, 2012, 556: 170–174.

15 Qiu X W, Zhang Y P, Liu C G. Effect of Ti content on structure and properties of $Al_2CrFeNiCoCuTi_x$ high-entropy alloy coatings. *Journal of Alloys and Compounds*, 2014, 585: 282–286.

16 Jones D A. Principles and Presentation of Corrosion, 2 edn. Prentice-Hall Reno, 1996.

17 Li B Y, Kun P E, Hu A P. Structure and properties of $FeCoNiCrCu_{0.5}Al_x$ high-entropy alloy. *Transactions of Nonferrous Metals Society of China*, 2013, 23(3): 735–741.

18 Hsu Y-J, Chiang W-C, Wu J-K. Corrosion behavior of $FeCoNiCrCu_x$ high-entropy alloys in 3.5% sodium chloride solution. *Materials Chemistry and Physics*, 2005, 92(1): 112–117.

19 Ren B, Liu Z X, Li D M, et al. Corrosion behavior of CuCrFeNiMn high entropy alloy system in 1 M sulfuric acid solution. *Materials and corrosion*, 2012, 63(9): 828–834.

20 Zhang C, Chen G J, Dai P Q. Evolution of the microstructure and properties of laser-clad FeCrNiCoBx high-entropy alloy coatings. *Materials Science and Technology*, 2016, 32(16): 1666–1672.

21 Zheng Z Y, Li X C, Zhang C, et al. Microstructure and corrosion behaviour of $FeCoNiCuSn_x$ high entropy alloys. *Materials science and technology*, 2024, 31(10): 1148–1152.

22 Kumar D, Maulik O, Sharma V K. Understanding the effect of tungsten on corrosion behavior of $AlCuCrFeMnW_x$ high-entropy alloys in 3.5 wt.% NaCl solution. *Journal of Materials Engineering and Performance*, 2018, 27(9): 4481–4488.

23 Zhang M, Zhang L, Liaw P K. Effect of Nb content on thermal stability, mechanical and corrosion behaviors of hypoeutectic CoCrFeNiNbχ high-entropy alloys. *Journal of Materials Research*, 2018, 33(19): 3276–3286.

24 Jayaraj J, Thinaharan C, Ningshen S, et al. Corrosion behavior and surface film characterization of TaNbHfZrTi high entropy alloy in aggressive nitric acid medium. *Intermetallics*, 2017, 89: 123–132.

25 Lin C-M, Tsai H-L. Evolution of microstructure, hardness, and corrosion properties of high-entropy $Al_{0.5}CoCrFeNi$ alloy. *Intermetallics*, 2011, 19(3): 288–294.

26 Lee C P, Chen Y Y, Hsu C Y. Enhancing pitting corrosion resistance of $Al_xCrFe_{1.5}MnNi_{0.5}$ high-entropy alloys by anodic treatment in sulfuric acid. *Thin Solid Films*, 2008, 517(3): 1301–1305.

27 Wu P, Gan K, Yan D. A non-equiatomic FeNiCoCr high-entropy alloy with excellent anti-corrosion performance and strength-ductility synergy. *Corrosion Science*, 2021, 183: 109341.

28 Birks N, Meier G H, Pettit F S. Introduction to the High-Temperature Oxidation of Metals. Cambridge University Press, 2006.

29 Wei Y, Fu Y, Pan Z M. Influencing factors and mechanism of high-temperature oxidation of high-entropy alloys: a review. *International Journal of Minerals, Metallurgy and Materials*, 2021, 28(6): 915–930.

30 Gorr B, Schellert S, Müller F. Current status of research on the oxidation behavior of refractory high entropy alloys. *Advanced Engineering Materials*, 2021, 23(5): 20011047.

31 Hong Y, Kivy M B, Zaeem M A. Competition between formation of Al_2O_3 and Cr_2O_3 in oxidation of $Al_{0.3}CoCrCuFeNi$ high entropy alloy: a first-principles study. *Scripta Materialia*, 2019, 168: 139–143.

32 Kai W, Li C C, Cheng F P. et al. Tsay, J.J. Kai, Air-oxidation of FeCoNiCr-based quinary high-entropy alloys at 700–900 °C. *Corrosion Science*, 2017, 121: 116–125.

33 Giggins C S, Pettit F S. Oxidation of Ni–Cr–Al alloys between 1000 °C and 1200 °C. *Journal of the Electrochemical Society*, 1971, 118(11): 1782.

34 Quastel A D, West A, Young G. Feeder pipe oxidation in the presence of steam during a nuclear reactor accident. *High Temperature Corrosion of Materials*, 2023, 99 (5–6): 399–413.

35 Shang X, Bo S, Guo Y. ZrC reinforced refractory-high-entropy-alloy coatings: compositional design, synthesis, interstitials, and microstructure evolution effects on wear, corrosion and oxidation behaviors+. *Applied Surface Science*, 2021, 564: 150466.

36 Huang D, Lu J, Zhuang Y, et al. The role of Nb on the high temperature oxidation behavior of $CoCrFeMnNb_xNi$ high-entropy alloys. *Corrosion Science*, 2019, 158: 108088.

37 Chen R Y, Yeun, W Y. Review of the high-temperature oxidation of iron and carbon steels in air or oxygen. *Oxidation of Metals*, 2003, 59: 433–468.

38 Voitovich V B, Lavrenko V A, Adejev V M, et al. High-temperature oxidation of tantalum of different purity. *Oxidation of Metals*, 1996, 43: 509–526.

39 Chen L, Zhou Z, Tan Z, et al. High temperature oxidation behavior of $Al_{0.6}CrFeCoNi$ and $Al_{0.6}CrFeCoNiSi_{0.3}$ high entropy alloys. *Journal of Alloys and Compounds*, 2018, 764: 845–852.

40 Lu S, Li X, Liang X, et al. Effect of Al content on the oxidation behavior of refractory high-entropy alloy $AlMo_{0.5}NbTa_{0.5}TiZr$ at elevated temperatures. *International Journal of Refractory Metals and Hard Materials*, 2022, 105: 105812.

41 Wang Y, Zhang M, Jin J. Oxidation behavior of CoCrFeMnNi high entropy alloy after plastic deformation. *Corrosion Science*, 2020, 163: 108285.

42 Kai W, Li C C, Cheng F P. The oxidation behavior of an equimolar FeCoNiCrMn high-entropy alloy at 950 °C in various oxygen-containing atmospheres. *Corrosion Science*, 2016, 108: 209–214.

43 Wang R, Tang Y, Li S, et al. Effect of lattice distortion on the diffusion behavior of high-entropy alloys. *Journal of Alloys and Compounds*, 2020, 825: 154099.

44 Atkinson, A. Diffusion along grain boundaries and dislocations in oxides, alkali halides and carbides. *Solid State Ionics*, 1984, 12: 309–320.

45 Li Y S, Niu Y, Gesmundo F, et al. Grain size effects on the oxidation of two-phase Cu–Fe alloys. *Corrosion science*, 2002, 44(7): 1457–1468.

46 Min S, Zhang H, Liu H, et al. Influence of defects on high-temperature oxidation performance of GH3536 superalloys fabricated by laser powder bed fusion. *Additive Manufacturing Letters*, 2022, 3: 100064.

47 Young D J. High Temperature Oxidation and Corrosion of Metals. Elsevier, 2008.

48 Tolpygo V K, Clarke, D R. Spalling failure of α-alumina films grown by oxidation: I. Dependence on cooling rate and metal thickness. *Materials Science and Engineering: A*, 2000, 278 (1–2): 142–150.

49 Li Y, Zhang P, Zhang J. Oxidation behavior of AlCoCrFeNiSi high-entropy alloys at 1100 °C. *Corrosion Science*, 2021, 190: 109633.

50 Liu Y Y, Chen Z, Chen Y Z. Effect of Al content on high temperature oxidation resistance of $Al_xCoCrCuFeNi$ high entropy alloys (x = 0, 0.5, 1, 1.5, 2). *Vacuum*, 2019, 169: 108837.

51 Gorr B, Müller F, Schellert S. A new strategy to intrinsically protect refractory metal based alloys at ultra high temperatures. *Corrosion Science*, 2020, 166: 108475.

52 Li L C, Li M X, Liu M. Enhanced oxidation resistance of MoTaTiCrAl high entropy alloys by removal of Al. *Science China Materials*, 2020, 64(1): 223–231.

53 Müller F, Gorr B, Christ H J, et al. On the oxidation mechanism of refractory high entropy alloys. *Corrosion Science*, 2019, 159: 108161.

54 Sheikh S, Bijaksana M K, Motallebzadeh A, et al. Accelerated oxidation in ductile refractory high-entropy alloys. *Intermetallics*, 2018, 97: 58–66.

55 Chen T, Shi Y, Ren Y. Effect of Cr content on synergistic effect of Cr-Al during oxidation of high-entropy AlCoCrxNiTi alloys. *Corrosion Science*, 2024, 227: 111690.

56 Chang C H, Titus M S, Yeh J W. Oxidation behavior between 700 and 1300 °C of refractory TiZrNbHfTa high-entropy alloys containing aluminum. *Advanced Engineering Materia*ls, 2018, 20(6): 1700948.

57 Butler T M, Chaput K J, Dietrich J R. High temperature oxidation behaviors of equimolar NbTiZrV and NbTiZrCr refractory complex concentrated alloys (RCCAs). *Journal of Alloys and Compounds*, 2017, 729: 1004–1019.

58 Tong Y, Zhao S, Bei H, et al. Severe local lattice distortion in Zr- and/or Hf-containing refractory multi-principal element alloys. *Acta Materialia*, 2020, 183: 172–181.

59 Laplanche G, Volkert U F, Eggeler G. Oxidation behavior of the CrMnFeCoNi high-entropy alloy. *Oxidation of Metals*, 2016, 85 (5–6): 629–645.

60 Kim Y K, Joo Y A, Kim H S, et al. High temperature oxidation behavior of Cr-Mn-Fe-Co-Ni high entropy alloy. *Intermetallics*, 2018, 98: 45–53.

61 Sakai H, Tsuji T, Naito K. Effect of oxygen partial pressure on oxidation of iron at 573 K *Journal of Nuclear Science and Technology*, 2012, 21(11): 844–852.

4

Additive Manufacturing and Phase Control

*Yan Liu, Bo Song**

School of Materials Science and Engineering, State Key Laboratory of Materials Processing and Die & Mould Technology, Huazhong University of Science and Technology, Wuhan, P.R. China

*Corresponding author: bosong@hust.edu.cn

4.1 Introduction to Additive Manufacturing Technologies

Additive manufacturing (AM), commonly known as 3D printing, stands at the forefront of advanced manufacturing technology. Since its emergence during the early 1980s, AM has rapidly evolved into a transformative force in global manufacturing, gaining significant traction throughout the 1990s. Unlike traditional subtractive machining, which removes material to achieve a desired shape, AM employs a groundbreaking layer-by-layer additive process. This method directly fabricates solid parts from three-dimensional digital models created using computer-aided design (CAD) software. The result is a manufacturing approach that not only significantly reduces production cycles but also enables the efficient creation of components with highly intricate and customized geometries. As a result, AM has become an indispensable tool in the modern manufacturing landscape, driving innovation across a diverse range of industries [1].

Over the past few decades, the versatility of AM has expanded to accommodate a wide variety of materials, including metals, polymers, composites, biomaterials, and even living tissues. This remarkable adaptability has established AM as a cornerstone of material science and engineering, enabling breakthroughs in both fundamental research and industrial applications. Among the myriad materials that benefit from AM, high-entropy alloys (HEAs) have emerged as a particularly promising area of focus. HEAs are a class of advanced materials characterized by their unique chemical complexity and exceptional mechanical, thermal, and corrosion-resistant properties. These materials offer significant potential for demanding applications, yet

Functional High Entropy Alloys and Compounds, First Edition. Edited by Yong Zhang, Yonggang Yao and Haojie Liu.

their complex composition often presents challenges for traditional manufacturing techniques.

The integration of AM into HEA research has provided a transformative solution to these challenges. The precision and flexibility inherent in AM allow for the tailored fabrication of HEA components, facilitating not only the exploration of novel compositions but also the optimization of microstructure and properties. By enabling rapid prototyping and the direct manufacturing of complex geometries, AM empowers researchers to overcome conventional processing limitations and unlock the full potential of HEAs. For instance, intricate lattice structures, internal channels, and porous designs—often impossible to achieve with traditional methods—can now be realized with precision and efficiency using AM.

Beyond its technical capabilities, the adoption of AM in HEA research aligns with the broader goals of sustainability and efficiency in manufacturing. The layer-by-layer process minimizes material waste, reduces energy consumption, and supports the development of lightweight, high-performance components. These attributes position AM as a key enabler for advancing HEA technologies in critical fields such as aerospace, biomedical engineering, and energy systems. AM represents a paradigm shift in the design, processing, and application of HEAs. Its ability to address the challenges posed by these chemically complex materials while offering unparalleled design freedom underscores its role as a pivotal tool in modern materials science. As the field continues to evolve, the synergy between AM and HEA research will undoubtedly drive innovation, enabling the development of next-generation materials and components that meet the demands of increasingly sophisticated applications. At present, the common energy sources are laser, electron beam, and electric arc, which have their own advantages and disadvantages in forming.

Laser AM. A manufacturing technology that uses a high-energy laser beam as a heat source and layers it into a two-dimensional plane through a three-dimensional model under the protection of an inert gas (Ar or N_2, etc.) melts and forms the powder or wire in layers according to a predetermined processing path, as shown in Figure 4.1a. Laser AM mainly uses powder as raw material, which can usually be divided into laser powder bed fusion (LPBF) and laser metal deposition (LMD) technology, also known as direct laser deposition (DLD) technology [2–4]. Among them, LPBF is often referred to as selective laser melting (SLM). SLM mainly uses a laser beam to melt the pre-laid two-dimensional metal powder according to a preset printing path, melt and solidify layer by layer from the bottom up, and finally form a three-dimensional solid part. SLM systems are usually composed of laser systems, powder spreading systems, 3D platforms, and vacuum systems [5]. The 3D platform moves on the two-dimensional plane according to a preset path. After forming a single-layer powder printing, the 3D platform moves in the printing direction/building direction, and the next layer of AM is carried out until the entire three-dimensional part of AM is completed. It is important that, before printing, the printing chamber must be filled with inert gases such as Ar or N_2 to prevent pollution and oxidation during the printing process. Usually, the diameter of the powder used in SLM is less than 50 μm, and the small size of the powder particles and the smaller laser radius (usually ~70–200 μm) eventually form

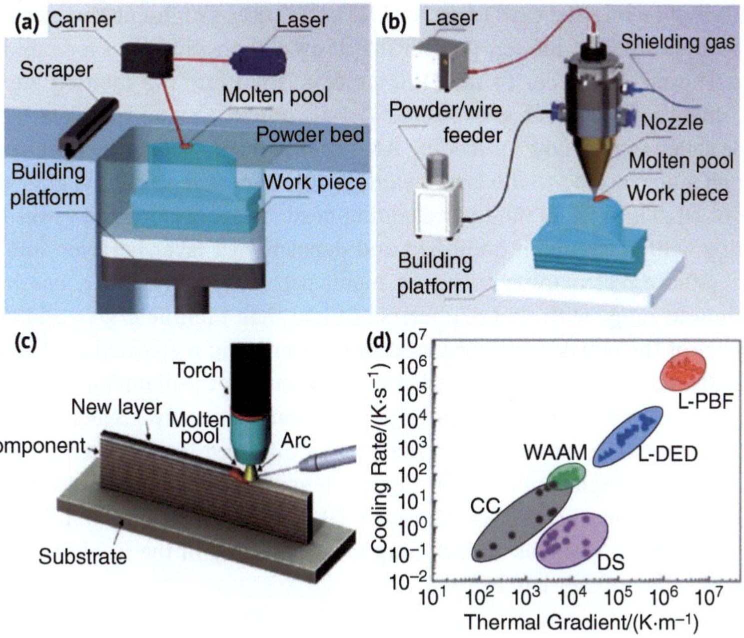

Figure 4.1 Comparison of different AM processes and cooling rates of HEA formation. (a) LPBF. (b) Laser-directed energy deposition (LDED). [6] / with permission of The American Association for the Advancement of Science. (c) Wire arc AM (WAAM). [7] / with permission of Elsevier. (d) Comparison of cooling rate and temperature gradient of different HEA-forming processes. [8] / permission of Springer Nature.

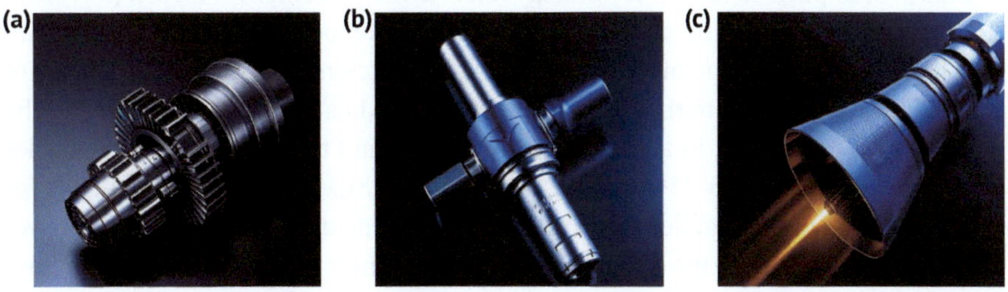

Figure 4.2 Engine parts printed by SLM. (a) High-pressure turbine blades. (b) Aviation engine fuel injector. (c) Rocket engine nozzle. Generated with AI using Doubao.

a thinner single layer. The parts printed out by SLM have extremely high dimensional accuracy and surface quality and can be processed without margin on some parts. As shown in Figure 4.2a–c, SLM technology can print complex structures of any shape that cannot be processed by traditional methods, such as spatial curved porous structures, complex cavity flow channel structures, and lightweight particle array sandwich structures. Therefore, SLM technology is suitable for the precision and rapid forming of small-size and complex-shaped metal components.

DLD is often referred to as LDED or LMD. DLD uses a high-energy laser as a heat source and creates a molten pool in the desired deposition area for rapid movement. The metal powder in the DLD process is fed into the melting area by an inert shielding gas, which is deposited directly onto the metal substrate. The DLD can be called coaxial powder feeding AM because the powder is fed synchronously through a nozzle next to the laser beam [9, 10]. Generally speaking, after the first layer of DLD is printed, the laser cladding head rises and starts the lower layer of printing, which is quickly solidified and depositioned layer by layer, and finally realizes the direct manufacturing of metal parts. Typically, DLDs use lasers in the kilowatt range with spot diameters of more than millimeters. In addition, the diameter of the powder particles used in DLD printing is selected in the range of ~25–200 μm; the smaller the spot diameter, the smaller the diameter of the selected particles and the larger the deposition chamber (generally greater than 1 m^3), which is conducive to the printing of parts. Compared to SLM technology, DLD technology has higher manufacturing efficiency and is often used for the repair of worn and damaged surfaces. However, DLD-prepared parts exhibit low dimensional accuracy, as well as surface roughness, resulting in the need for secondary machining before they can be used.

Electron Beam AM. According to the state of the material, it can be divided into Electron Beam Directed Manufacturing (EBDM) and electron beam selective melting (EBSM) technology. EBDM refers to the electron beam bombarding the metal surface to form a molten pool, and the wire raw material is fed into the molten pool through the wire feeding device and melted to realize the solidification and accumulation of metal materials layer by layer and finally form 3D metal parts or blanks [11–13]. EBDM is characterized by very high forming efficiency and vacuum printing to avoid the harm of impurities. EBSM technology refers to the movement of an electron beam according to a set scanning path, and the pre-laid metal powder is melted and solidified layer by layer, and the metal part is formed by stacking and forming it. The principle of EBSM is similar to that of SLM. However, the heat source is different and the processing is carried out in a vacuum environment, and the diameter of the powder particles is generally ~45–150 μm [13, 14]. Similar to SLM, EBSM is also suitable for manufacturing more complex and smaller- to medium-sized parts but with a lower surface roughness. In terms of formability, EBSM is more suitable for printing alloys with high tough-brittle transition temperatures because the chamber temperature can be heated to more than 1000 °C before printing.

Arc AM. It is a manufacturing technology that uses the arc of the plasma welding power source as a heat source to form 3D metal parts by adding wire. Compared to lasers and electron beams, arc forming is more efficient and can often be used with multiaxis robotic positioners to achieve high degrees of freedom and fabrication of arbitrary complex paths [15–17]. Wire arc AM (WAAM) is suitable for the molding of large and complex parts, which are usually open and do not limit the size of the part, and the technology is easy to stack and repair any local position of the large part. WAAM uses an electric arc to melt the filament, which can melt up to 3–6 kg/h per filament, resulting in a high deposition rate. In addition, WAAM is insensitive to metals, has low equipment costs and short manufacturing cycles, and can be used to expand arc additives of dissimilar metals such as aluminum and titanium.

In addition, there are two main types of metal AM according to the feeding method: one is powder bed fusion, such as SLM and electron beam melting, and the other is direct energy deposition (DED) technology based on powder feeding and wire feeding processes, such as laser engineering net shape (LENS), electron beam freeform fabrication, and WAAM. The main technologies used in the research of AM of HEAs are SLM, EBSM, and LMD, and a few researchers have explored ink extrusion and binder jetting.

4.2 3D Printing of High-Entropy Materials

4.2.1 Introduction of HEAs for 3D Printing

HEAs have emerged as a promising material class for 3D printing due to their unique composition and properties. These alloys typically consist of five or more principal elements in near-equiatomic or high concentrations, forming a single-phase solid solution with remarkable stability [18–21]. The multielement nature of HEAs imparts exceptional mechanical strength, high-temperature stability, and corrosion resistance, making them highly attractive for applications requiring superior performance and durability [21–27].

In 3D printing, HEAs offer several advantages. First, their compositional flexibility allows for the tuning of material properties to meet specific application demands. For example, by adjusting the constituent elements, HEAs can be optimized for use in industries such as aerospace, automotive, and medical systems, where high strength and reliability are crucial. Additionally, HEAs typically exhibit an excellent strength-to-weight ratio, which is particularly valuable for printed components that need to withstand heavy loads and demanding conditions [28–39].

Moreover, HEAs are known for their excellent thermal stability, which makes them suitable for high-temperature 3D printing applications [39–47]. Many HEAs maintain their structural integrity even at elevated temperatures, making them ideal for sectors like aerospace and energy. The multielement composition also often provides superior corrosion resistance, expanding their applicability in harsh environments such as marine or chemical industries.

However, challenges remain in the 3D printing of HEAs. Their complex solidification behavior can lead to issues such as cracking or phase segregation during the printing process, requiring precise control of printing parameters. Additionally, the quality and flowability of the alloy powder are critical factors in achieving successful prints, as uniform particle size and good flowability are essential for ensuring the structural integrity and performance of the printed parts. Postprocessing treatments, including heat treatment, are often necessary to optimize the mechanical properties and reduce residual stresses in the printed components [47–54]. Despite these challenges, research into the use of HEAs in AM continues to advance, with ongoing efforts to improve processing techniques and material performance, unlocking their full potential for a wide range of applications.

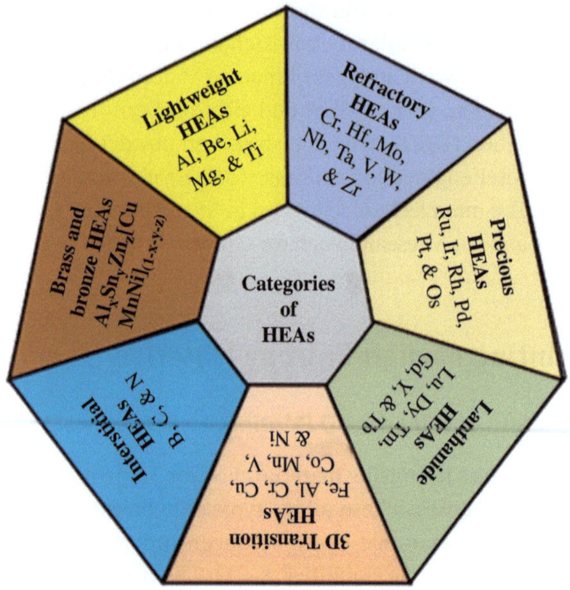

Figure 4.3 Categories of HEAs.

4.2.2 Classification of HEAs for 3D Printing

Although there are many types of elements in HEAs, their structure is a simple solid-solution phase structure, such as common face-centered cubic (FCC) and body-centered cubic (BCC). Hexagonal closed stacked (HCP) phases are also present. According to the metal elements used and the design purpose, HEAs are divided into seven categories, as shown in Figure 4.3.

3d Transition Metal HEAs. The 3D transition group metal HEAs are the earliest discovered family of HEAs, including the famous Cantor alloy (CoCrFeMnNi) and the remaining 345 alloys formed by the permutations and combinations of five to six elements in Al, Co, Cr, Cu, Fe, Ni, Mn, Ti, and V [51, 55–59]. This type of HEA is an extension of stainless steel and nickel-based superalloy materials, but, in terms of strength, none of them can surpass nickel-based superalloys and precipitation-reinforced stainless steel.

Refractory HEAs. Refractory metal HEAs are often formed by the arrangement and combination of Cr, Hf, Mo, Nb, Ta, Ti, V, W, Zr, and Al [60–63]. It can be seen as an alloy family composed of four quaternary alloys such as MoNbTaW, HfNbTaZr, CrMoNbTa, and CrNbVZr, with other elements added as the alloy base. Most of the selected elements are high-melting-point metals, and low-melting-point metals are added to reduce alloy density and improve alloy properties.

Lightweight HEAs. Light metal HEAs are lightweight structural metals designed for space transportation, and the elements are mostly selected from Al, Be, Li, Mg,

Sc, Si, Sn, Ti, and Zn [64, 65]. Due to the wide range of melting and boiling temperatures of the selected elements, their development is limited.

Lanthanide HEAs. Rare-earth metal HEAs are pure solid-solution alloys for the preparation of single-phase HCP. In this type of alloy, there are at least four rare metals. Two alloys have been reported: DyGdLuTbY and DyGdLuTbTm [66].

High-Entropy Brasses and Bronzes. The design of cyan and brass HEAs is to extend the existing family of cyan and brass alloys and improve the properties of traditional cyan and brass alloys by forming solid solutions. The commonly used elements are Al, Cu, Mn, Ni, Sn, and Zn [67], and the reported alloy is $Al_xSn_yZn_z[CuMnNi]_{(1-x-y-z)}$.

Precious Group of HEAs. Noble metal HEAs are designed as catalysts, and at least four of Ag, Au, Co, Cr, Ni, Cu, Pd, Pt, Rn, and Ru are selected as the constituent elements of the alloys, including MoPdRnRu HEAs with HCP structure [68].

HEAs Containing Boron, Carbon, and Nitrogen Gap Compounds. HEAs containing boron, carbon, and nitrogen gap compounds exist in both the 3D transition metal HEA family and the refractory metal HEA family [69, 70]. The most reported are HEAs containing nitrides, which are usually made into thin flakes and have high oxygen content.

4.3 Shape Design and Phase Control in High-Entropy Materials by 3D Printing

4.3.1 Phase Structure of HEAs

The number of components of HEAs is relatively large, and the structure of HEAs is often simple solid-solution phases, including simple FCC, BCC, and HCP solid-solution phases, due to the close mole ratio of each component, as shown in Figure 4.4. The valence electron concentration (VEC) can predict the phase structure of HEA, and when $VEC \geq 8$, the HEA has an FCC-phase structure. When $6.87 < VEC < 8$, the HEA has a mixed-phase structure of FCC and BCC; when $VEC \leq 6.87$, the HEA has a BCC-phase structure. The FCC phase is mainly composed of 3d transition metal elements in near-equiatomic ratios. In general, FCC-phase HEAs have better plasticity than

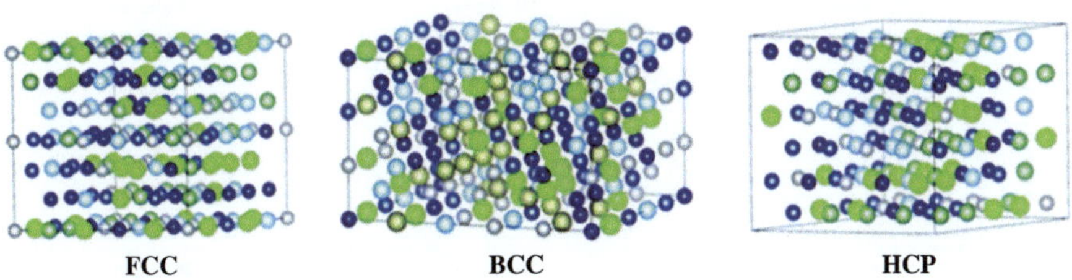

Figure 4.4 Schematic structure of HEA FCC, BCC, and HCP phases. [70] / with permission of Frontiers Media S.A./ CC BY 4.0.

BCC-phase HEAs, but their strength is weaker. Typical FCC-phase HEA CrMnFeCoNi has excellent thermodynamic stability and plasticity, while BCC-phase HEA is mostly composed of refractory elements with high thermal stability and hardness, and the typical representative is TaNbMoW. There are relatively few HCP-phase HEAs, mainly rare-earth element-based HEAs, and the representative one is CoFeReRu HEA.

4.3.2 The Influence of Printing Parameters on the Phase Structure

These parameters affect the cooling speed, which in turn affects the phase structure of the alloy. Using SLM to construct CoCrFeMnNi HEA with 5–15% TiC, the results show that at 100 and 400 W laser power, the sample is a single-phase BCC structure; when the laser power rises to 700 W, due to the high laser power, the bath temperature will increase, resulting in a high-temperature reaction, and the second phase of MnC appears in the sample (Figure 4.5) [71]. AlCoCrFeNiNiCu and AlTiCrFeCoNi HEA samples with 600–800 W laser power and 4–12 mm/s scanning speed are used (Figure 4.6) [72]. The BCC phase with a small amount of FCC structure was observed at a scanning speed of 4 mm/s. The high volume fraction of the BCC phase manifests as the scanning speed increases. This is because the high scanning speed leads to a high cooling rate, which moves the FCC to the BCC-phase structure.

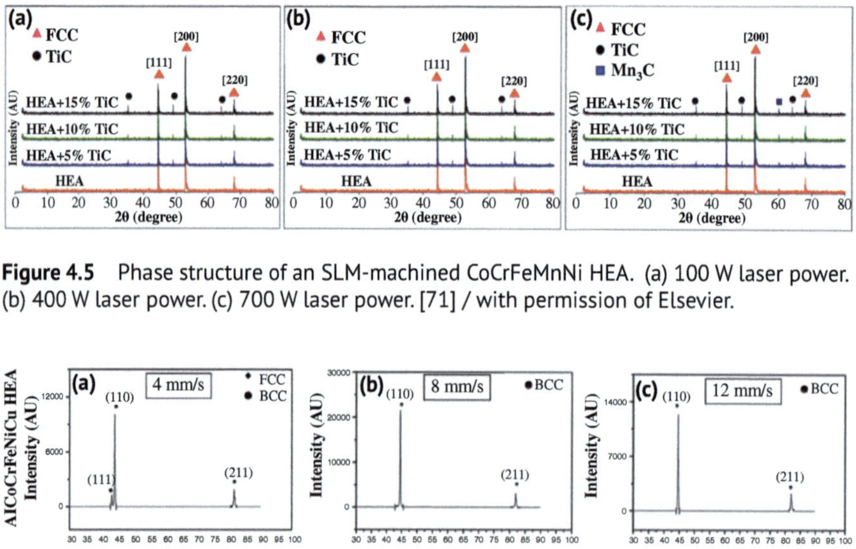

Figure 4.5 Phase structure of an SLM-machined CoCrFeMnNi HEA. (a) 100 W laser power. (b) 400 W laser power. (c) 700 W laser power. [71] / with permission of Elsevier.

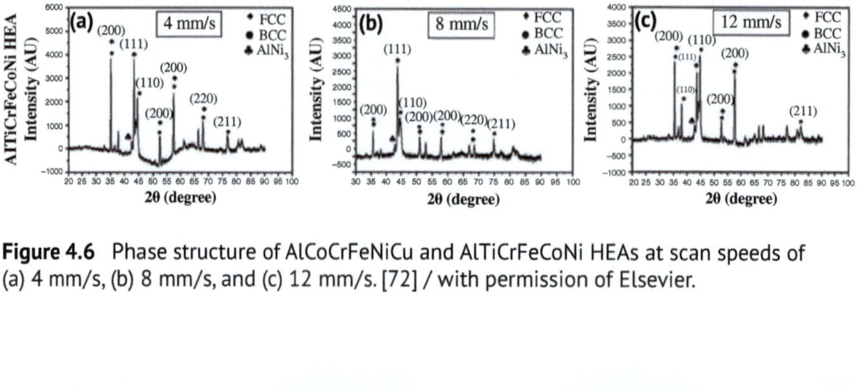

Figure 4.6 Phase structure of AlCoCrFeNiCu and AlTiCrFeCoNi HEAs at scan speeds of (a) 4 mm/s, (b) 8 mm/s, and (c) 12 mm/s. [72] / with permission of Elsevier.

4.3.3 Effect of Alloying Elements on Phase Structure

The addition of specific elements to the HEA system can significantly change the structure and precipitated phase of the HEA, thereby affecting the mechanical properties of the alloy. The effects of different alloying elements on the phase structure are as follows:

Addition of Al Element. Borkar et al. [73] studied the evolution process of Al content on the solidified tissue of Al_xCoCrFeNi blocks under LMD. With the increase of Al content, the phase structure of the solidified tissue changes from FCC (containing the sequence L12 phase) to FCC + BCC, and finally, when $x = 0.85$, it all changes to the BCC-phase structure (containing the sequence B2 phase) and the FCC is the first solidified phase, and when the Al content is high, the BCC phase solidifies first [73]. In the study on the preparation of Al_xCrCuFeNi by SLM [74], with the increase of aluminum content, the crystal structure also changed from FCC + BCC/B2, accompanied by the transition of columnar crystals to isometric crystals. When $x = 0.75$ and $x = 1.0$, the alloy is formed by a sheet/cell-like FCC matrix and inlay. Chen et al. [75] prepared $Al_x(CoFeNiCu)_{1-x}$ HEA by LMD and found that, with the increase of Al content and the decrease of Cu content, the crystal structure changed from mainly FCC to a combination of FCC and BCC.

Addition of Mn and Ni. Mn and Ni have a high diffusion coefficient and tend to be isolated in the interdendrite region [1] or at the boundary of the cell structure [77]. CrMnFeCoNi may form secondary phases during the printing process, such as σ phase and HCP martensitic phase. Compared with traditional methods, AM has fast cooling speed, grain refinement, and high grain boundary density, which is conducive to atomic diffusion and therefore conducive to the formation of the tetragonal σ phase [77]. In addition, the high dislocation density forms a martensitic phase, causing the Shockley dislocation to propagate along the alternating slip surface [78].

Addition of Three-Dimensional Transition Metals. Other elements added to three-dimensional transition metal HEAs include Si [79, 80], Cu [81], Ti [82], and Mo [80]. The addition of 1.5 at.% Si to CrFeCoNi HEA results in a single FCC phase [81], while the addition of a higher Si content (~9 at.%) to CrCoNi HEA results in a Cr- and Si-rich σ phase, a Ni-rich precipitate, and CrS inclusions [83]. In addition, the addition of Si has caused new dislocation structures in the printing material, including planar defects [81] and dislocation loops [81], which are attributed to the reduction of lamination fault energy.

The Addition of Cu Elements. In the LPBF process, the uneven mixing of the CrFeCoNi alloy and Cu powder can lead to unmelted Cu-rich particles due to the easy segregation of Cu elements [84]. However, successful mixing of powders yields HEAs with uniform elemental distribution [81], suggesting that the aerosolized CrFeCoNiCu powder will further promote the formation of a single FCC phase.

The Addition of Ti Elements. $CoCrMoNbTi_x$ ($x = 0, 0.2, 0.4, 0.5$, and 1.0) prepared by arc smelting is a refractory HEA. When the alloy does not contain Ti, in addition to the main phase BCC solid solution, there is also a small amount of Laves phase in the alloy. When a small amount of Ti element ($x = 0.2$) is added, the alloy is still dominated by the BCC solid-solution phase; when the Ti element content is increased to 0.4, the alloy is a single-phase BCC solid solution; when the Ti element content is increased to 0.5, 1.0, except for the main BCC solid-solution phase, the alloy is a single-phase BCC solid solution. A small amount of Cr_2Nb and Co_2Ti type Laves phases have been re-precipitated in the alloy. However, after Ti is added, its high melting point may cause the powder to be mixed in situ, and the Ti particles are not melted when the gas atomization. The powder elements are evenly distributed, and there is no secondary phase [84]. Adding 4.8 (%Mo) to CrFeCoNi will cause the Mo-rich phase to precipitate near the boundary of the bath, while there is still no precipitated phase in the center of the bath [80].

4.3.4 Research Progress on Additive Production of HEAs

In recent years, AM has emerged as an effective technique for manufacturing metallic materials. This technique enables the fabrication of complex components based on three-dimensional CAD models. AM technology is gaining prominence in materials science and is widely used to manufacture products with complex geometries. To date, various AM-based methods have been developed and are now commonly used in the manufacturing process of a variety of engineering materials, including titanium alloys [85], aluminum alloys [86], steels [87], magnesium alloys [88], Co alloys [89], and composite materials [90] (Figure 4.7).

The preparation method of HEAs is mainly based on vacuum arc smelting and casting technologies. However, the cooling speed of HEAs prepared by arc smelting is slow, and the prepared alloys have macroscopic defects such as component segregation, pores, and incisions; at the same time, the use of traditional techniques to prepare HEAs have a long production cycle and low accuracy, especially the preparation of refractory HEAs is prone to low raw material utilization and high cost and the preparation process of complex structural parts is limited, which is not conducive to the practical application needs of HEAs. The use of 3D printing technology to form HEAs can effectively improve production efficiency and reduce costs while meeting the integrated forming of complex components. At present, a series of HEA components with small tissues and excellent properties have been successfully prepared at home and abroad. Currently, the major AM techniques used for HEAs include SLM, Selective Laser Sintering (SLS), SEBM, DED, Binder Jetting Technology (BJT), direct ink writing (DIW), and additive friction stir deposition (AFSD) [98–104]. Studying the microstructural evolution, property enhancement, and defect control of HEAs fabricated via these AM methods is of paramount importance and has attracted considerable attention in the academic community.

SLM Technology. A schematic diagram of SLM is illustrated in Figure 4.8a. SLM, also known as LPBF, is an advanced AM technology that employs high-power lasers to selectively melt and solidify specific areas of metal powder, conforming to the

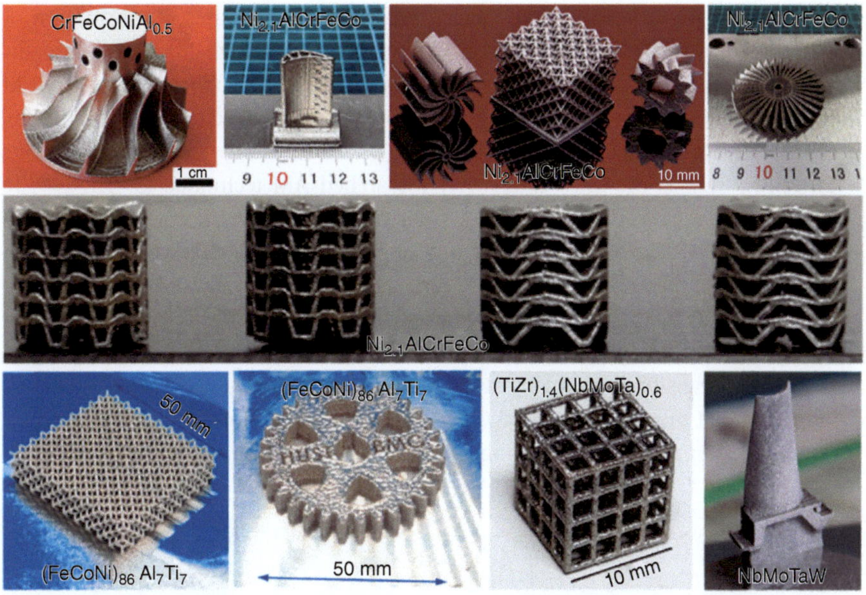

Figure 4.7 Industrial components consisting of AM-fabricated HEAs, including a CrFeCoNi $Al_{0.5}$ 3d transition metal HEA [91]. Reproduced with permission of Elsevier. A $Ni_{2.1}AlCrFeCo$ eutectic HEA [8, 92–94]. Reproduced with permission of Elsevier. An $(FeCoNi)_{86}Al_7Ti_7$ precipitation-strengthened HEA [95]. Reproduced with permission of Elsevier; $(TiZr)_{1.4}$ $(NbMoTa)_{0.6}$ and NbMoTaW refractory HEAs [96, 97]. Reproduced with permission of Elsevier and Springer Nature.

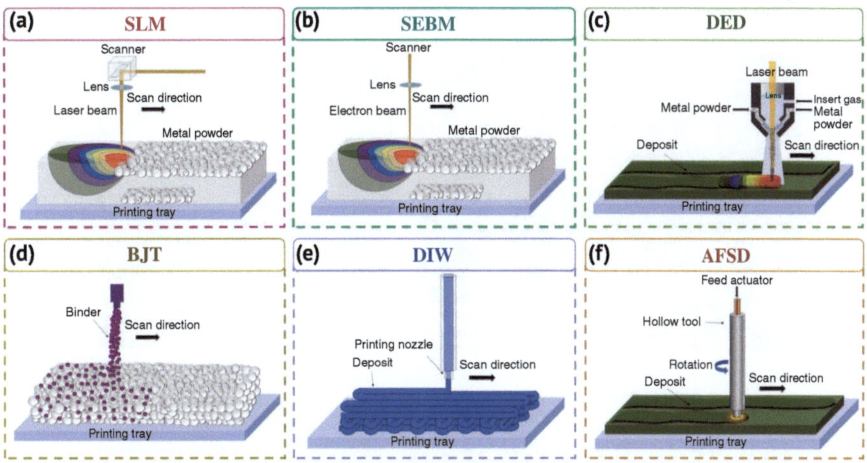

Figure 4.8 Schematic of typical (a) SLM process; (b) SEBM process; (c) DED process; (d) BJT process; (e) DIW process; (f) AFSD process [105]. / MDPI / CC BY 4.0.

cross-section data of a 3D CAD model. This layer-by-layer approach makes it possible to fabricate complex geometries that are often impossible or challenging using traditional manufacturing methods. SLM has several unique advantages over SEBM and DED. It uses a lower laser power and a thinner, high-beam-quality laser beam.

This promotes rapid heating and cooling rates, which facilitates the formation of nonequilibrium phases and fine-grained microstructures with innovative properties. A wide range of microstructures and properties can be customized using SLM by adjusting processing parameters such as laser traversal speed, power, beam size, precursor composition, and feed rate.

Currently, SLM is the most extensively researched AM technique for HEAs. The Cantor alloy (i.e., CoCrFeMnNi) and its derivatives (CoCrFeNi, AlCoCrFeNi, etc.) produced by SLM have been thoroughly investigated. Brif et al. [106] conducted a pioneering study on the application of SLM to HEAs. Their work showed that a uniform single-phase FCC solid-solution scheme in isotropic FeCoCrNi HEAs in a range of processing parameters using beam traversing rates of 0.33, 0.36, 0.38, 0.44, 0.48, and 0.5 m^{-1}, with no phase transitions occurring at the melting point of the alloy (1414 °C). Li et al. [1] reported that the SLM processed CoCrFeMnNi HEA with unique microstructural features including nanotwin and tetragonal σ-phase (Figure 4.9a–e), which significantly

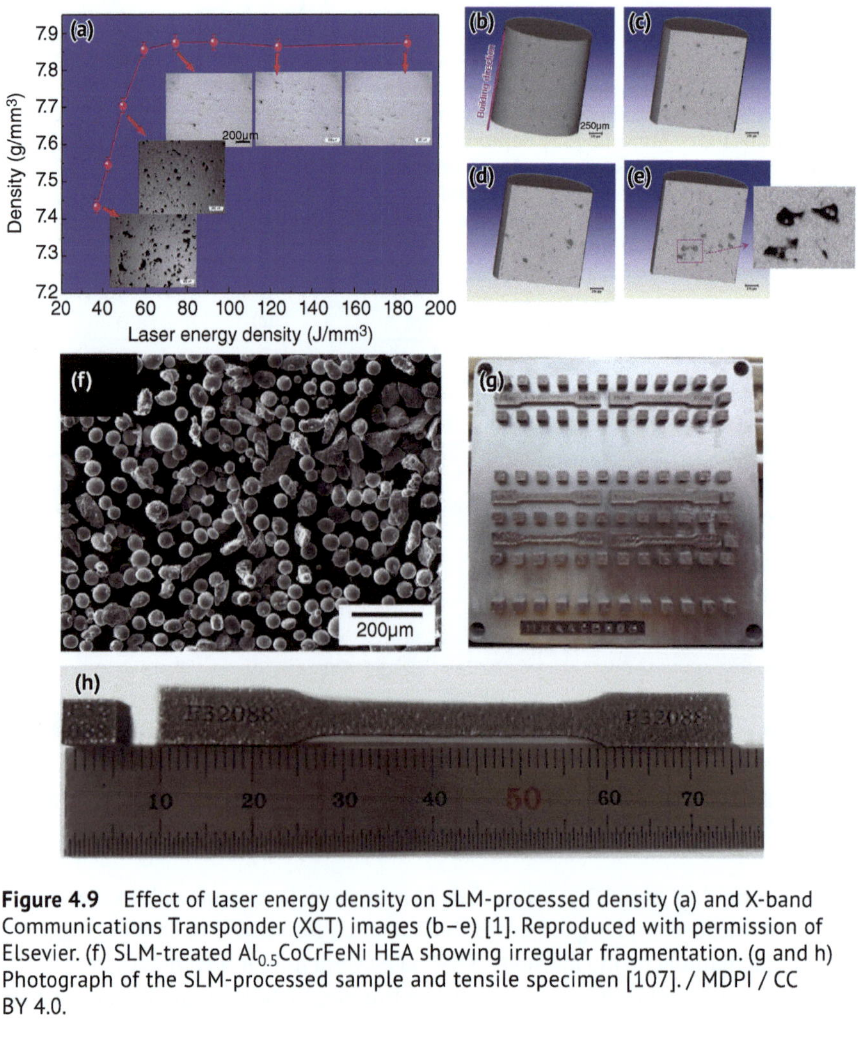

Figure 4.9 Effect of laser energy density on SLM-processed density (a) and X-band Communications Transponder (XCT) images (b–e) [1]. Reproduced with permission of Elsevier. (f) SLM-treated $Al_{0.5}CoCrFeNi$ HEA showing irregular fragmentation. (g and h) Photograph of the SLM-processed sample and tensile specimen [107]. / MDPI / CC BY 4.0.

improved the mechanical properties of the alloy through grain refinement and synergistic strengthening effects. In addition, Sun et al. [107] found that the microstructure of SLM-treated $Al_{0.5}$CoCrFeNi HEA showed to consist of FCC and BCC phases without complex intermetallic compounds (Figure 4.9f–h).

SLS Technology. SLS is similar to SLM, but it uses a laser to sinter the powder bed instead of melting it. This allows the production of lower-density materials for specialized applications. In the field of HEAs, the available literature on SLS is quite sparse. Only one paper was found to be related to HEAs fabricated by SLS. Yan et al. prepared $Ni_{30}Cr_{25}Al_{15}Co_{15}Mo_5Ti_5Y_5$ HEA coatings by SLS under different pulsed/continuous wave laser operating modes to improve the surface properties of pressurized light-transmitting tubes. The effect of laser-induced plasma on the SLS process was investigated by analyzing the micromechanics, friction, and wear behavior of the HEA coatings.

SEBM Technology. SEBM, also known as electron powder bed fusion, is an alternative AM technology based on a powder bed system that utilizes an electron beam instead of a laser beam, as shown in Figure 4.8b. The focused electron beam sequentially melts points established in the slice data obtained from the 3D CAD file. Therefore, this fundamental difference requires SEBM to operate under a high vacuum system (over 10^{-4} mbar) to ensure a high-quality electron beam and a contamination-free processing environment. In addition, SEBM employs larger beam sizes and thicker layer thicknesses, from 50 to 200 μm, compared to SLM, requiring the use of larger sizes of spherical metal powders for printing. The cooling rate of SEBM is typically two orders of magnitude lower than that of SLM, resulting in slightly poorer dimensional accuracy and surface quality of the built parts compared to those produced by SLM. In particular, the SEBM process starts with a high-temperature preheating of the powder bed (up to 1100 °C) by means of a diffuse-focused electron beam, high beam current, and velocity values. This step causes a slight sintering of the powder compared to SLM, resulting in a lower temperature gradient and cooling rate (103–105 K s^{-1}) for SEBM. As a result, thermal cracks and structural failures due to thermal stresses during printing can be avoided to a certain extent.

Shiratori et al. [108] investigated the microstructure of SEBM-prepared equiatomic AlCoCrFeNi HEAs (Figure 4.10a), revealing a predominantly nanolamellar mixture of disordered BCC and ordered BCC (B2) phases, with the precipitation of the BCC phases at the grain boundaries being affected by a unique preheating process of the SEBM (Figure 4.10b, c). Fujieda et al. [109] found that the as-cast AlCoCrFeNi specimens and as-prepared powders had a BCC single-phase structure, whereas the AlCoCrFeNi specimens of SEBM had both BCC and FCC structures (Figure 4.10d, e), while Wang et al. [110] found that the CoCrFeNiMn HEA assemblies constructed by SEBM exhibited hierarchical microstructures with long columnar particles and intraparticle cellular structures (dendrites) aligned along the build direction (Figure 4.10f, g).

DED Technology. DED is the use of focused thermal energy to fuse materials by melting during the deposition process. It is an advanced manufacturing

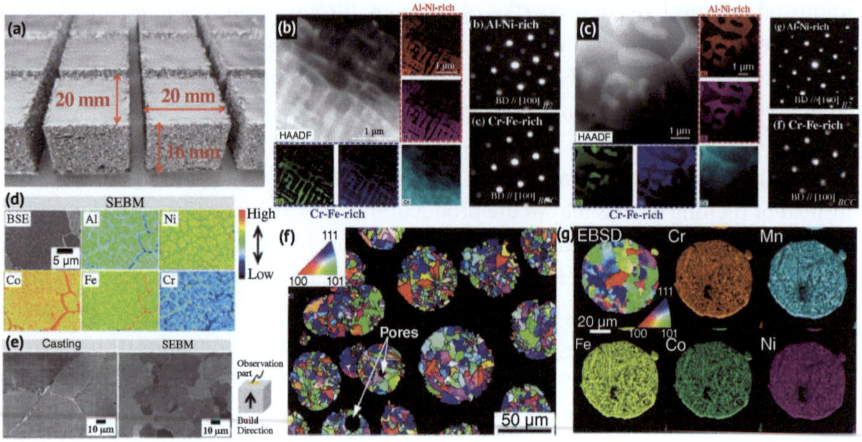

Figure 4.10 (a) AlCoCrFeNi alloy specimens fabricated by SEBM. HAADF–STEM image and Energy Dispersive Spectrometer (EDS) elemental maps for the (b) top and (c) bottom parts of the SEBM specimens [108]. Reproduced with permission from / Elsevier. (d) High-magnification backscattered electron (BSE) micrographs of SEBM specimens and corresponding Electron Probe X-ray Micro-Analyzer (EPMA) elemental mapping. (e) BSE micrographs of cast and SEBM specimens [109]. Reproduced with permission from / Elsevier. (f) Electron Back Scatter Diffraction (EBSD) showing polycrystalline of HEA powder. (g) EBSD image and the corresponding elemental distribution of the as-received powder [112]. Reproduced with permission from [112] / Elsevier.

technology based on AM and laser cladding technology. In the DED process, raw metal materials are melted using various energy sources such as laser beams, electron beams, or electric arcs. The raw material for the process can be in the form of powder, using blown powder deposition techniques, or wire, using arc and laser melting as heat sources. Figure 4.8c illustrates the setup of the DED technique. This technique can overcome the limitations of low deposition rates, limited sample size constrained by the size of the powder bed, and substantial material waste.

The most widely studied alloy using the DED method is CrMnFeCoNi HEA. Most studies reported that DED-treated CrMnFeCoNi HEA has a single-phase FCC structure. In a study by Tong et al. [111], DED-produced FeCrCoMnNi HEA maintained a single FCC solid solution after heat treatment (Figure 4.11a–c). However, Gao and Lu [112] found that DED-prepared CoCrFeMnNi HEA exhibited a fine distribution of BCC phases at the grain boundaries of the FCC substrate. Li et al. [113] used DED to fabricate a single-track CoCrFeMnNi HEA (Figure 4.11d–f) and found that the lower temperature gradient to solidification rate ratio was associated with the transformation of columnar to equiaxed morphology. Li et al. [114] investigated the effect of thermal confinement, thermal cycling, and temperature gradient along the deposition direction on the residual stress distribution of CoCrFeMnNi HEA and found that the thermal gradient during the lowering process reduces the level of residual stresses, which affects the deformation and the functional properties of the alloy. DED-prepared HEAs have a microstructure of directional solidification from the dendritic columnar grains close to the melting pool boundary to those far away from the equiaxed grains.

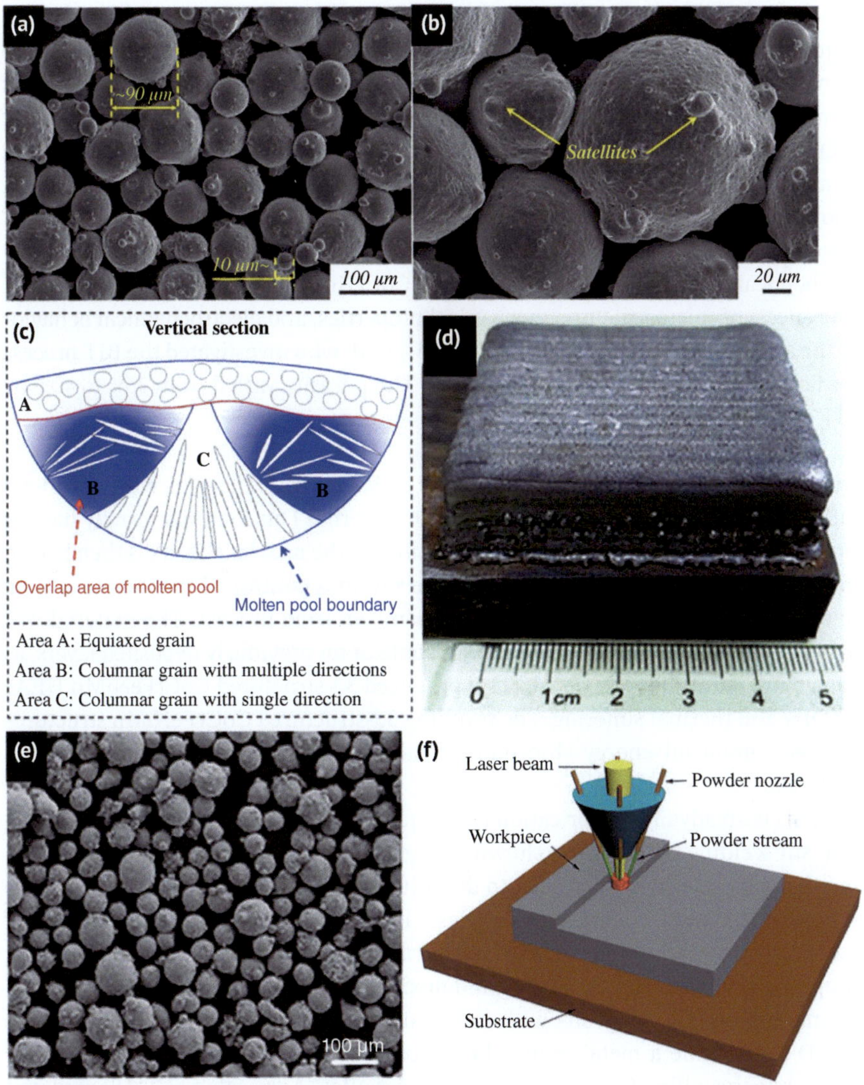

Figure 4.11 (a, b) SEM morphology of prealloyed FeCrCoMnNi alloy powders. Reproduced with permission from [111] / Elsevier. (c) Schematic illustration of the microstructural features in the vertical section of a single molten pool. [111] / with permission of Elsevier. (d) Appearance image of the printed CoCrFeMnNi HEA sample. (e) SEM morphology of CoCrFeMnNi HEA powder prepared by DED. Reproduced with permission from [112] / Elsevier. (f) Schematic diagram of the coaxial powder feeding laser 3D printing system. [112] / Elsevier / CC BY 4.0.

BJT Technology. BJT is an AM process in which a liquid binder is selectively deposited to bind powder materials together. It utilizes a powder bed system and uses a liquid binder instead of a heat source, as shown in Figure 4.8d. This method consists of applying the binder material to the powder bed and fabricating the components layer by layer to form a "green body" that needs to be sintered to achieve sufficient density. Therefore, postprocessing steps, such as sintering or infiltration, are often necessary to obtain the final material properties. Key features of this technique include its ability to create complex geometries, accommodate a wide range of materials (e.g., metals, ceramics, and polymers), and generally build faster than some other AM methods.

Only three papers on HEAs fabricated using the binder-based AM technique were found. Karlsson et al. [102] prepared high-density compositions of AlCoCrFeNi alloys by BJT and subsequent sintering. Chen et al. [115] explored the effects of sintering temperature and hot isostatic pressure on the microstructure and electrochemical behavior of the isoatomic FeNiCoCrHeNi HEAs fabricated by the BJT. The effects of microstructure, mechanical properties, and electrochemical behavior of the samples were investigated by Xu et al. [116], who investigated the BJT process for fabricating porous CoCrFeMnNi HEA. The porosity varied from 35% to 40% by BJT and subsequent sintering process.

Other Technologies. DIW is another type of AM process. It begins with the preparation of a specially formulated ink or paste material. This material usually has the right viscosity to allow it to pass smoothly through the nozzle. The ink is then loaded into a dispensing system that extrudes the ink in a controlled manner according to a preprogrammed pattern during the manufacturing process. The material was deposited layer by layer, either on the substrate or on previously deposited layers, as shown in Figure 4.8e. Peng et al. [117] produced 3d-structured CoCrFeNiMn HEAs by DIW and thermal sintering. The resulting 3d-structured CoCrFeNiMn structures showed significant energy absorption capabilities over conventionally structured materials.

AFSD is an advanced fabrication technique that combines the principles of friction stir welding with AM, as shown in Figure 4.8f. In the AFSD process, a rotating tool generates frictional heat to deposit material layer by layer, which makes it possible to create complex geometries and structures, especially those made of metal. This method offers significant advantages, including enhanced material properties due to localized heating and mechanical agitation, which helps to form strong, high-quality deposits with minimal porosity. Karthik et al. [104] utilized AFSD to fabricate a metal-to-metal composite containing an AA5083 reinforced with nanocrystalline CoCrFeNi HEA particles alloy. Agrawal et al. [118] explored AFSD as a solid-state AM technique capable of producing defect-free 3D components with a uniform, equiaxed microstructure. AFSD-treated $Fe_{40}Mn_{20}Co_{20}$ $Cr_{15}Si_5$ HEA exhibited transformation and twinning-induced plasticity during deformation.

The Influence of Process Parameters on Forming Quality. Kunce et al. [119] of the Poland Military Technical Institute prepared AlCoCrFeNi HEAs using LENS technology and found that the alloys had a simple BCC-ordered solid-solution structure. As the laser scanning rate increased, from 2.5 to 40 mm/s, the average grain size decreased from 108.3 ± 32.4 to 30.6 ± 9.2 μ0. Compared to the hardness of as-cast alloys, the hardness of the alloy formed by LENS is about 13% higher. Zhou et al. [120] found that the density of the FeCoCrNi bulk alloy formed by SLM increased with the increase of scanning power and velocity, the yield strength reached 656 MPa, and the elongation was 13% at the highest density. Niu et al. [121, 122] used SLM technology to fabricate AlCoCrFeNi HEAs, and the density of SLM samples gradually increased with the increase of volumetric energy density (VED), but there were still a small number of pores in the alloy as VED was maximum. Hiroshi Shiratori et al. studied the effect of the preheating process on the phase structure of AlCoCrFeNi HEAs formed by SEBM. The preheating process leads to the precipitation of the FCC phase at the BCC/B2 grain boundary, and the content of FCC precipitation at the bottom of the molded part is significantly higher than that at the top due to the different preheating durations of the upper and lower parts of the molded part [108].

Analysis of the Texture and Performance of the Molded Parts. In 2015, Yevgeni Brif et al. [106] confirmed for the first time the feasibility of SLM-forming HEAs. The FeCoCrNi HEA prepared by SLM has a single-phase FCC solid-solution structure, and the composition is uniform and nonpolarized. The yield strength of the alloy is as high as 600 MPa, which is significantly better than the cast FeCoCrNi alloy formed by traditional arc smelting technology. Fujieda et al. [109], using AlCoCrFeNi prealloy powder as raw material, SEBM was used for the first time to successfully prepare AlCoCrFeNi HEA. Compared with cast alloys, SEBM-formed AlCoCrFeNi alloys have fine dendrite structure and better compression properties, and their compression breaking strength is 1400 MPa.

The Ev value significantly affects the grain size and grain orientation of CrMnFeCoNi HEA. As the Ev increases, the grain size increases, and the texture is sequentially constructed in the direction of <233>, <001>, <203>, <101>; this is related to the shape of the bath [123]. In addition, the grain texture will be affected by the scanning strategy [124, 125], the rotation angle [126], and the position in the printed part [127]. This phenomenon can be understood by recognizing that the grain structure of the alloy manufactured by LPBF is aligned with the largest thermal gradient, and different thermal gradients occur as the laser rotates, stops, or continues to advance.

The Impact of Post-Treatment on the Organization and Performance of the Formed Parts. The AlCoCrFeNi HEAs prepared by Direct Laser Fabrication (DLF) were 600, 800, 1000, and 1200 °C. After maintaining the constant temperature for 168 h, the tissue properties were observed, and it was found that, as the temperature of the aging treatment increased, the content of the FCC phase first increased and then decreased. The tissue was adjusted by the amplitude of the DLF sample. The decomposition and precipitation are transformed into tissues with Al-rich and Ni-rich phases as the substrate and FCC and σ-rich phases as the precipitation. When the aging temperature is low, due to the increase of the metal phase, the compressive strength of the alloy decreases, and the plasticity increases.

After heat treatment of the $Al_{0.3}CoCrFeNi$ HEA manufactured by LENS, the yield strength was increased to 750 MPa, and the tensile ductility reached about 10%. The increase in alloy strength is due to the increase in dislocation density under fast cooling and fast heating, the formation of subgrain boundaries, and the formation of Al–Ni atomic clusters under repeated laser heating. The plasticity of the alloy stems from the excellent plasticity of the FCC phase. It can be seen from this that, to achieve the combination of strength and toughness of the alloy, a large volume of the active phase and suitable strengthening factors are required.

Joseph et al. [128, 129] also studied the deformation mechanism of $Al_xCoCrFeNi$ ($x = 0.3, 0.6, 0.85$) HEA prepared by DLF under tensile and compression, as well as the effects of hot isostatic pressing (HIP) on their mechanical properties and grain structure. The study found that, for $Al_{0.3}$, after HIP, the segregation of Al-rich in the grain boundary was reduced, and the alloy was more uniform, thereby improving the toughness of the alloy. For $Al_{0.6}CoCrFeNi$, the Widmanstätten structure in the DLF sample did not disappear after HIP treatment but became coarser. Also roughened is the B2 phase of the intergranular region, which leads to improved compressive performance but decreased tensile performance. For $Al_{0.85}CoCrFeNi$, the σ phase appeared after HIP treatment, and the microstructure changed into an Al–Ni-rich B2 matrix and Fe–Cr-rich BCC precipitate. Although the spinodal decomposition disappeared due to heat treatment, the compressive properties of the σ phase formed are somewhat reduced because it is a brittle phase.

4.4 Applications of 3D Printed High-Entropy Materials

Compared with traditional manufacturing, 3D printed high-entropy materials (HEM) has potential applications in a variety of industries, including energy, aerospace, and biomedicine, due to its increased design freedom, reduced material waste, and unique microstructure control.

4.4.1 Energy Industry

The combination of HEM and AM is unlocking innovative applications across the energy sector. In power generation, 3D-printed HEAs are being used for turbine components, such as blades and nozzles, where their high-temperature strength and oxidation resistance ensure durability and efficiency. Similarly, in nuclear energy, HEMs excel in radiation resistance and corrosion resistance, making them ideal for reactor components like heat exchangers and fuel cladding. AM enables precise customization, optimizing performance under extreme conditions. HEMs also hold promise in renewable energy and storage technologies. For example, they are used in wind turbines for durable gearboxes and bearings and in hydrogen production systems for efficient catalysts and storage materials. Additionally, high-entropy ceramics (HECs) are advancing battery and fuel cell technologies by improving electrode designs and enhancing charge transport.

In applications like heat exchangers, oil and gas pipelines, and thermal management systems, the combination of HEMs' material properties and the design freedom of AM allows for more efficient and robust solutions. The ability to produce lightweight, corrosion-resistant, and high-strength components is helping to address critical challenges in energy efficiency and sustainability.

Replacing fossil fuels with renewable resources is an important cornerstone for sustainable development. The development of efficient fuel cells, hydrogen storage, and sustainable hydrogen production methods is an important way to develop renewable resources. Among them, the bipolar flow plate in the fuel cell is an important component that converts the stored hydrogen into electrical energy. Bipolar flow plates have complex flow path patterns and are subjected to extremely corrosive environments with high potential differences, high operating temperatures, and low pH during service [129, 130]. The combination of 3D printing and HEAs is expected to optimize flow channels and have excellent service performance at the same time. It has been reported that CrMnFeCoNi alloys prepared by LPBF can form a stable antioxidation layer in the simulated cathodic environment after heat treatment at 800 °C, which is expected to be a candidate material for bipolar flow plates [129]. The TiZrNbMoV HEA powders by the LENS method showed that a stable BCC structure could be obtained after laser triple remelting, and its hydrogen storage capacity could reach 0.59 wt%, which could be increased to 0.61 wt% after heat treatment [130]. The equimolar prealloyed ZrTiVCrFeNi HEA powder-forming sample has a hydrogen storage capacity of 1.81 wt% [119]. These results fully demonstrate the application potential of 3D-printed HEA in hydrogen storage.

In addition, 3D-printed magnetic HEAs are expected to solve the problems of high cost, poor mechanical properties, and complex manufacturing processes of traditional soft magnetic materials such as super alloys, silicon steel, Permendur, Molybdenum Permalloy Powder cores, and high magnetic fluxes. Overall, the synergy between 3D printing and HEMs is driving significant advancements in energy systems, offering customized, durable, and high-performance solutions that are shaping the future of the energy industry.

4.4.2 Aerospace Industry

The aerospace industry demands materials with exceptional strength, thermal stability, and resistance to extreme environments. HEMs, including HEAs and HECs, have emerged as ideal candidates to meet these challenges. The integration of AM techniques, such as SLM and LMD, allows for the precise fabrication of complex aerospace components, further enhancing their applicability.

One major application of 3D-printed HEMs in aerospace is in high-temperature engine components, such as turbine blades, combustors, and nozzles. HEAs exhibit excellent high-temperature strength, oxidation resistance, and thermal fatigue properties, making them suitable for use in jet engines and rocket propulsion systems. AM enables intricate cooling channels and lightweight structures within these components, improving efficiency and durability.

Structural components in aircraft and spacecraft also benefit from 3D-printed HEMs. The ability to tailor HEM compositions for enhanced strength-to-weight ratios makes them ideal for load-bearing parts, such as airframes, brackets, and fasteners. AM techniques provide the design flexibility needed to reduce weight without compromising structural integrity, contributing to fuel efficiency and payload capacity. In addition to mechanical applications, HEMs are used for thermal protection systems in spacecraft. Their superior thermal stability and resistance to thermal shock ensure reliable performance during extreme temperature fluctuations, such as those experienced during reentry. AM facilitates the fabrication of multifunctional heat shields and protective layers with optimized geometries.

Moreover, HECs are finding applications in aerospace for advanced coatings and ablative materials. These materials protect components from erosion, oxidation, and high-speed aerodynamic heating. The precision of AM processes allows for customized coatings that enhance the lifespan and performance of critical parts.

Three-dimensionally printed HEAs offer great application potential for lightweight devices in the aerospace field due to their excellent strength, corrosion resistance, wear resistance, and high hardness and fatigue properties. The specific strength of TiBeAlSiFeHEA is more than twice that of $TiAl_6V_4$, the most widely used commercial alloy in aerospace applications [88]. AlSiMnFe [131] and AlTiVCr [132] have excellent corrosion resistance while maintaining low density and low cost. The high-entropy matrix composite $(CrCoNi)_{96.9}W_{0.95}Al_{0.65}Nb_{0.47}Re_{0.47}Ti_{0.30}C_{0.24}–Y_2O_3$ has significantly better creep resistance at high temperatures than other state-of-the-art LPBF-fabricated alloys [133]. Therefore, the use of 3D printing of HEAs to prepare valve trains, exhaust nozzles, combustion chambers, compressor blades, and aeroengine gas turbines [134] and other components is expected to further increase the operating temperature of the engine, increase the thrust-to-weight ratio of the engine, and promote the development of the aviation industry.

The combination of AM and HEMs is revolutionizing aerospace engineering. By enabling the production of lightweight, durable, and high-performance components, this synergy is addressing key challenges in efficiency, safety, and sustainability, paving the way for the next generation of aerospace innovations.

4.4.3 Nuclear Application

In nuclear reactors, 3D-printed HEAs are used for critical components such as fuel cladding, heat exchangers, and structural supports. Their ability to resist neutron irradiation reduces material degradation, such as swelling and embrittlement, under prolonged radiation exposure. AM enhances the design flexibility of these components, allowing for features like optimized cooling channels that improve thermal efficiency and operational safety.

HEMs also play a key role in nuclear waste management. Their outstanding corrosion resistance ensures the long-term integrity of storage containers for radioactive waste. AM further allows for the creation of durable, customized containment systems that maximize safety while minimizing material waste.

Additionally, HEMs are being explored for advanced nuclear fuel cladding materials and fusion energy systems. Their resistance to oxidation and mechanical degradation at high temperatures makes them suitable for demanding applications such as plasma-facing components in fusion reactors. AM enables precise control over the material's composition and microstructure, improving performance and extending the lifespan of these components.

By combining the unique properties of HEMs with the precision and flexibility of AM, the nuclear industry is advancing toward safer, more efficient, and sustainable energy solutions. This synergy is critical to addressing the challenges of modern nuclear systems and supporting the development of next-generation energy technologies.

The excellent low activation energy, creep resistance, corrosion resistance, and oxidation resistance of HEAs make them ideal for thermonuclear reactor materials. At present, equimolar AlCoCrFeNi HEA prepared by SLM has a single-phase BCC structure, and the gap of the layered structure can be adjusted by adjusting the SLM parameters, which is suitable for next-generation filters in the extreme environment of nuclear reaction and oxidation [135].

4.4.4 Medical Applications

One key application of 3D-printed HEAs in medicine is the development of orthopedic implants and prosthetics. HEAs possess excellent mechanical properties, such as high strength and toughness, which make them suitable for load-bearing implants like hip and knee replacements. Their inherent corrosion resistance ensures long-term stability in the human body, while AM allows for the creation of patient-specific designs, such as porous structures that promote bone ingrowth and osseointegration.

HEMs are also being explored for use in dental implants and surgical instruments. The wear resistance and biocompatibility of these materials ensure durability and safety in demanding environments. AM further enables the production of intricate geometries, such as customized dental crowns and bridges, enhancing patient comfort and treatment outcomes. In addition to implants, 3D-printed HEAs are being utilized in medical devices and components for diagnostic and therapeutic applications. For example, the corrosion resistance and mechanical precision of HEAs make them ideal for devices operating in harsh environments, such as fluid delivery systems and components in imaging equipment.

For example, TiTaNbMoZr HEA can be used for knee implants using microplasma arc AM. The hardness of TiFeCrCoNi HEA produced by the LPBF process is as high as 776.3 HV, which is 2.8 times that of pure Ti. The strengthening effect due to grain refinement, as well as the increased hardness of the alloy, enhances the wear resistance of the alloy and is expected to be suitable for dental and orthopedic surgery [107]. In addition, LPBF-prepared CrFeCoNiCu exhibits antimicrobial properties against *Escherichia coli* and *Staphylococcus aureus* [136]. By replacing Cu with Ti, HEA with better cytocompatibility can be obtained than pure Ti. Similarly, a variety of novel bio-HEAs can be designed by utilizing other biocompatible metals as the main elements in HEAs, such as zinc [137] and magnesium [82].

The synergy between HEMs and AM is driving advancements in the medical industry, enabling the production of innovative, patient-specific, and high-performance solutions. This combination not only enhances the functionality and durability of medical devices but also opens new avenues for personalized healthcare and improved patient outcomes.

4.4.5 Coating Application

One major application of HEM coatings is in the protection of components exposed to harsh environments, such as those in the aerospace, energy, and marine industries. For example, HEM coatings can withstand extreme temperatures and resist oxidation, making them ideal for turbine blades, heat exchangers, and pipelines. AM allows for uniform and defect-free coatings, ensuring reliable performance under demanding conditions.

In the industrial sector, HEM coatings are used to enhance wear and erosion resistance for tools and machinery components. Applications include cutting tools, molds, and rollers, where HEM coatings significantly improve durability and reduce maintenance costs. AM enables the precise deposition of these coatings, even on complex geometries, ensuring optimal protection without compromising the underlying material.

HEM coatings are also being utilized in biomedical applications, such as on implants and surgical instruments. The biocompatibility and corrosion resistance of HEAs ensures safe and long-lasting performance in the human body. AM further enables the production of functional coatings with tailored porosities and surface features that enhance biological integration and reduce the risk of rejection.

The combination of HEMs and AM is transforming the field of coatings by offering advanced solutions that enhance durability, efficiency, and functionality. This synergy is enabling the development of coatings with superior properties, catering to the specific needs of various industries and applications. The excellent interface and adhesive properties of HEAs can be effectively used for base metal coatings. $MoFeCrTiWAlNb_x$ ($x = 1.5$, 2, 2.5, and 3) HEA coating was performed on M_2 tool steel by AM laser cladding. The addition of Nb improves the microhardness and wear behavior of the coated samples, and AlCrFeCoCuNiTi HEA is applied to Q235 steel by an AM laser cladding process to enhance the surface and corrosion resistance [138]. In addition, performance can be further improved by applying a high-entropy thermal barrier coating on the blade, such as $Al_{30}Cr_{23}$–$Ni_{23}Co_{22}Si_2$ [139] or $(Sm_{0.2}Eu_{0.2}Tb_{0.2}Dy_{0.2}Lu_{0.2})_2Zr_2O_7$ [140].

References

1 Li R, Niu P, Yuan T, et al. Selective laser melting of an equiatomic CoCrFeMnNi high-entropy alloy: processability, non-equilibrium microstructure and mechanical property. *Journal of Alloys and Compounds*, 2018, 746: 125–134.

2 Svetlizky D, Zheng B, Vyatskikh A, et al. Laser-based directed energy deposition (DED-LB) of advanced materials. *Materials Science and Engineering*: A, 2022, 840: 142967.

3 Yan L, Chen Y, Liou F. Additive manufacturing of functionally graded metallic materials using laser metal deposition. *Additive Manufacturing*, 2019, 31(4): 100901.

4 DebRoy T, Wei H L, Zuback J S, et al. Additive manufacturing or metallic components—process, structure and properties. *Progress in Materials Science*, 2018, 92: 112–224.

5 Hooyar A, Shima E H, Damon K, et al. Recent developments and opportunities in additive manufacturing of titanium-based matrix composites: a review. *International Journal of Machine Tools & Manufacture*, 2018, 133: 85–102.

6 Gu D, Shi X, Poprawe R, et al. Material-structure performance integrated laser-metal additive manufacturing. *Science*, 2021, 372(6545): aebg1487.

7 Mcandew A R, Rosales M A, Colegrove P A, et al. Interpass rolling of Ti-6Al-4V wire+ arc additively manufactured features for microstructural refinement. *Additive Manufacturing*, 2018, 21: 340–349.

8 Ren J, Zhang Y, Zhao D, et al. Strong yet ductile nanolamellar high-entropy alloys by additive manufacturing. *Nature*, 2022, 608(7921): 62–68.

9 Wei H L, Mukherjee T, Zhang W, et al. Mechanistic models for additive manufacturing of metallic components. *Progress in Materials Science*, 2021, 116: 100703.

10 Yue H, Chen Y, Wang X, et al. Effect of beam current on microstructure, phase, grain characteristic and mechanical properties of Ti-47Al-2Cr-2Nb alloy fabricated by selective electron beam melting. *Journal of Alloys and Compounds*, 2018, 750:617–625.

11 Terner M, Biamino S, Baudana G, et al. Initial oxidation behavior in air of TiAl-2Nb and TiAl-8Nb alloys produced by electron beam melting. *Journal of Materials Engineering and Performance*, 2015, 24(10):3982–3988.

12 Tebaldo V, Faga M G. Influence of the heat treatment on the microstructure and machinability of titanium aluminides produced by electron beam melting. *Journal of Materials Processing Technology*, 2017, 244: 289–0303.

13 Guo C, Ge W, Lin F. Effects of scanning parameters on material deposition during electron beam selective melting of Ti-6Al-4V powder. *Journal of Materials Processing Technology*, 2015, 217: 148–157.

14 Xin J, Wu D, Shen C, et al. Multi-physical modelling of alloy element transportation in wire arc additive manufacturing of a y-TiAl alloy. *International Journal of Thermal Sciences*, 2022, 179: 107641.

15 Wang L, Shen C, Zhang Y, et al. Effect of Al content on the microstructure and mechanical properties of γ-TiAl alloy fabricated by twin-wire plasma arc additive manufacturing system. *Materials Science and Engineering: A*, 2021, 826: 142008.

16 Wang L, Zhang Y, Hua X, et al. Fabrication of γ-TiAl intermetallic alloy using the twin-wire plasma arc additive manufacturing process: microstructure evolution and mechanical properties. *Materials Science and Engineering: A*, 2021, 812: 141056.

17 张勇, 周云军, 陈国良, 快速发展中的高熵溶体合金. 物理, 2008, 37(08): 600–605.

18 Li W, Xie D, Li D, et al. Mechanical behavior of high-entropy alloys. *Progress in Materials Science*, 2021, 118: 100777.

19 Tong C J, Chen Y L, Chen S K, et al. Microstructure characterization of AlCoCrCuFeNi high-entropy alloy system with multiprincipal elements. *Metallurgical and Materials Transactions A: Physical Metallurgy and Materials Science*, 2005, 36(4): 881–893.

20 Chang H W. Huang PK, Yeh J W, et al. Influence of substrate bias, deposition temperature and post-deposition annealing on the structure and properties of multi-principal-component (AlCrMoSiTi)N coatings. *Surface and Coatings Technology*, 2008, 202(14): 3360–3366.

21 Ng C, Guo S, Luan J, et al. Entropy-driven phase stability and slow diffusion kinetics in an $Al_{0.5}$CoCrCuFeNi high entropy alloy. *Intermetallics*, 2012, 31: 165–172.

22 Tsai M H, Yeh J W, Gan J Y. Diffusion barrier properties of AlMoNbSiTaTiVZr high-entropy alloy layer between copper and silicon. *Thin Solid Films*, 2008, 516(16): 5527–5530.

23 Ranganathan S. Alloyed pleasures: multimetallic cocktails. *Current Science*, 2003, 85(10): 1404–1406.

24 Feng R, Zhang C, Gao M C, et al. High-throughput design of high-performance lightweight high-entropy alloys. *Nature Communications*, 2021, 12(1): 6–15.

25 Butler T M, Chaput K J. Native oxidation resistance of $Al_{20}Nb_{30}Ta_{10}Ti_{30}Zr_{10}$ refractory complex concentrated alloy (RCCA). *Journal of Alloys and Compounds*, 2019, 787: 606–617.

26 Wang Z Q, Wu H, Wu Y, et al. Solving oxygen embrittlement of refractory high-entropy alloy via grain boundary engineering. *Materials Today*, 2022, 54: 83–89.

27 Wang M, Lu Y, Wang T, et al. A novel bulk eutectic high-entropy alloy with outstanding as-cast specific yield strengths at elevated temperatures. *Scripta Materialia*, 2021, 204: 114132.

28 Tong Y, Chen D, Han B, et al. Outstanding tensile properties of a precipitation-strengthened $FeCoNiCrTi_{0.2}$ high-entropy alloy at room and cryogenic temperatures. *Acta Materialia*, 2019, 165: 228–240.

29 Han L, Maccari F, Souza Filho I R, et al. A mechanically strong and ductile soft magnet with extremely low coercivity. *Nature*, 2022, 608(7922): 310–316.

30 Yao Y, Huang Z, Xie P, et al. Carbothermal shock synthesis of high-entropy-alloy nanoparticles. *Science*, 2018, 359(6383): 1489–1494.

31 Gorbatov O I, Stroev A Y, Gornostyrev Y N, et al. Effective cluster interactions and pre-precipitate morphology in binary Al-based alloys. *Acta Materialia*, 2019, 179: 70–84.

32 Zhao Y Y. Chen H W, Lu Z P, et al. Thermal stability and coarsening of coherent particles in a precipitation-hardened $(NiCoFeCr)_{94}Ti_2Al_4$ high-entropy alloy. *Acta Materialia*, 2018, 147: 184–194.

33 Niu M C, Yin L C, Yang K, et al. Synergistic alloying effects on nanoscale precipitation and mechanical properties of ultrahigh-strength steels strengthened by NisTi, Mo-enriched, and Cr-rich co-precipitates. *Acta Materialia*, 2021, 209: 116788.

34 Zhao Y L, Yang T. Tong Y, et al. Heterogeneous precipitation behavior and stacking-fault-mediated deformation in a CoCrNi-based medium-entropy alloy. *Acta Materialia*, 2017, 138: 72–82.

35 Laplanche G, Berglund S, Reinhart C, et al. Phase stability and kinetics of σ-phase precipitation in CrMnFeCoNi high-entropy alloys. *Acta Materialia*, 2018, 161:338–351.

36 Yang T, Zhao Y L, Fan L, et al. Control of nanoscale precipitation and elimination of intermediate-temperature embrittlement in multicomponent high-entropy alloys. *Acta Materialia*, 2020, 189: 47–59.

37 Mantri S A, Dasari S, Sharma A, et al. Effect of micro-segregation of alloying elements on the precipitation behaviour in laser surface engineered Alloy 718. *Acta Materialia*, 2021, 210: 116844.

38 Wang L, Wang L, Zhou S, et al. Precipitation and micromechanical behavior of the coherent ordered nanoprecipitation strengthened Al-Cr-Fe-Ni-V high entropy alloy. *Acta Materialia*, 2021, 216: 117121.

39 Daniel R, Zalesak J, Matko I, et al. Microstructure-dependent phase stability and precipitation kineties in equiatomic CrMnFeCoNi high-entropy alloy: role of grain boundaries. *Acta Materialia*, 2022, 223: 117470.

40 Qin S, Yang M, Jiang P, et al. Designing structures with combined gradients of grain size and precipitation in high entropy alloys for simultaneous improvement of strength and ductility. *Acta Materialia*, 2022, 230: 117847.

41 Cao P P, Huang H L, Jiang S H, et al. Microstructural stability and aging behavior of refractory high entropy alloys at intermediate temperatures. *Journal of Materials Science and Technology*, 2022, 122: 243–254.

42 Peng J, Li Z, Fu L, et al. Carbide precipitation strengthening in fine-grained carbon-doped FeCoCrNiMn high entropy alloy. *Journal of Alloys and Compounds*, 2019, 803: 491–498.

43 Man J, Wu B, Duan G, et al. The synergistic addition of Al, Ti, Mo and W to strengthen the equimolar CoCrFeNi high-entropy alloy via thermal-mechanical processing. *Journal of Alloys and Compounds*, 2022, 902: 163774.

44 Mehranpour M S, Shahmir H, Derakhshandeh A, et al. Significance of Ti addition on precipitation in CoCrFeNiMn high-entropy alloy. *Journal of Alloys and Compounds*, 2021, 888: 2–5.

45 Chen M, Lan L, Shi X, et al. The tribological properties of $Al_{0.6}$CoCrFeNihigh-entropy alloy with the σ phase precipitation at elevated temperature. *Journal of Alloys and Compounds*, 2019, 777: 180–189.

46 Fan L, Yang T, Luan J H, et al. Control of discontinuous and continuous precipitation of γ′-strengthened high-entropy alloys through nanoscale Nb segregation and partitioning. *Journal of Alloys and Compounds*, 2020, 832: 154903.

47 Gwalani B, Torgerson T, Dasari S, et al. Influence of fine-scale B2 precipitation on dynamic compression and wear properties in hypo-eutectic $Al_{0.5}$CoCrFeNi high-entropy alloy. *Journal of Alloys and Compounds*, 2021, 853: 157126.

48 Nandal V, Sarvesha R. Singh S S, et al. Influence of pre-deformation on the precipitation characteristics of aged non-equiatomic $Co_{1.5}CrFeNi_{1.5}$ high entropy alloys with Ti and Al additions. *Journal of Alloys and Compounds*, 2021, 855: 157521.

49 Joseph J, Senadeera M, Chao Q, et al. Computational design of thermally stable and precipitation-hardened Al-Co-Cr-Fe-Ni-Ti high entropy alloys. *Journal of Alloys and Compounds*, 2021, 888: 161496.

50 Yeh J W, Chen S K, Lin S J, et al. Nanostructured high-entropy alloys with multiple principal elements: Novel alloy design concepts and outcomes. *Advanced Engineering Materials*, 2004, 6(5): 299–303.

51 Chen X F, Wang Q, Cheng Z Y, et al. Direct observation of chemical short-range order in a medium-entropy alloy. *Nature*, 2021, 592: 712–716.

52 Zhang R, Zhao S, Ding J, et al. Short-range order and its impact on the CrCoNi medium-entropy alloy. *Nature*, 2020, 581(7808): 283–287.

53 Wang S D, Liu X J, Lei Z F, et al. Chemical short-range ordering and its strengthening effect in refractory high-entropy alloys. *Physical Review B*, 2021, 103(10): 3–10.

54 Cantor B, Chang I T H., Knight P, et al, Microstructural development in equiatomic multicomponent alloys. *Materials Science and Engineering*: A, 2004, (375): 213–218.
55 Zhang Y, Zuo T T., Tang, Z, et al. Microstructures and properties of high-entropy alloys. *Progress in Materials Science*, 2014, 61: 1–93.
56 Tong C J, Chen M R, Chen S K, et al. Mechanical performance of the $Al_xCoCrCuFeNi$ high-entropy alloy system with multiprincipal elements. *Metallurgical and Materials Transactions: A*, 2005, 36: 1263–1271.
57 Tsai M H, Yeh J W. High-entropy alloys: a critical review. *Material Research Letters*, 2014, 2: 107–123.
58 Couzinie J P, Dirras G., Perriere L, et al. Microstructure of a near-equimolar refractory high-entropy alloy. *Materials Letters*, 2014, 126: 285–287.
59 Lilensten L, Couzinie J P, Perriere L, et al. New structure in refractory high-entropy alloys. *Materials Letters*, 2014, 132: 123–125.
60 Liu C M, Wang H M, Zhang S Q, et al. Microstructure and oxidation behavior of new refractory high entropy alloys. *Journal of Alloys and Compounds*, 2014, 583: 162–169.
61 Senkov O N, Senkova S V, Dimiduk D M, et al. Oxidation behavior of a refractory $NbCrMo_{0.5}Ta_{0.5}TiZr$ alloy. *Journal of Materials Science*, 2012, 47: 6522–6534.
62 Senkov O N, Senkova S V, Woodward, C, et al. Effect of aluminum on the microstructure and properties of two refractory high entropy alloys. *Acta Materialia*, 2014, 68: 214–228.
63 Hammond V H, Atwater M A, Darling K A, et al. Equalchannel angular extrusion of a low-density high-entropy alloy produced by high-energy cryogenic mechanical alloying. *Jom*, 2014, 66: 2021–2029.
64 Yang X, Chen S Y, Cotton J D, et al. Phase stability of low-density, multiprincipal component alloys containing aluminum, magnesium, and lithium. *Jom*, 2014, 66: 2009–2020.
65 Takeuchi A, Amiya K, Wada T, et al. High-entropy alloys with a hexagonal close-packed structure designed by equi-atomic alloy strategy and binary phase diagrams. *Jom*, 2014, 66: 1984–1992.
66 Laws K J, Crosby, C. Sridhar A. et al. High entropy brasses and bronzes-microstructure, phase evolution and properties. *Journal of Alloys and Compounds*. 2015, 650: 949–961.
67 Paschoal J O A, Kleykamp H, Thummler F. Phase equilibria in the quaternary molybdenum-ruthenium-rhodium-palladium system. *Zeitschrift fur Metallkunde*, 1983, 74: 652.
68 Chen T K, Shun T T, Yeh J W, et al. Nanostructured nitride films of multi-element high-entropy alloys by reactive DC sputtering. *Surface and Coating Technology*, 2004, 188: 193–200.
69 Braic V. Vladescu A, Balaceanu M, et al. Nanostructured multi-element (TiZrNbHfTa)N and (TiZrNbHfTa)C hard coatings. *Surface & Coating Technology*, 2012, 211: 117–121.
70 Tian F Y. A review of solid-solution models of high-entropy alloys based on ab initio calculations. *Frontiers in Materials*, 2017, 4: 36.
71 Akinwande A A, Balogun O A, Adediran A A, et al. Experimental analysis, statistical modeling, and parametric optimization of quinary-$(CoCrFeMnNi)_{100-x}$/

TiC_x high-entropy-alloy (HEA) manufactured by laser additive manufacturing. *Results in Engineering*, 2023, 17: 100802.

72 Dada M, Popoola P, Mathe N, et al. Effect of laser parameters on the properties of high entropy alloys: A preliminary study. *Materials Today: Proceedings*, 2021, 38: 756–761.

73 Choudhuri D, Gwalani B, Gorsse S, et al. Change in the primary solidification phase from FCC to BCC-based B2 in high entropy or complex concentrated alloys. *Scripta Materialia*, 2017, 127: 186–190.

74 Su Y, Luo S, Wang Z. Microstructure evolution and cracking behaviors of additively manufactured $Al_xCrCuFeNi_2$ high entropy alloys via selective laser melting. *Journal of Alloys and Compounds*, 2020, 842: 155823.

75 Chen X, Yan L, Karnati S, et al. Fabrication and characterization of $Al_xCoFeNiCu_{1-x}$ high entropy alloys by laser metal deposition. *Coatings*, 2017, 7(4): 47.

76 Guan S, Wan D, Solberg K, et al. Additive manufacturing of fine-grained and dislocation-populated CrMnFeCoNi high entropy alloy by laser engineered net shaping. *Materials Science and Engineering: A*, 2019, 761: 138056.

77 Wang H, Zhu Z, Chen H, et al. Effect of cyclic rapid thermal loadings on the microstructural evolution of a CrMnFeCoNi high-entropy alloy manufactured by selective laser melting. *Acta Materialia*, 2020, 196: 609–625.

78 Lin D, Xu L, Li X, et al. A Si-containing FeCoCrNi high entropy alloy with high strength and ductility synthesized in situ via selective laser melting. *Additive Manufacturing*, 2020, 35: 101340.

79 Jung H, Lee J, Gu G H. et al. Multiscale defects enable synergetic improvement in yield strength of CrCoNi-based medium entropy alloy fabricated via laser-powder bed fusion. *Additive Manufacturing*, 2022, 61: 103360.

80 Gao J, Jin Y, Fan Y, et al. Fabricating antibacterial CoCrCuFeNi high-entropy alloy via selective laser melting and in-situ alloying. *Journal of Materials Science & Technology*, 2022, 102: 159–165.

81 Fujieda T, Chen M, Shiratori H, et al. Mechanical and corrosion properties of CoCrFeNiTi-based high-entropy alloy additive manufactured using selective laser melting. *Additive Manufacturing*, 2019, 25: 412–420.

82 Huang Y, Xie Z, Li W. et al, Dynamic mechanical properties of the selective laser melting $NiCrFeCoMo_{0.2}$ high entropy alloy and the microstructure of molten pool. *Journal of Alloys and Compounds*, 2022, 927: 167011.

83 Farquhar L, Maddison G, Hardwick L, et al. In-situ alloying of CoCrFeNiX high entropy alloys by selective laser melting. *Metals*, 2022, 12(3): 456.

84 Kunce I, Polanski M, Bystrzycki J. Structure and hydrogen storage properties of a high entropy ZrTiVCrFeNi alloy synthesized using laser engineered net shaping (LENS). *International Journal of Hydrogen Energy*, 2013, 38(27): 12180–12189.

85 Dutta B, Froes F S. The additive manufacturing (AM) of titanium alloys. *Metal Powder Report*. 2017, 72: 96–106.

86 Kotadia H, Gibbons G, Das A, et al. A review of Laser Powder Bed Fusion Additive Manufacturing of aluminium alloys: Microstructure and properties. *Additive Manufacturing*, 2021, 46: 102155.

87 Bajaj P, Hariharan A, Kini A, et al. Steels in additive manufacturing: a review of their microstructure and properties. *Materials Science and Engineering: A*, 2020, 772: 138633.

88 Chen J, Chen B. Progress in additive manufacturing of magnesium alloys: a review. *Materials*, 2024, 17: 3851.

89 Chen B, Chen R, Fan Y, et al. Microstructural characterization of Stellite 6 alloy processed by electron beam melting. *Vacuum*, 2024, 229: 113591.

90 Rajendran, S, Palani, G, Kanakaraj, A, et al. Metal and polymer based composites manufactured using additive manufacturing—a brief review. *Polymers*, 2023, 15: 2564.

91 Zhou P, Xiao D, Wu Z, et al. $Al_{0.5}$FeCoCrNi high entropy alloy prepared by selective laser melting with gas-atomized pre-alloy powders. *Materials Science and Engineering A*, 2019, 739: 86–89.

92 Guo Y, Su H, Yang P, et al. New insight into tailorable eutectic high entropy alloys with remarkable strength–ductility synergy and ample shaping freedom fabricated using laser powder bed fusion. *Additive Manufacturing*, 2022, 60: 103257.

93 Guo, Y, Su H, Zhou H, et al. Unique strength-ductility balance of $AlCoCrFeNi_{2.1}$ eutectic high entropy alloy with ultra-fine duplex microstructure prepared by selective laser melting. *Journal of Materials Science and Technology*, 2022, 111: 298–306.

94 Peng X, Cao X, Wang Y, et al. Design, microstructure, and mechanical property of negative Poisson's ratio porous structure fabricated by LPBF of AlCoCrFeNi2. 1 eutectic high-entropy-alloy, *Journal of Materials Science and Technology*, (2023), 24: 2028–2040.

95 Kwak K, Okamura Y, Mine Y, et al. Micromechanical characterisation of slip behaviour and precipitation strengthening in CoCrFeNiTiMo alloy additively manufactured by laser powder bed fusion. *Materials Science and Engineering A*, 2022, 840: 142970.

96 Ishimoto T, Ozasa R, Nakano, K, et al. Development of TiNbTaZrMo bio-high entropy alloy (BioHEA) super-solid solution by selective laser melting, and its improved mechanical property and biocompatibility. *Scripta Materialia*, 2021, 194: 113658.

97 Zhang H, Xu W, Xu Y, et al. The thermal-mechanical behavior of WTaMoNb high-entropy alloy via selective laser melting (SLM): experiment and simulation. *The International Journal of Advanced Manufacturing Technology*, 2018, 96: 461–474.

98 Wang W-R, Wang W-L, Yeh J-W. Phases, microstructure and mechanical properties of Al_xCoCrFeNi high-entropy alloys at elevated temperatures. *Journal of Alloys and Compounds*, 2014 589: 143.

99 Song M, Zhou R, Gu J, et al. Nitrogen induced heterogeneous structures overcome strength-ductility trade-off in an additively manufactured high-entropy alloy. *Applied Materials Today*, 2020, 18: 100498.

100 Lin W-C, Chang Y-J, Hsu T-H, et al. Microstructure and tensile property of a precipitation strengthened high entropy alloy processed by selective laser melting and post heat treatment. *Additive Manufacturing*, 2020, 36: 101601.

101 Xue Y, Xu J-Y, Guan X-H, et al. Optimization of microstructure and properties of laser sintered $Ni_{30}Cr_{25}Al_{15}Co_{15}Mo_5Ti_5Y_5$ high-entropy alloy coatings via controlling plasma. *Transactions of Nonferrous Metals Society of China*, 2023, 33: 168–188.

102 Karlsson D, Lindwall G, Lundbäck A, et al. Binder jetting of the AlCoCrFeNi alloy. *Additive Manufacturing*, 2019, 27: 72–79.

103 Kenel C, Casati N P M, Dunand, D C. 3D ink-extrusion additive manufacturing of CoCrFeNi high-entropy alloy micro-lattices. *Nature Communications*, 2019, 10: 904.

104 Karthik G, Panikar S, Ram G J, et al. Additive manufacturing of an aluminum matrix composite reinforced with nanocrystalline high-entropy alloy particles. *Materials Science and Engineering A*, 2017, 679: 193–203.

105 Chen B. Progress in additive manufacturing of high-entropy alloys. *Materials*, 2024, 17(23): 5917.

106 Brif Y, Thomas M, Todd I. The use of high-entropy alloys in additive manufacturing. *Scripta Materialia*, 2015, 99: 93–96.

107 Sun K, Peng W, Yang L, et al. Effect of SLM processing parameters on microstructures and mechanical properties of $Al_{0.5}CoCrFeNi$ high entropy alloys. *Metals*, 2020, 10: 292.

108 Shiratori H, Fujieda T, Yamanaka K, et al. Relationship between the microstructure and mechanical properties of an equiatomic AlCoCrFeNi high-entropy alloy fabricated by selective electron beam melting. *Materials Science and Engineering A*, 2016, 656: 39–46.

109 Fujieda T, Shiratori H, Kuwabara K. First demonstration of promising selective electron beam melting method for utilizing high-entropy alloys as engineering materials. *Materials Letters*, 2015, 159: 12–15.

110 Wang P, Huang P, Ng F L, et al. Additively manufactured CoCrFeNiMn high-entropy alloy via pre-alloyed powder. *Materials & Design*. 2019, 168: 107576.

111 Tong Z, Ren X, Jiao J, et al. Laser additive manufacturing of FeCrCoMnNi high-entropy alloy: Effect of heat treatment on microstructure, residual stress and mechanical property. *Journal of Alloys and Compounds*, 2019, 785: 1144–1159.

112 Gao X, Lu Y. Laser 3D printing of CoCrFeMnNi high-entropy alloy. *Materials Letters*, 2019, 236: 77–80.

113 Li H, Huang Y, Jiang S, et al. Columnar to equiaxed transition in additively manufactured CoCrFeMnNi high entropy alloy. *Materials & Design*, 2021, 197: 109262.

114 Li H, Lee T, Zheng W, et al. Characterization of residual stress in laser melting deposited CoCrFeMnNi high entropy alloy by neutron diffraction. *Materials Letters*, 2020, 263: 127247.

115 Chen L, Fu Z, Chen W, et al. Enhancing mechanical properties and electrochemical behavior of equiatomic FeNiCoCr high-entropy alloy through sintering and hot isostatic pressing for binder jet 3D printing. *Additive Manufacturing*, 2024, 81: 103999.

116 Xu Z, Zhu Z, Wang P, et al. Fabrication of porous CoCrFeMnNi high entropy alloy using binder jetting additive manufacturing. *Additive Manufacturing*, 2020, 35: 101441.

117 Peng S, Mooraj S, Feng R, et al. Additive manufacturing of three-dimensional (3D)-architected CoCrFeNiMn high-entropy alloy with great energy absorption. *Scripta Materialia*, 2021, 190: 46–51.

118 Agrawal P, Haridas R S, Agrawal P, et al. Deformation based additive manufacturing of a metastable high entropy alloy via Additive friction stir deposition. *Additive Manufacturing*, 2022, 60: 103282.

119 Kunce I, Polanski M, Bystrzycki J. Microstructure and hydrogen storage properties of a TiZrNbMoV high entropy alloy synthesized using laser engineered net shaping (LENS). *International Journal of Hydrogen Energy*, 2014, 39: 9904–9910.

120 Zhou R, Liu Y, Zhou C, et al. Microstructures and mechanical properties of C-containing FeCoCrNi high-entropy alloy fabricated by selective laser melting. *Intermetallics*, 2018, 94: 165–171.

121 Niu P D, Li R D, Yuan T C, et al. Microstructures and properties of an equimolar AlCoCrFeNi high entropy alloy printed by selective laser melting. *Intermetallics*, 2019, 104: 24–32.

122 Niu P, Li R, Zhu S, et al. Hot cracking, crystal orientation and compressive strength of an equimolar CoCrFeMnNi high-entropy alloy printed by selective laser melting. *Optics & Laser Technology*, 2020, 127: 106147.

123 Jin M, Piglione A, Dovgyy, B, et al. Cyclic plasticity and fatigue damage of CrMnFeCoNi high entropy alloy fabricated by laser powder-bed fusion. *Additive Manufacturing*, 2020, 36: 101584.

124 Xu Z, Zhang H, Li W, et al. Microstructure and nanoindentation creep behavior of CoCrFeMnNi high-entropy alloy fabricated by selective laser melting. *Additive Manufacturing*, 2019, 28: 766–771.

125 Zhang C, Feng K, Kokawa H, et al. Cracking mechanism and mechanical properties of selective laser melted CoCrFeMnNi high entropy alloy using different scanning strategies. *Materials Science and Engineering A*, 2020, 789: 139672.

126 He X, Wang H, Zhu Z, et al., Texture evolution in a CrMnFeCoNi high-entropy alloy manufactured by laser powder bed fusion. *Journal of Materials Science*. 2022, 57(21): 1–12.

127 Wang R, Zhang K, Davies C, et al. Evolution of microstructure, mechanical and corrosion properties of AlCoCrFeNi high-entropy alloy prepared by direct laser fabrication. *Journal of Alloys & Compounds*, 2017, 694: 971–981.

128 Joseph J, Jarvis T, Wu X, et al. Comparative study of the microstructures and mechanical properties of direct laser fabricated and arc-melted Al_xCoCrFeNi high entropy alloys. *Materials Science and Engineering: A*, 2015, 633(1): 184–193.

129 Joseph J, Hodgson P, Jarvis T, et al. Effect of hot isostatic pressing on the microstructure and mechanical properties of additive manufactured AlCoCrFeNi high entropy alloys. *Materials Science and Engineering: A*, 2018, 733(22): 59–70.

130 Cheng H, Liu Z, Luo H, et al., Tuning the microstructure to improve corrosion resistance of additive manufacturing high entropy alloy in proton exchange membrane fuel cells environment. *Corrosion Science*, 2023, 213: 110969.

131 Tseng K. A light-weight high-entropy alloy $Al_{20}Be_{20}Fe_{10}Si_{15}Ti_{35}$. *Science China Technological Sciences*, 2018, 61(2): 184–188.

132 O'Brien S. Christudasjustus J, Esteves L, et al. A low-cost, low-density, and corrosion resistant AlFeMnSi compositionally complex alloy. *NPJ Materials Degradation*, 2021, 5(1): 1–10.

133 Qiu Y, Thomas S, Gibson M, et al. Microstructure and corrosion properties of the low-density single-phase compositionally complex alloy AlTiVCr. *Corrosion Science*, 2018, 133: 386–396.

134 Smith T M, Kantzos C A, Zarkevich N A, et al. A 3D printable alloy designed for extreme environments. *Nature*, 2023, 617: 513–518.

135 Tepylo N, Huang X, Patnaik P C. Laser-based additive manufacturing technologies for aerospace applications. *Advanced Engineering Materials*, 2019, 21: 1900617.

136 Karimi J, Ma P, Jia Y D, et al. Linear patterning of high entropy alloy by additive manufacturing. *Manufacturing Letters*, 2020, 24: 9–13.

137 Shuai C, Xie J, Jiang X, et al. Additively manufactured high entropy alloy with high wear resistance for biomedical implant. *Vacuum*, 2024, 221: 112939.

138 Dong Z, Han C, Zhao Y, et al. Role of heterogenous microstructure and deformation behavior in achieving superior strength-ductility synergy in zinc fabricated via laser powder bed fusion. *International Journal of Extreme Manufacturing*, 2024, 6: 045003.

139 Li M, Benn F, Derra T, et al. Microstructure, mechanical properties, corrosion resistance and cytocompatibility of WE43 Mg alloy scaffolds fabricated by laser powder bed fusion for biomedical applications. *Materials Science and Engineering: C*, 2021, 119: 111623.

140 Qiu X. Microstructure and corrosion properties of $Al_2CrFeCo_xCuNiTi$ high entropy alloys prepared by additive manufacturing. *Journal of Alloys and Compounds*, 2021, 887: 161422.

141 Mohan S K K Y N, Alam T, Dasari S, et al. Enhanced tensile yield strength in laser additively manufactured $Al_{0.3}CoCrFeNi$ high entropy alloy. *Materialia*, 2020, 9: 100522.

5

Functional Properties: Thermal, Electric, Magnetic, and Optics

Hang Xiang[2], *Yong Zhang*[1,2,*]

[1] *Key Laboratory of Silicon-based Materials, The Ministry of Education, and School of Materials Science and Engineering, Fuyao University of Science and Technology, Fuzhou, China*
[2] *University of Science and Technology Beijing, State Key Laboratory for Advanced Metal Materials, Beijing, China*

*Corresponding author: drzhangy@ustb.edu.cn/yongzhang@fyust.edu.cn

5.1 Introduction to Functional Properties and Applications

Since ancient times, materials have been witnesses and promoters of the progress of human civilization, with metallic materials playing a central role. Alloys can be found in both the national economy and the national defense and military. In today's era of rapid development of science and technology, the application environment of alloys has become more complex and demanding. In addition to meeting the basic mechanical properties, the functional properties put forward higher and higher requirements. Traditional alloy design and development has encountered a bottleneck, and it is difficult to meet the growing demand for comprehensive performance.

Starting in 2004, the research on high-entropy alloys (HEAs) has become a new field and soon expanded from HEAs to middle-entropy alloys, as well as ceramics, polymers, and composites. High-entropy materials (HEMs) have four unique effects, which can form a unique microstructure and organization. These properties not only endow HEMs with excellent mechanical properties but also make them have special functional properties. However, most of the research on HEMs focuses on their potential applications as structural materials, and the development of their functional properties is still relatively underdeveloped. At present, HEMs have excellent performance in properties such as catalytic properties, irradiation resistance, hydrogen storage properties, biocompatibility, and superconducting properties and have great potential for applications in catalysis, nuclear energy, energy and power, biomedicine, aerospace, and other fields.

Functional High Entropy Alloys and Compounds, First Edition. Edited by Yong Zhang, Yonggang Yao and Haojie Liu.

In the thermal field, HEMs exhibit excellent thermal stability and low thermal conductivity, making them suitable for use as thermal barrier coatings and high-temperature structural materials. Electrically, they have high electrical conductivity and thermoelectric properties, making them suitable for thermoelectric conversion and superconducting materials. Magnetically, the complex magnetic behavior of HEMs gives them potential for applications in magnetic refrigeration. In optics, the tunable bandgap structure of HEMs makes them show promise for applications in fields such as photoelectric conversion and optical detection. Therefore, HEMs have great potential in aeroengines, high-speed rail transit, and photovoltaic power generation. High-throughput synthesis and characterization of these materials is a challenge for current research, but through a multidisciplinary approach and scientific exchange, it is expected that customized high-performance materials will be achieved, driving technological progress and sustainable development.

5.2 Thermal and Electronic Properties in HEMs

Unlike conventional alloys, which are dominated by only one element, HEMs are composed of multiple elements, each of which plays a major role. The presence of multiple elements increases the conformational entropy of the material, avoids the formation of intermetallic compounds, and facilitates the formation of simple structures. Thus, atomic disorder is a key factor in determining the properties of HEMs, while differences in atomic size lead to severe lattice distortions that affect the transport of electrically and thermally conductive media concerned in HEMs. The huge design space of components greatly improves the development potential of high-performance materials in the field of thermology and acoustics, and there have been several researches on the thermal and superconducting properties of HEMs, which show a great prospect of development. At the same time, the low lattice thermal conductivity of HEMs has also become a thermoelectric material with great development potential.

5.2.1 Thermal Conduction in HEMs

Predicting the lattice thermal conductivity of HEAs is challenging, and traditional experimental methods are time-consuming and costly. Molecular dynamics (MD) simulations and first-principles calculations are also limited by computational resources. Therefore, it is crucial to explore a fast and effective prediction method. Machine learning (ML) techniques enable fast and accurate prediction of thermal conductivity by analyzing the complex structure of HEA. Lu et al. [1] used the ML method to predict the lattice thermal conductivity of AlCoCrNiFe HEAs. The study utilized MD simulations to train an ML model, achieving high accuracy with an R^2 value of 0.91 for the test set. The findings indicated that aluminum content significantly influences thermal conductivity. The model successfully predicted both low and high lattice thermal conductivity samples, aligning well with MD simulation results (e.g., Figure 5.1). The study also explored the phonon density of states and elastic modulus to understand the thermal conductivity mechanisms. This work

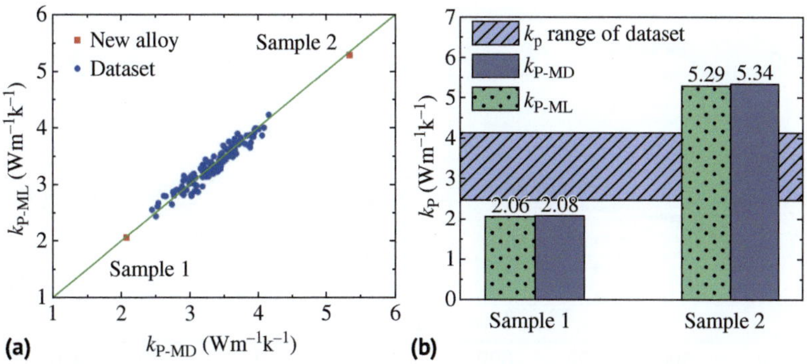

Figure 5.1 (a) Comparison between the new alloys selected by the SVR (Support Vector Regression) model and other alloys of the dataset. (b) Comparison between the kP–ML predicted by the SVR model and the kP–MD calculated by NEMD (Non-Equilibrium Molecular Dynamics) for new alloys. [1] / AIP Publishing / CC BY 4.0.

offers a powerful tool for developing HEAs with tailored properties, advancing materials science by combining ML with traditional simulation methods.

The effect of alloying on the transport properties of Ti—Cr—Mo—W—V—Nb—Ta refractory HEAs (RHEAs), including electrical conductivity, thermal conductivity, and Seebeck coefficient, has been systematically investigated using density functional theory (DFT) by Singh et al. (2024) [2]. The relaxation time of charge carriers is a key fundamental parameter in determining the thermal conductivity, which is challenging to predict using first principles alone. Therefore, the authors follow the approach of Mukherjee, Satsangi, and Singh to optimize the relaxation time of RHEAs. The experimental results validate the thermal conductivity predictions for monolithic solids, binary and ternary alloys, and RHEAs and compare these predictions with experimental results, which show a good correlation. To understand the observed trends in thermal conductivity, the authors evaluated the phase stability, electronic structure, phonons, and intrinsic and tensile strengths of selected RHEAs, which correlate with the observed compositional trends in RHEA thermal transport. These DFT evaluations and thermodynamic predictions provide a unique design guideline for RHEAs with tailored thermal conductivity.

Yang et al. [3]. have probed the magnetic and thermal properties of a CrMnFeCoNi HEA using physical property measurements and neutron scattering. This alloy consists of a variety of elements in near-iso-atomic proportions, leading to extensive atomic disorder on a highly symmetric lattice. It has been shown that the thermal conductivity of CrMnFeCoNi is significantly lower than that of nickel (Ni), with a total thermal conductivity of about 10 W m^{-1} K^{-1} at room temperature (e.g., Figure 5.2). Both electron and lattice contributions to the thermal conductivity are suppressed. The lattice thermal conductivity shows an unusual temperature dependence, with a glass-like plateau in the 50–150 K region. This behavior suggests the possible existence of two thermal conduction channels: normal phonon transport and diffusive conduction. Inelastic neutron scattering measurements are used to probe the lattice dynamics of CrMnFeCoNi. The results show that the anharmonicity of the lattice is weak, indicating that defect scattering dominates the heat transport. Significant phonon

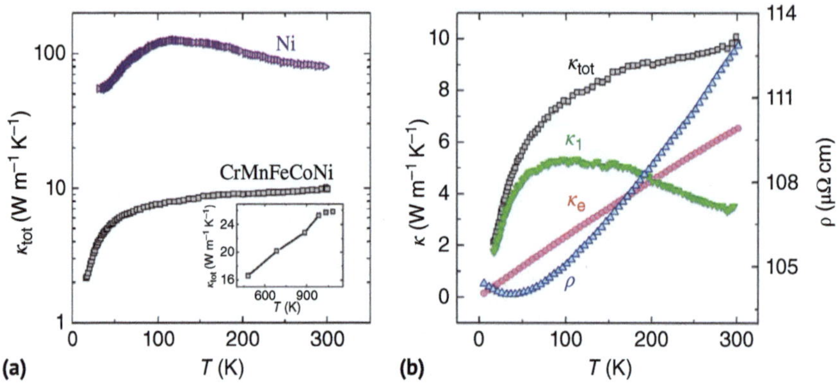

Figure 5.2 (a) The temperature dependencies of thermal conductivity of CrMnFeCoNi and Ni. The inset shows the high-temperature thermal conductivity of CrMnFeCoNi. (b) The temperature dependencies of the low-temperature total, electronic, and lattice thermal conductivity of CrMnFeCoNi as well as electrical resistivity (ρ). [3] / with permission of Elsevier.

softening is observed compared to pure nickel, which may be related to the expansion of the unit cell volume and the lower melting point of the alloy. Atomic disorder in CrMnFeCoNi leads to complex magnetic behavior and low thermal conductivity.

5.2.2 Superconducting in High-Entropy Materials

The phenomenon of superconductivity, originally discovered by Dutch scientist Heike Kammerling-Onnes in 1911, is a physical property that manifests itself in specific materials, whereby the resistance of a superconductor decreases significantly with decreasing temperature until it virtually disappears as it approaches absolute zero. The superconducting properties of conventional metals and alloys have been widely explored. However, since the beginning of the 21st century, the introduction of HEAs has brought new opportunities for the study of metal-based superconductors. Not only have they greatly expanded the family of superconductors but also opened up research directions for the further development of superconductivity theory.

Ta–Nb–Hf–Zr–Ti system, with composition $[TaNb]_{0.67}[HfZrTi]_{0.33}$, is the first reported HEA superconductor (body-centered cubic [BCC] structure) with $Tc = 7.3$ K and shows an upper critical field $\mu_0 H_{c2}(0) = 8.2$ T [4]. Quite remarkably, this compound shows an extraordinarily robust zero-resistance superconductivity at pressures up to 190 GPa. The alloy, with a BCC structure, exhibited an increase in its superconducting transition temperature (T_c) from 7.7 K at ambient pressure to 10 K at approximately 60 GPa and then gradually decreased to 9 K at 190.6 GPa (e.g., Figure 5.3). This high pressure is comparable to that found in the Earth's outer core. The study suggests that the alloy's unique electronic and mechanical properties allow it to maintain zero-resistance superconductivity across a wide pressure range. The findings indicate that this HEA superconductor could have significant potential for applications in extreme conditions and challenges existing quantum physics theories.

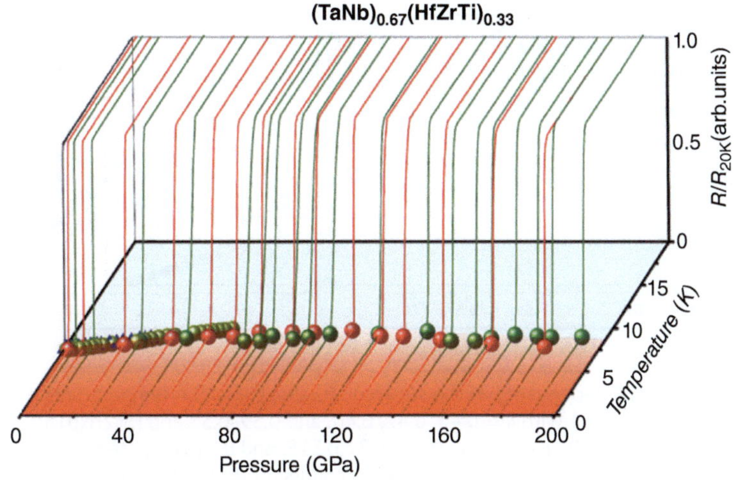

Figure 5.3 Phase diagram of superconducting transition temperature versus applied pressure up to 190.6 GPa for the HEA, combined with plots of the corresponding resistance versus temperature. Blue triangles and dark-yellow balls in the diagram represent the zero-resistance superconducting transition temperature T_c obtained from measurements at a pressure below 60 GPa, while the red and olive balls stand for zero-resistance T_c from the measurements at pressures up to 179.2 and 190.6 GPa, respectively. [4] / with permission of the National Academy of Sciences.

Stolze and other researchers [5] created new pentanary $(ScZrNb)_{1-x}(RhPd)_x$ and hexanary $(ScZrNb)_{1-x}(RhPd)_x$ HEAs that have CsCl-type structures and exhibit mixed-site occupancies. These alloys are unique in that they are type-II superconductors, with critical temperatures (T_c) that vary significantly based on the valence electron count (VEC). The (T_c) increases as the VEC decreases, diverging from trends observed in other superconducting materials (e.g., Figure 5.4a). Notably, the $(ScZrNb)_{0.65}[RhPd]_{0.35}$ HEA demonstrated a high T_c of about 9.3 K and a substantial upper critical field $\mu_0 H_{c2}(0)$ of 10.7 T (e.g., Figure 5.4b). The study also found that the presence of niobium positively affects T_c, and even the niobium-free $(ScZr)_{0.50}[RhPd]_{0.50}$ alloy showed superconductivity, indicating an intrinsic superconducting feature of the material. The findings highlight the potential of these HEAs for practical applications due to their high T_c, strong upper critical fields, and ductile mechanical properties.

Marik et al. [6] investigated the superconducting properties of a new HEA with a nominal composition of $Nb_{20}Re_{20}Zr_{20}Hf_{20}Ti_{20}$. The alloy, which is the first equimolar BCC-structured HEA superconductor, was characterized using various techniques including powder X-ray diffraction (XRD), energy-dispersive X-ray spectroscopy (EDX), magnetization, transport, and thermodynamic measurements. The results indicate that the alloy exhibits superconductivity at a critical temperature (T_c) of 5.3 K (e.g., Figure 5.5a). It was found to be a type-II superconductor with a lower critical field (H_{c1}) of 33 mT (e.g., Figure 5.5b) and an upper critical field (H_{c2}) of 8.88 T. The specific heat measurement suggests that the sample is a moderately coupled superconductor, fitting well with the single-gap BCS (Bardeen-Cooper-Schrieffer) model. The study provides insights

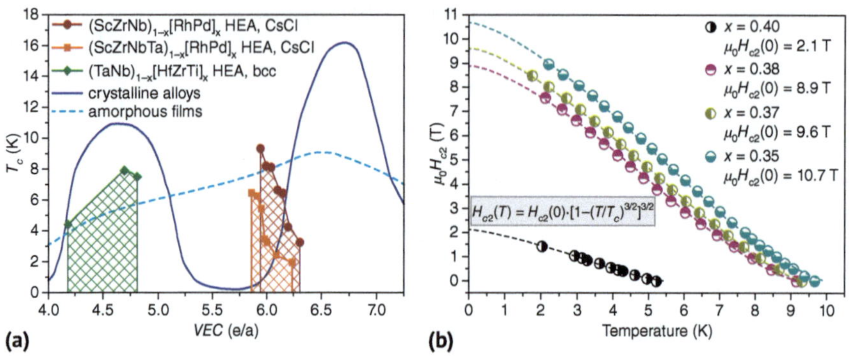

Figure 5.4 (a) VEC dependency of the superconducting transition temperatures of $(ScZrNb)_{1-x}[RhPd]_x$ HEA with $x = 0.45$, 0.42, 0.40, and 0.39–0.35 and $(ScZrNbTa)_{1-x}[RhPd]_x$ with $x = 0.43$, 0.38,0.35,0.33,0.328,0.32,0.315, and 0.31 compared to the T_c versus VEC trend lines of $[TaNb]_{1-x}(HfZrTi)_x$ HEA superconductor (BCC), crystalline binary alloys, and amorphous alloy films. (b) Temperature-dependent upper critical fields $\mu_0 H_{c2}(T)$ of $(ScZrNb)_{1-x}[RhPd]_x$ HEA with $x = 0.40, 0.38, 0.37$, and 0.39. The open circles are the 50% values obtained from the magnetic-field dependent $\rho(T)$ plots; the dotted lines show the fits for the determination of $\mu_0 H_{c2}(0)$, according to the equation depicted in the gray frame. [5] / with permission of the American Chemical Society.

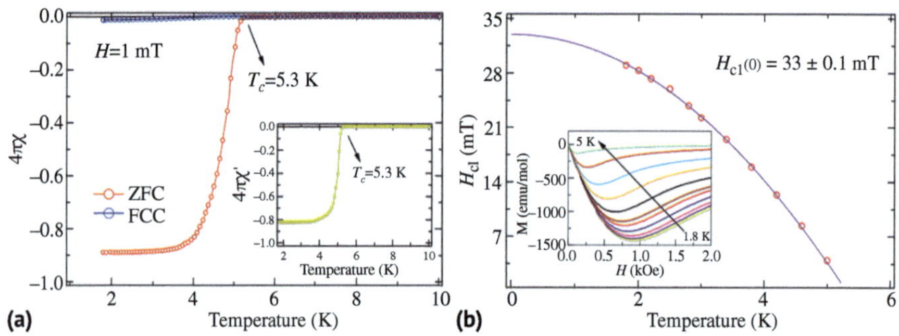

Figure 5.5 (a) The temperature variation of DC susceptibility for $Nb_{21}Re_{16}Zr_{20}Hf_{23}Ti_{20}$ taken in 1 mT field shows the superconducting transition at $T_c = 5.3$ K. Inset shows the AC susceptibility measurement. (b) Temperature dependence of the lower critical field H_{c1} and the fitting using Ginzburg–Landau relation gives $H_{c1}(0) = 33 \pm 1$ mT . Inset: The magnetization curves as a function of the applied magnetic field at various temperatures. [6] / with permission Elsevier.

into the potential of HEA superconductors for future applications, particularly under extreme conditions, and highlights the importance of exploring the role of spin-orbit coupling in enhancing the upper critical field of such materials.

5.2.3 High-Entropy Materials for Thermoelectric Applications

The chemical complexity of HEAs, such as significant atomic size disparities, coupled with high configurational entropy, leads to substantial lattice distortion and enhanced high-temperature phase stability. This distortion significantly increases phonon scattering, thereby reducing lattice thermal conductivity. Consequently, it is plausible to hypothesize that HEAs inherently exhibit low lattice thermal conductivity across a broad temperature spectrum, making them a promising candidate for thermoelectric materials.

Fan et al. [7] designed and investigated the microstructure and thermoelectric properties of the PbSnTeSe HEA. It was found that the PbSnTeSe HEA has a simple face-centered cubic (FCC) structure and possesses quite low lattice thermal conductivity at low temperatures, which could be ascribed to the strong phonon scattering due to its severe lattice distortion (e.g., Figure 5.6). The study found that adding small amounts of lanthanum (La) can enhance the Seebeck coefficient and electrical conductivity at high temperatures while suppressing the bipolar effect (e.g., Figure 5.7). The dimensionless figure of merit (zT), a key parameter for thermoelectric materials, was improved by the addition of La, with the highest zT value reaching approximately 0.8 at 873 K for the alloy with 1.5% La. The research suggests that HEAs could be a promising material for thermoelectric applications if challenges related to the bipolar effect are addressed. The findings highlight the potential of the HEA concept in designing novel thermoelectric materials.

Shi et al. [8] have proposed a novel method that enhances thermoelectric power generation performance by achieving global energy band convergence in high-entropy $Yb_xCa_{1-x}Mg_yZn_{2-y}Sb_2$ materials. The researchers designed a series of compositions that achieve simultaneous band convergence, leading to larger power factors and lower thermal conductivities. One specific composition, $Yb_{0.7}Ca_{0.3}Mg_{0.55}Zn_{1.45}Sb_2$, demonstrated a high thermoelectric figure of merit (zT) value of ~1.2 at 773 K, among the highest for p-type Zintl materials (e.g., Figure 5.8). The material exhibited excellent thermal and temporal stability, making it a promising candidate for waste heat recovery and thermoelectric power generation.

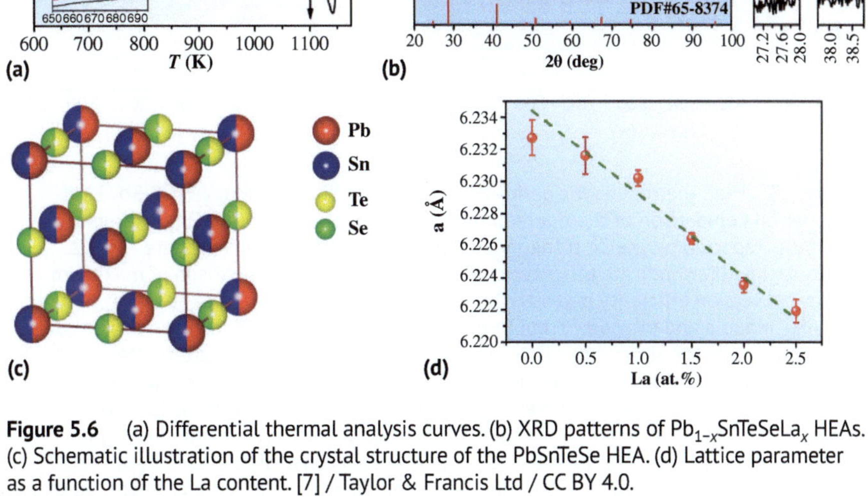

Figure 5.6 (a) Differential thermal analysis curves. (b) XRD patterns of $Pb_{1-x}SnTeSeLa_x$ HEAs. (c) Schematic illustration of the crystal structure of the PbSnTeSe HEA. (d) Lattice parameter as a function of the La content. [7] / Taylor & Francis Ltd / CC BY 4.0.

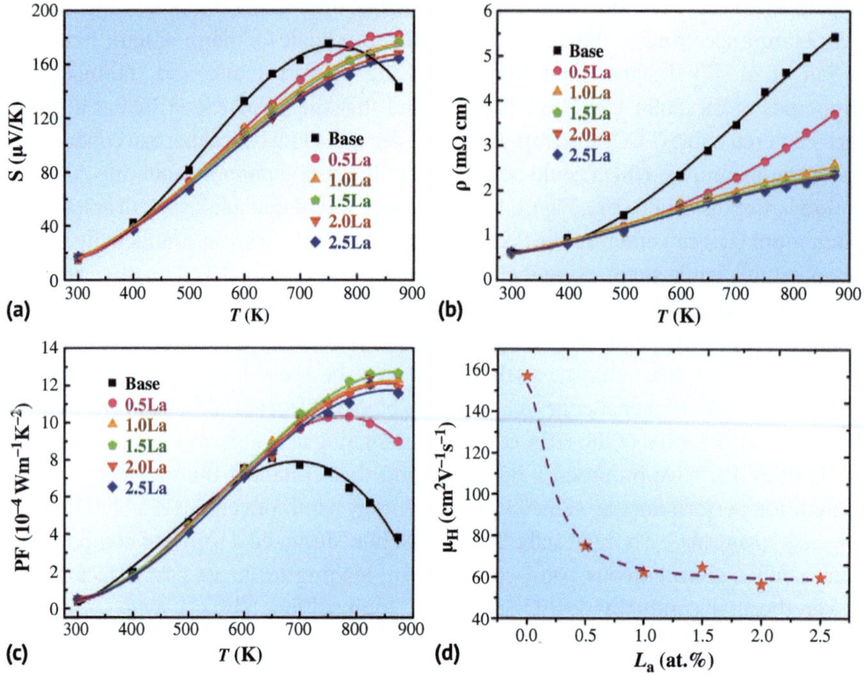

Figure 5.7 (a) Seebeck coefficient S, (b) electrical resistivity ρ, (c) power factor PF as a function of temperature for all the HEAs, and (d) Hall mobility μ_H at room temperature as a function of the La content. [7] / Taylor & Francis Ltd / CC BY 4.0.

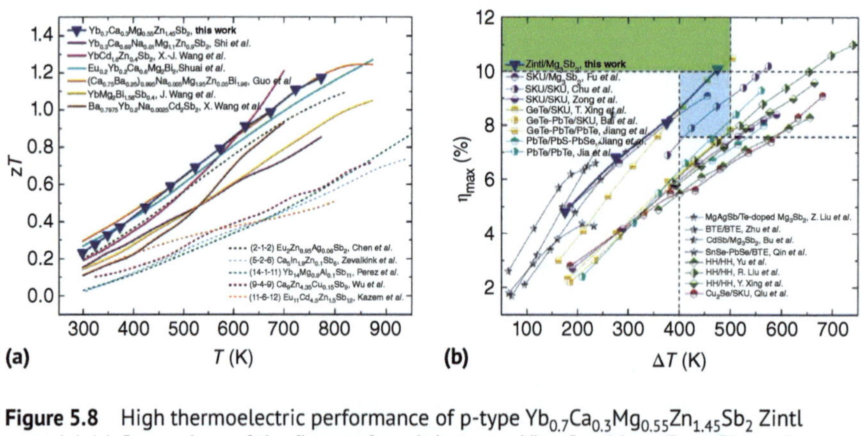

Figure 5.8 High thermoelectric performance of p-type $Yb_{0.7}Ca_{0.3}Mg_{0.55}Zn_{1.45}Sb_2$ Zintl material. (a) Comparison of the figure of merit between $Yb_{0.7}Ca_{0.3}Mg_{0.55}Zn_{1.45}Sb_2$ and previously reported p-type Zintl thermoelectric materials. Solid lines indicate AM2X2 Zintls; dashed lines indicate other types of Zintl materials. (b) Comparison of maximum energy conversion efficiency between the all-Zintl $Yb_{0.7}Ca_{0.3}Mg_{0.55}Zn_{1.45}Sb_2/Mg_{3+\delta}Y_{0.02}Sb_{1.5}Bi_{0.5}$ module and previously reported single-stage thermoelectric modules. The cold-side temperature for all modules is around room temperature. SKU, skutterudite-based materials; BTE, Bi_2Te_3-based materials; HH, half-Heusler-based materials. The green area represents the region in which the energy conversion efficiency is at least 10%, and the temperature difference is not higher than 500 K. The cyan area represents the region in which mid-to-high-temperature modules exhibit energy conversion efficiency at least 75% that of ours under $DT = 400$ to 500 K. [8] / with permission of The American Association for the Advancement of Science.

The researchers synthesized a series of $Yb_xCa_{1-x}Mg_yZn_{2-y}Sb_2$ alloys and discovered that the assumption of a linear relationship between the crystal-field splitting energy (D) of the alloy and its parent compounds was valid. This assumption was fundamental to their compositional design strategy. The stability of the material was confirmed through thermal cycling and long-term exposure to air, showing minimal degradation in performance (e.g., Figure 5.9). The study also successfully assembled a thermoelectric module using the p-type $Yb_{0.7}Ca_{0.3}Mg_{0.55}Zn_{1.45}Sb_2$ and an n-type $Mg_{3+\delta}$ $Y_{0.02}Sb_{1.5}Bi_{0.5}$-based material. This all-Zintl module showed exceptional heat-to-electricity conversion efficiency exceeding 10% at a temperature difference of 475 K (e.g., Figure 5.10b). The module's performance, combined with its nontoxicity and lack of tellurium, positions it as a strong contender for next-generation thermoelectric power generation.

In conclusion, the research introduces a groundbreaking method for designing high-performance thermoelectric materials with global band convergence.

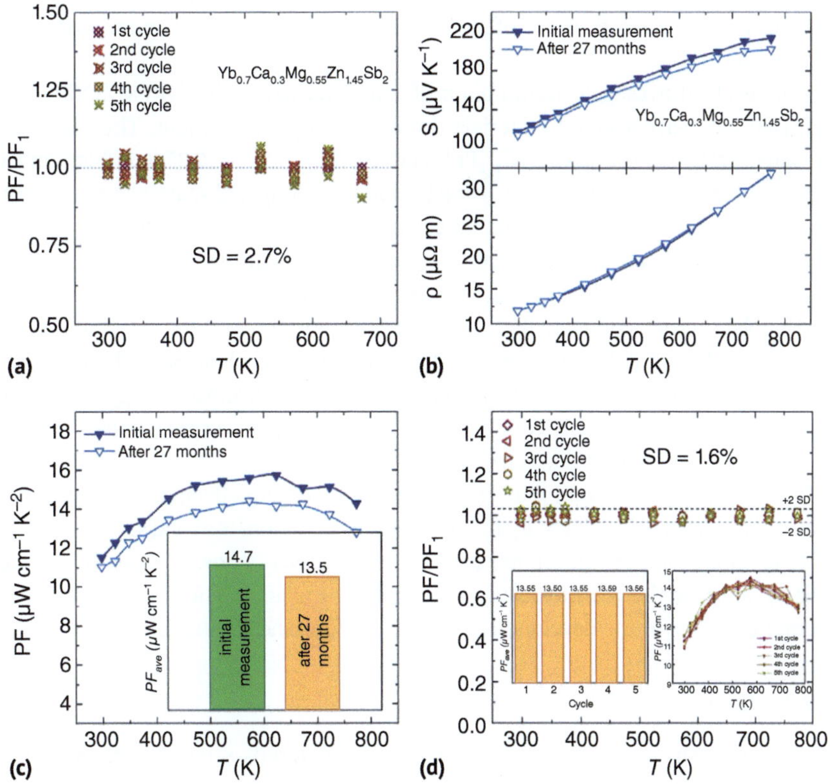

Figure 5.9 Thermal and temporal stability of $Yb_{0.7}Ca_{0.3}Mg_{0.55}Zn_{1.45}Sb_2$. (a) Power-factor deviation over five thermal cycles from 298 to 673 K and from 673 back to 298 K. (b) Seebeck coefficient (top) and electrical resistivity (bottom) results from the initial measurement and after 27 months of exposure to air. (c) Power-factor results from the initial measurement and after 27 months of exposure to air. (Inset) Average power-factor values from 298 to 773 K for the initial measurement and after 27 months of exposure to air. (d) Power-factor deviation over five thermal cycles from 298 to 773 K and from 773 back to 298 K for the sample measured after 27 months of exposure to air. (Left inset) Average power-factor values from 298 to 773 K for the five thermal cycles. (Right inset) Power-factor results from the five thermal cycles. [8] / with permission of The American Association for the Advancement of Science.

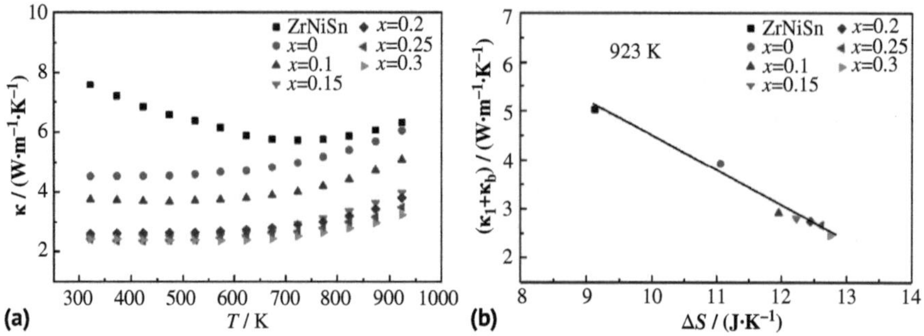

Figure 5.10 (a) Temperature dependence of total thermal conductivity. (b) Change of lattice thermal conductivity with configuration entropy at 923 K. [9] / with permission of Science Press.

The findings could significantly impact the field of thermoelectric power generation, offering a path to more efficient energy conversion and more sustainable materials.

Wang et al. [9] investigate the inhibition of lattice thermal conductivity in ZrNiSn-based half-Heusler thermoelectric materials through entropy adjustment. The researchers prepared ZrNiSn and $Zr_{0.5}Hf_{0.5}Ni_{1-x}Pt_xSn$ ($x = 0$, 0.1, 0.15, 0.2, 0.25, 0.3) high-entropy half-Heusler thermoelectric alloys using levitation melting and spark plasma sintering. They manipulated the configurational entropy of the alloys by substituting Hf for Zr and Pt for Ni and studied the effects of configurational entropy on the thermoelectric properties. The results showed that the minimum sum of lattice thermal conductivity and bipolar thermal conductivity ($\kappa_l + \kappa_b$) at 673 K for $Zr_{0.5}Hf_{0.5}Ni_{0.85}Pt_{0.15}Sn$ was optimized at 2.1 $W \cdot m^{-1} K^{-1}$, which was significantly reduced by about 58% when compared with ZrNiSn (e.g., Figure 5.10). This finding provides an effective strategy for reducing the lattice thermal conductivity of ZrNiSn-based alloys, offering great potential for further improvement of thermoelectrics.

Wang et al. [9] also found that the HEA design idea can effectively solve the problem of high thermal conductivity in HH alloys. By changing the solid solution amount of Pt to adjust the configurational entropy of the ZrNiSn alloy, the researchers were able to optimize the thermoelectric properties of the alloy. The research provides a theoretical basis for the development of high-performance thermoelectric materials.

5.2.4 Conclusion

HEMs represent a significant advancement in materials science, offering unique thermal and electronic properties distinct from conventional alloys. The presence of multiple major elements in HEMs increases configurational entropy, preventing the formation of intermetallic compounds while enhancing hardness, corrosion resistance, and thermal stability. The atomic disorder is a key determinant of HEM properties, and variations in atomic size lead to severe lattice distortions that affect the transport of electrically and thermally conductive media.

1. Predicting lattice thermal conductivity in HEAs is challenging due to complex structures, but ML offers a fast and accurate prediction method. Studies show that aluminum content significantly influences thermal conductivity in AlCoCrNiFe HEAs,

and alloying affects transport properties in Ti—Cr—Mo—W—V—Nb—Ta RHEAs [2]. CrMnFeCoNi HEA exhibits lower thermal conductivity than nickel, with unusual temperature dependence and dominant defect scattering in heat transport [3].

2. HEAs have expanded the family of superconductors and opened new research directions. The Ta—Nb—Hf—Zr—Ti system shows robust zero-resistance superconductivity up to 190 GPa, indicating potential for extreme condition applications [4]. New HEAs like $(ScZrNb)_{1-x}{}^{[RhPd]}{}_x$ and $(ScZrNb)_{1-x}{}^{[RhPd]}{}_x$ exhibit type-II superconductivity with high T_c and strong upper critical fields, suitable for practical applications [5].
3. HEAs with high mixing entropy often form simple crystal structures, leading to substantial lattice distortion and reduced lattice thermal conductivity, making them promising for thermoelectric materials. PbSnTeSe HEA shows low lattice thermal conductivity and improved thermoelectric performance with La addition [7]. $YbxCa_{1-x}Mg_yZn_{2-y}Sb_2$ materials achieve global energy band convergence, demonstrating high thermoelectric performance and stability, suitable for waste heat recovery and power generation [8].

In summary, HEMs offer unique opportunities in thermal and electron properties, with significant potential in superconductivity and thermoelectric applications, pushing the boundaries of materials science.

5.3 Magnetic and Optical Properties in HEMs

5.3.1 Magnetic Properties in High-Entropy Materials

In addition to the excellent mechanical properties of HEAs, the magnetic properties of HEAs have attracted much attention. Some transition group HEAs have ferromagnetism due to the high content of ferromagnetic elements (Fe, Co, Ni) and have the possibility to be used as hard or soft magnetic functional materials. The rapid development of high-entropy soft magnetic materials is primarily based on ferromagnetic elements Fe, Co, and Ni, with minor additions of non-ferromagnetic elements such as Cr, Cu, Si, Ga, Al, Ti, and Mn. These additions help enhance both magnetic and mechanical properties, leading to materials with excellent overall performance. In soft magnetic materials, usually, these materials must have high saturation magnetization strength (M_s) and low coercivity (H_c). It has been shown that high saturation magnetization strength (M_s), low coercivity (H_c), and good mechanical properties can be obtained by carefully designing the composition of alloys. Meanwhile, the magnetic properties of HEAs are related to the processing and preparation of the alloys and are also affected by the heat treatment process.

Zhang et al. [10] designed and investigated a new soft magnetic material known as $FeCoNi(AlSi)_x$ HEAs, which exhibit exceptional magnetic, electrical, and mechanical properties. The optimal balance of these properties was found at a molar ratio of $x = 0.2$, where the alloy demonstrated high saturation magnetization (1.15 T), low coercivity (1400 A/m), desirable electrical resistivity (69.5 $\mu\Omega \cdot$cm), and good ductility with 50% strain without fracture. The study utilized ab initio calculations to elucidate the high magnetic saturation and the impact of composition on the magnetic characteristics of HEAs. The findings pave the way for new applications of HEAs.

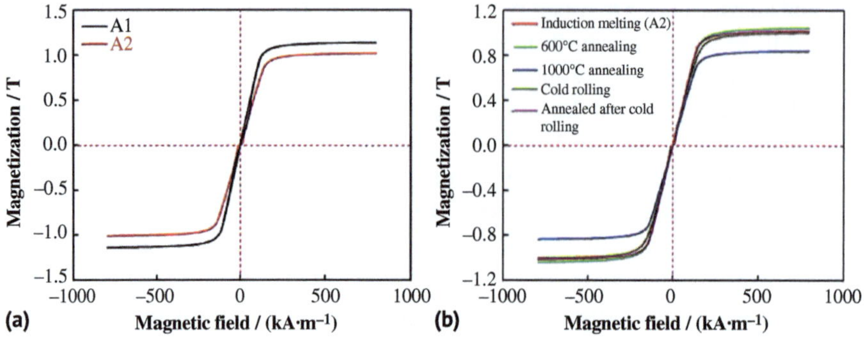

Figure 5.11 Magnetic properties of the $FeCoNiAl_{0.2}Si_{0.2}$ alloys. (a) Hysteresis loops of the alloy prepared by two methods. (b) Hysteresis loops of A2 samples after different processes. [11] / with permission of Springer Nature.

Zuo et al. [11] investigated the effect of processing methods on the magnetic and mechanical properties of $FeCoNiAl_{0.2}Si_{0.2}$ HEAs. The HEA was prepared by arc melting and induction melting, resulting in two samples denoted A1 and A2, respectively. Both samples exhibited an FCC structure and dendritic morphology. The A2 sample showed slightly higher tensile yield strength and elongation compared to A1. The magnetic properties of both alloys were similar, with saturation magnetization and coercivity values of 1.151 T and 1400 A/m for A1 and 1.015 T and 1431 A/m for A2 (e.g., Figure 5.11). Heat treatment at various temperatures did not significantly change the phases, except for the A2 sample heat-treated at 600 °C, which showed new phase precipitation and a decrease in coercivity. Cold rolling and subsequent heat treatment did not significantly improve the magnetic properties but did enhance the strength of the A2 sample. The study concludes that the $FeCoNiAl_{0.2}Si_{0.2}$ HEAs have excellent overall properties, including good plasticity and high magnetic saturation, with slight variations between the two preparation methods.

Zuo et al. [12] investigated how the structure and properties of CoFeNi HEAs are affected by the additions of aluminum (Al) and silicon (Si). The research focuses on $Al_xCoFeNi$ and $CoFeNiSi_x$ alloys with varying Al and Si molar ratios. It was discovered that Al and Si additions significantly influence the phase structures; high Al ratios can transform the FCC structure to BCC, while increased Si leads to new compound formations. Both elements enhance yield strength and hardness but at the expense of plasticity, with Si having a more pronounced effect. Magnetic property characterizations reveal that all alloys exhibit ferromagnetic behavior, with saturation magnetization decreasing with increased Al and Si content. Si addition reduces magnetization more significantly than Al (e.g., Figure 5.12). Electrical resistivity increases with Al and Si additions, with Si causing a more substantial increase. The alloys also display minimal magnetostriction, crucial for maintaining material integrity under magnetic fields. The findings suggest that a careful balance of Al and Si ratios is necessary to optimize the alloys' mechanical and magnetic properties.

Zuo et al. [13] conducted a study on the impact of metal doping on the magnetic characteristics of HEAs CoFeMnNiX, where X represents Al, Cr, Ga, and Sn.

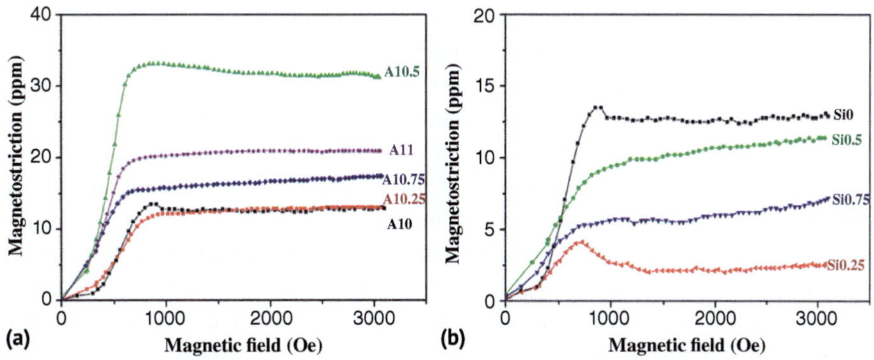

Figure 5.12 The magnetostrictive curves of (a) $Al_xCoFeNi$ and (b) $CoFeNiSi_x$ alloys. [12] / with permission of Elsevier.

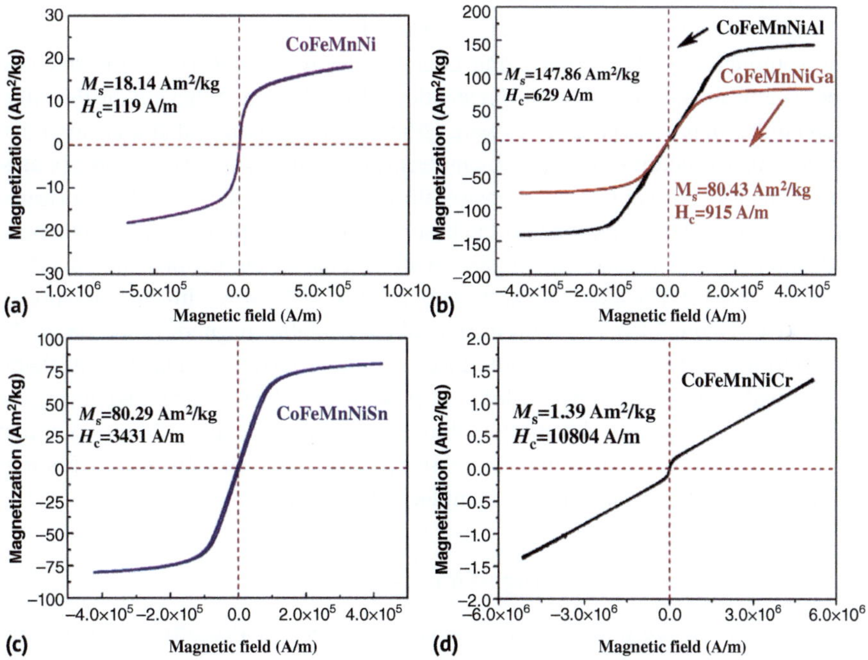

Figure 5.13 Hysteresis loops of (a) CoFeMnNiMn, (b) CoFeMnNiAl and CoFeMnNiGa, (c) CoFeMnNiSn, and (d) CoFeMnNiCr alloys at room temperature. [13] / with permission of Elsevier.

The study finds that the addition of Al, Ga, and Sn induces a phase transition from FCC to ordered BCC structures, enhancing the saturation magnetization significantly. For instance, the CoFeMnNiAl alloy achieves a high M_S of 147.86 Am^2/kg (e.g., Figure 5.13). DFT calculations reveal that doping elements like Al can suppress the anti-ferromagnetism of Mn atoms, promoting ferromagnetism. The research also discusses the potential for further optimization of these alloys for soft magnetic applications, suggesting future work on $CoFe(Cr,Mn)_xNi(Al,Ga)_y$ compositions and

L21-structured alloys. The findings provide valuable insights into tailoring magnetic properties in HEAs, crucial for advancing their functional applications.

Zhang et al. [14] concentrated on the $(Fe_{0.3}Co_{0.5}Ni_{0.2})_{100-x}(Al_{1/3}Si_{2/3})_x$ system for the development of HEAs with excellent soft magnetic properties, in which the base alloy composition $Fe_{0.3}Co_{0.5}Ni_{0.2}$ was chosen due to its near-zero saturated magnetostriction coefficient. The $(Fe_{0.3}Co_{0.5}Ni_{0.2})_{95}(Al_{1/3}Si_{2/3})_5$ alloy demonstrated superior soft magnetic characteristics, including a coercivity of 96 A/m, saturation magnetization of 1.49 T, and good ductility with 38% elongation after annealing at 1000 °C. The findings suggest this alloy could be a promising candidate for next-generation soft magnets, offering advantages in fabrication, high thermal stability, and a balance of soft magnetic and mechanical properties over traditional materials like silicon steel and amorphous soft magnetic materials. However, the study is limited to room temperature analyses, and further research is needed to explore the alloy's performance under different temperature conditions and practical applications.

Song et al. [15] successfully prepared $Co_{47.5}Fe_{28.5}Ni_{19}Si_{3.4}Al_{1.6}$ HEA with excellent comprehensive properties using laser powder bed fusion (LPBF) technology. The organizational evolution and magnetic and mechanical behaviors of this alloy under different process parameters were systematically investigated. It was found that unique microstructures, including high-density dislocations and Al-rich precipitates, contributed to excellent comprehensive attributes. Specifically, the alloy fabricated at 200 W and 800 mm/s exhibited the best performance with a saturation magnetic induction of 1.479 T, coercivity of 188.3 A/m, and maximum permeability of 1171.8 (e.g., Figure 5.14). The tensile yield strength reached 417.0 MPa with an elongation of 33.9% (e.g., Figure 5.15). The study attributed the favorable magnetic properties to the high ferromagnetic element concentration and uniform composition. The mechanical strength was mainly due to dislocation-precipitate frameworks formed during the LPBF process. However, the study is limited to the specific alloy composition and LPBF parameters explored, and further research is needed to generalize these findings to other compositions and additive manufacturing processes.

5.3.2 Optical Properties in High-Entropy Materials

In recent years, the optical performance of HEMs has also attracted considerable interest, particularly for applications in solar energy absorption coatings. HEA nitride films, an important subset of HEMs, typically exhibit simpler and denser structures compared to traditional binary or ternary nitride films, and they demonstrate enhanced performance. Extensive research has indicated that these HEA nitride films hold broad application prospects in hard coatings and solar thermoelectric conversion coatings. Moreover, the thermal stability of HEA nitride films is particularly remarkable. Thus, due to their excellent optical properties, thermal stability, and the ability to optimize performance by adjusting composition and microstructure, HEA nitride films show significant application potential in the field of solar energy absorption coatings. As research progresses, the optical properties of HEMs are expected to find applications in an increasingly wider array of fields.

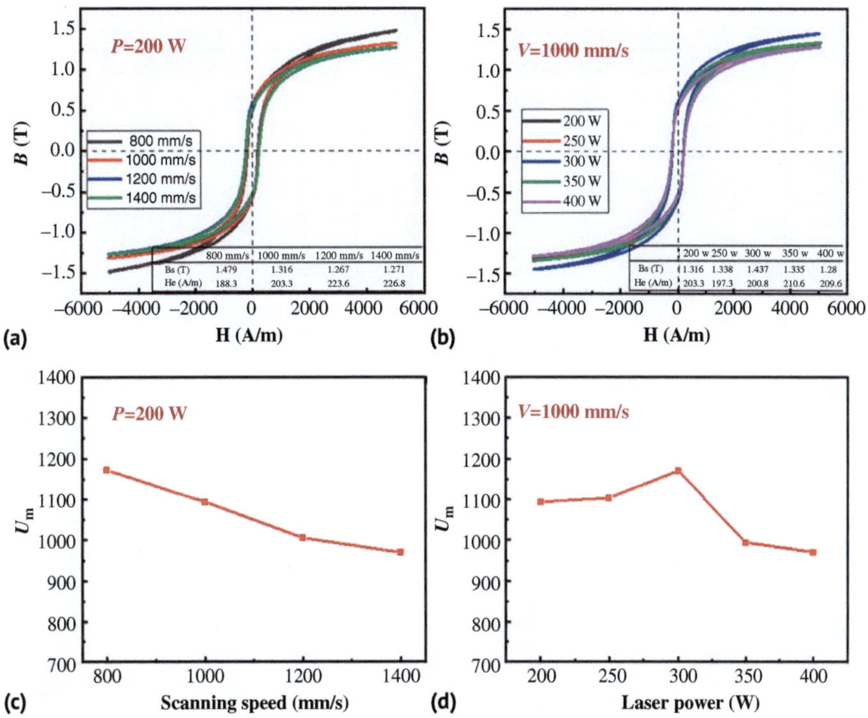

Figure 5.14 Hysteresis loops of $Co_{47.5}Fe_{28.5}Ni_{19}Si_{3.4}Al_{1.6}$ HEA under different LPBF process parameters. (a), (c) 200 W, ~800–1400 mm/s. (b), (d) 1000 mm/s, ~200–400 W. Note: As can be seen in the figure, with the increases in the scanning speed, the magnetic properties of the samples severely deteriorated and the influencing effects of the changes in laser power on the magnetic properties were limited. [15] / with permission of Elsevier.

Zhang et al. [16] revealed several key findings in the study of the optical properties of $FeCrVTa_{0.4}W_{0.4}$ HEA nitride films. The study on the optical performance of $FeCrVTa_{0.4}W_{0.4}$ HEA nitride films reveals several key findings. The films deposited on monocrystalline silicon substrates were investigated for their solar spectral selective absorption. It was found that the films exhibit selective absorbance, with lower reflectivity for wavelengths between 300 and 400 nm and higher reflectivity for the 400–800 nm range. The absorption rate was highest when three layers of nitride films were applied, suggesting an interaction between the layers and the film's surface roughness, which contributes to light absorption.

The research also indicated that the films' optical properties can be tailored by adjusting the nitrogen content. An increase in nitrogen flow led to a transformation from an amorphous structure to an FCC structure, which enhanced the crystallinity and, in turn, affected the optical performance. Notably, a specific nitrogen flow rate of 15 mL min^{-1} yielded the optimal mechanical properties, which coincided with the best optical performance in terms of solar spectrum selective absorptivity. In summary, the study provides insights into how HEA nitride films can be engineered for specific optical performance characteristics, making them promising candidates for applications in solar energy absorption and other photonic technologies.

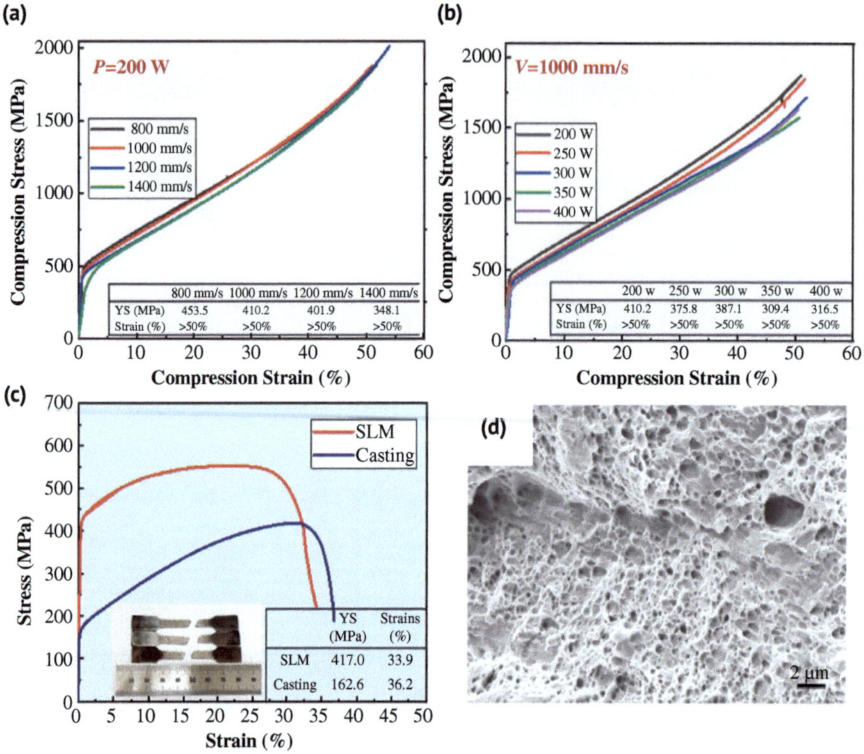

Figure 5.15 Compressive properties of $Co_{47.5}Fe_{28.5}Ni_{19}Si_{3.4}Al_{1.6}$ HEA at room temperature. (a) 200 W, ~800–1400 mm/s. (b) 1000 mm/s, ~200–400 W. (c) Engineering tensile stress–strain curves of the LPBF sample with 200 W, 800 mm/s and a cast sample of $Co_{47.5}Fe_{28.5}Ni_{19}Si_{3.4}Al_{1.6}$ HEA. (d) Tensile fracture micromorphology of LPBF sample with 200 W, 800 mm/s in LPBF. [15] / with permission of Elsevier.

Zhang et al. [17] investigated the optical properties of high-entropy films (HEFs) in the pseudobinary system (Cr, Fe, V)–(Ta, W). The research investigates how varying the atomic content of elements affects the structure and solar absorptivity of the films. The films exhibited an amorphous structure when the compositional parameter x ranged from 86.9 to 32.5 and a BCC structure formed when x was less than 24. The solar absorptivity was found to be highest at the transitional area from amorphous to BCC structure, peaking when the composition was $Cr_6Fe_6V_6Ta_{42}W_{40}$, with an absorptivity of 81.94% (e.g., Figure 5.16). The study suggests that combinatorial strategies are effective for identifying high-performance HEFs with optimal solar absorptivity for potential use in photothermal conversion applications.

Gao et al. [18] have worked on the design and preparation of high-entropy nitride films and photothermal conversion performance. However, the photothermal mechanism of high-entropy spectrally selective absorption coating is still unclear. Based on previous research results on high-entropy nitride light-absorbing materials, and to meet the application requirements of concentrating solar power systems under high-temperature working conditions, the researchers chose the constituent elements and explored the effect of high-entropy

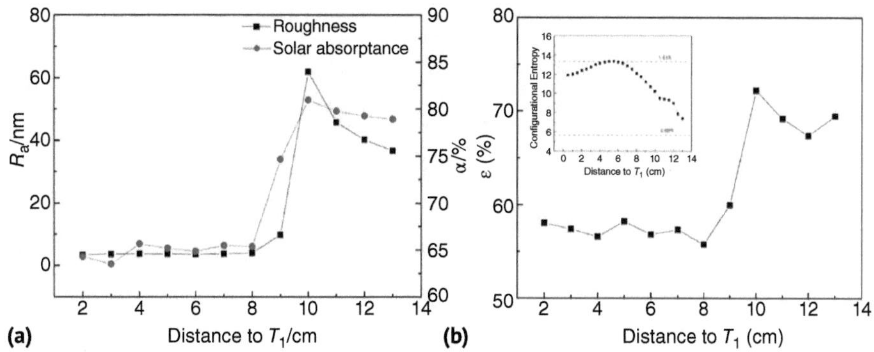

Figure 5.16 (a) Variations in the toughness and light absorption ratio of the $(Cr_{0.33}Fe_{0.33}V_{0.33})_x(Ta_{0.5}W_{0.5})_{100-x}$ HEFs according to composition. (b) Variations in toughness and light absorption ratio of $(Cr_{0.33}Fe_{0.33}V_{0.33})_x(Ta_{0.5}W_{0.5})_{100-x}$ HEFs. Inset illustrates variations in the mixing entropy of the film according to composition. [17] / with permission of American Chemical Society.

design on the electronic structure of the materials. The study shows that the complex energy band structure of high-entropy nitrides increases the density of states near the Fermi energy level. Meanwhile, the flat energy band structure broadens the solar energy absorption range of the material, enhances the interband jump effect, and improves the solar energy absorption performance of the material. The well-designed refractive index gradient enables destructive interference effects, offering a roughness-insensitive high solar absorptance (92.8%) and low thermal emittance (39.2%) (e.g., Figure 5.17). Impressively, the solar absorptance is 9.1% higher than the reported best-performing selective nanofilm-based textile. As a result, such a textile achieves a record-high photothermal conversion efficiency (82.2% under 0.6 suns, at 0 °C). This textile yields a 3.5 °C drop in the set point of indoor air-conditioner temperature. Besides, on a winter morning with an air temperature of 7.5 °C, it warms up the human skin by as much as 12 °C under weak sunlight (350 W m^{-2}) (e.g., Figure 5.18). The coating is of great significance in enhancing the efficiency of the trough concentrator system by its photothermal conversion capability, thermal stability, and double-layer structure that is easy to produce on a large scale.

5.3.3 High-Entropy Materials for Optoelectronic Applications

Metal nanoparticle co-catalysts have been extensively studied for their ability to enhance the performance of photocatalytic materials, but their inherent instability under light irradiation often limits their practical applications. This challenge has stimulated interest in exploring HEAs, which are composed of five or more elements in near-equimolar ratios and exhibit low Gibbs free energies due to a large increase in entropy (ΔS_{mix}), resulting in excellent intrinsic stability. In addition, variations in atomic size among the constituent elements induce significant lattice distortions, impose higher energy barriers to atomic diffusion, and contribute to the formation and stability of HEA NPs. Furthermore, the local variability and tunability of the electronic structure facilitate the

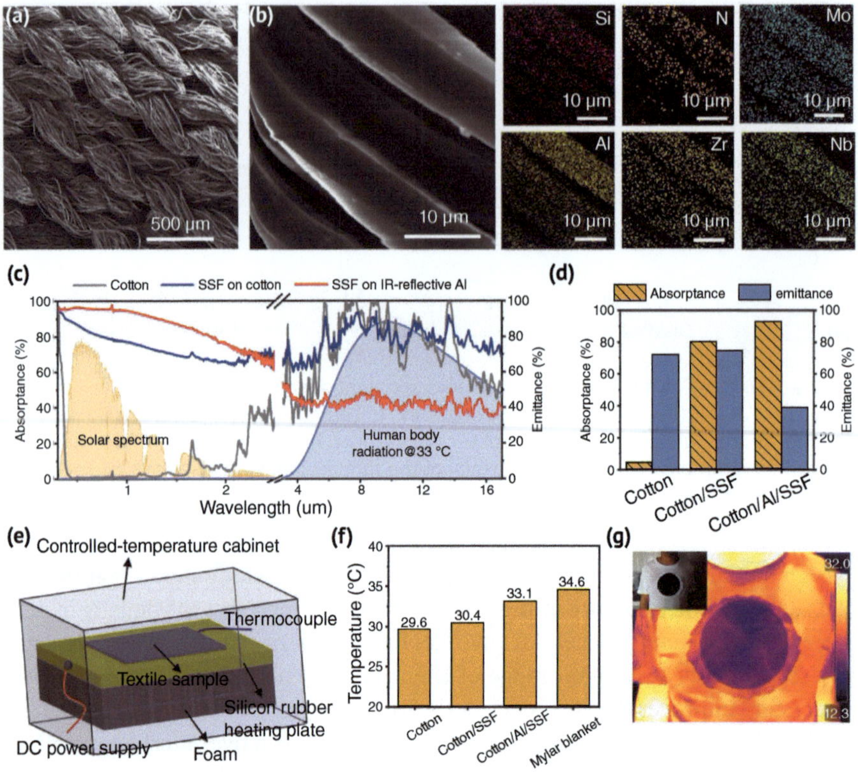

Figure 5.17 Microstructure, optical properties, and indoor application of the SSF-decorated cotton. (a) Surface SEM topography of the SSF-decorated cotton. (b) The corresponding amplifying region and elemental mappings of Si, N, Mo, Al, Zr, and Nb. Reproduced with permission from [18] / John Wiley & Sons / CC BY 4.0.
(c) Reflectance spectra of cotton, SSF/cotton, and SSF/Al/cotton. (d) The corresponding solar absorptance and thermal emittance. (e) Schematic of a homemade experimental setup for indoor passive radiative heating performance tests—the temperature is 20 °C. (f) Final temperatures of skin simulator and the ones covered with cotton, cotton/Al/SSF, and Mylar blanket. (g) Visible passive radiative heating evaluation of a human body covered with SSF/Al/cotton in indoor ambient. [18] / John Wiley & Sons / CC BY 4.0.

surface activation of adsorbed substances. Therefore, HEA nanoparticles are expected to be used as catalysts in the field of photovoltaics.

Huang et al. [19] developed and investigated noble metal-free HEA nanoparticles for efficient solar-driven photocatalytic CO_2 reduction. The study presents a significant enhancement in photocatalytic CO_2 reduction when these HEA nanoparticles are immobilized on TiO_2, achieving a 23-fold increase over pristine TiO_2. The CO and CH_4 production rates reached 235.2 and 19.9 $\mu mol\ g^{-1}\ h^{-1}$, respectively (e.g., Figure 5.19). Notably, the HEA nanoparticles demonstrated excellent stability under simulated solar irradiation, as well as high-energy X-ray irradiation, highlighting their potential to revolutionize the field of photocatalysis with their superior stability and performance.

Ge et al. [20] presented in the document investigates the synthesis and optical properties of high-entropy oxide (HEO) sub-1 nm nanosheets (SNSs) induced by

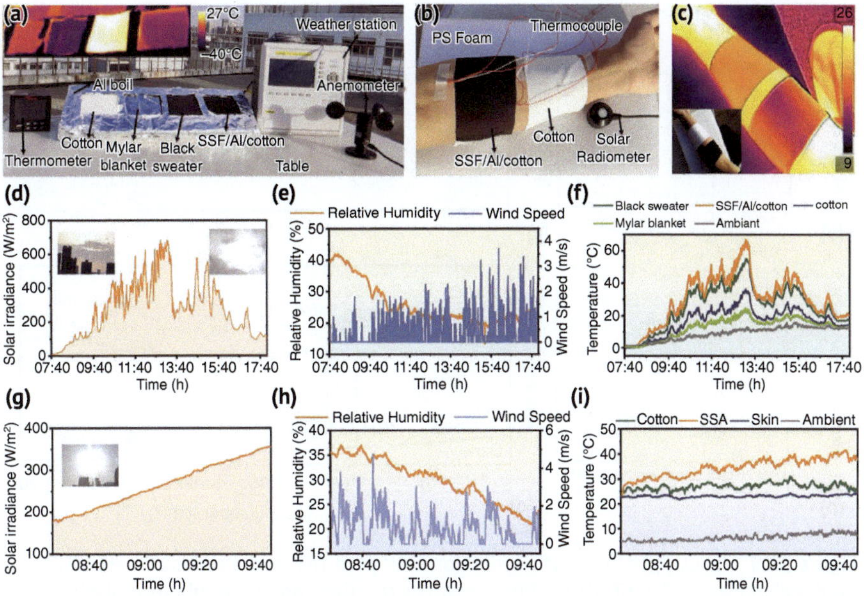

Figure 5.18 Solar heating performance of the SSF-decorated cotton. (a) Digital photograph of a homemade solar heating apparatus placed on the rooftop (February 18, 2021; Lanzhou China, 103:50 E, 36:03 N) to evaluate the solar radiative ability; the inset is the corresponding IR image. (b) Optical image of solar radiative warming performance evaluation for the cotton and SSF-decorated cotton on a real human. (c) IR and optical photographs of the cotton and SSF-decorated cotton on a real human. (d) Recorded solar irradiance intensity. (e) Relative humidity and wind speed during the daytime experiment, and (f) corresponding real-time temperatures of the skin simulator covered with different textiles. (g) Recorded solar irradiance intensity and (h) relative humidity and wind speed during the measurement on real human skin. (i) Corresponding real-time temperatures of the skin covered with different textiles. [18] / John Wiley & Sons / CC BY 4.0.

polyoxometalate (POM) clusters for use in photoelectrocatalysts for Zn$^-$ air batteries. The study reveals that these 2D HEOs-POM SNSs exhibit strong light absorption across the ultraviolet to near-infrared (NIR) spectrum, with the CFNMCZ-PM A(CoFeNiMnCuZnO$_x$-phosphomolybdic acid) SNSs showing the most potent absorption, particularly in the visible light range. The bandgap energies were calculated to be 2.12 eV for CFNMCZ-PMA, 2.32 eV for NMCZ (NiMnCuZnO$_x$)-PMA, and 2.18 eV for FNMCZ (FeNiMnCuZnO$_x$)-PMA SNSs, indicating the potential for efficient light utilization and charge carrier dynamics. The photocurrent responses were also highest for the CFNMCZ-PMA SNSs, underscoring their effectiveness in separating and transferring photogenerated charge carriers (e.g., Figure 5.20). These findings suggest that the 2D HEOs-POM SNSs could significantly enhance the photoelectrocatalytic performance for oxygen evolution and reduction reactions in Zn$^-$ air batteries under light irradiation.

Folgueras et al. [21] have made a breakthrough in their research on the synthesis of high-entropy halide chalcogenide single crystals, which is very promising for photovoltaic applications. Conventional HEMs need to be synthesized at temperatures above 1000 °C. However, researchers such as Yang have developed a method to create these

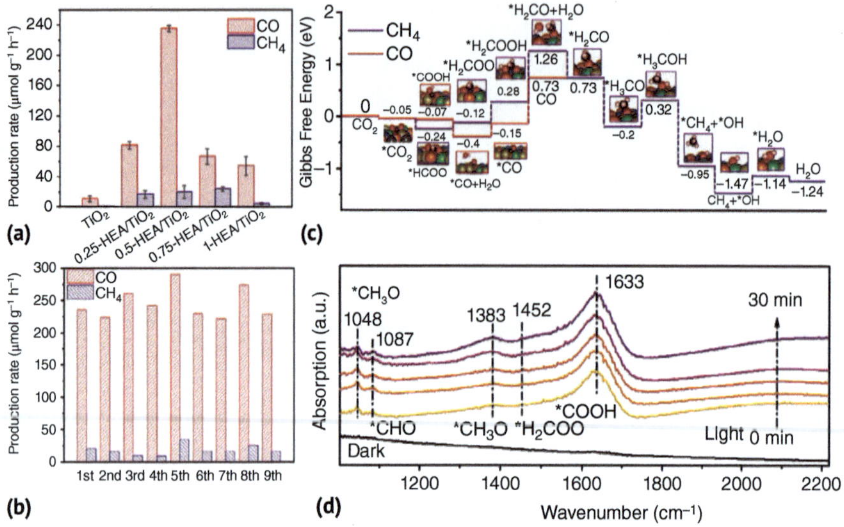

Figure 5.19 (a) CO_2 photocatalytic reduction over pure TiO_2 and a series of x-HEA/TiO_2. (b) Recycle tests for 0.5-HEA/TiO_2. Photocatalytic reaction conditions: photocatalysts (0.015 g), 1 atm CO_2, 10 μL H_2O, AM1.5G simulated light irradiation, 100 mW cm^{-2}, and irradiation time 4 h. (c) The DFT calculated potential energy diagram of CO_2 reduction on HEA (111) and the corresponding adsorption configurations of CO_2 and reactive intermediates of CO and CH_4 generation. (d) In situ FTIR (Fourier Transform Infrared Spectroscopy) spectra for coadsorption of a mixture of CO_2 and H_2O vapor on 0.5-HEA/TiO_2. [19] / with permission of John Wiley & Sons.

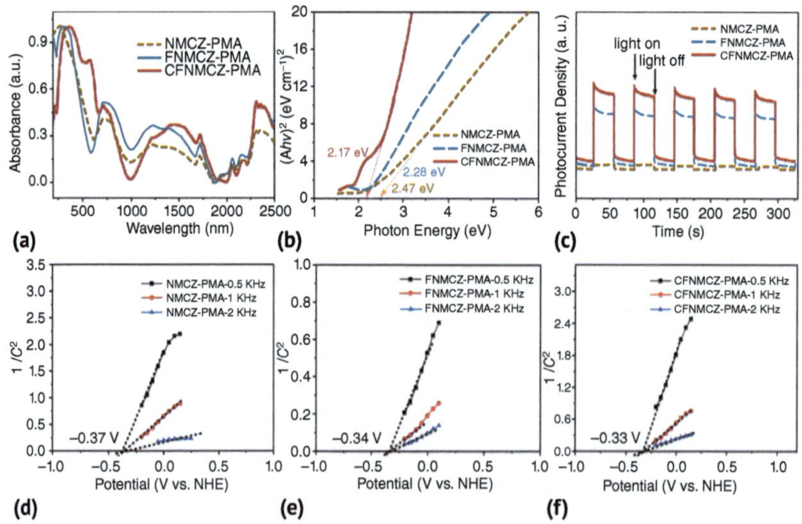

Figure 5.20 Bandgap and band structure of 2D HEOs-PMA SNSs. (a) UV–vis–NIR absorbance spectra of the NMCZ-PMA, FNMCZ-PMA, and CFNMCZ-PMA SNSs. (b) Tauc plots of the three sorts of SNSs derived from the absorbance spectra. (c) Transient photocurrent density of the SNSs versus time under the full arc of an Xe lamp. Mott–Schottky plots of the (d) NMCZ-PMA SNSs, (e) FNMCZ-PMA SNSs, and (f) CFNMCZ-PMA SNSs. [20] / with permission of the American Chemical Society.

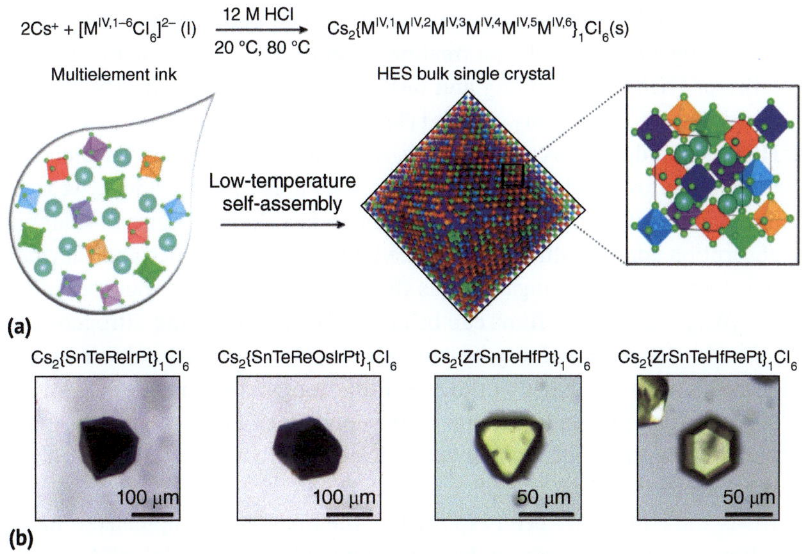

Figure 5.21 Synthetic design of high-entropy five- and six-element Cs_2MX_6 single crystals. (a) Schematic representation of the room- and low-temperature-driven self-assembly processes in solution from multielement inks to high-entropy semiconductor perovskite single crystals. The use of 12 M HCl as the solvent for multielement inks was a specific design choice because an excess chloride environment promotes the formation of six coordinated $[MCl_6]^{2-}$ 2-octahedral molecules in solution. (b) Optical microscope images of high-entropy five-element $Cs_2\{SnTeReIrPt\}c_1Cl_6$, six-element $Cs_2\{SnTeReOsIrPt\}_1Cl_6$, five-element $Cs_2\{ZrSnTeHfPt\}_1Cl_6$, and six-element $Cs_2\{ZrSnTeHfRePt\}_1Cl_6$ single crystals. [21] / with permission of Springer Nature.

crystals at room temperature (20 °C) or low temperature (80 °C) (e.g., Figure 5.21). These crystals are based on a cubic Cs_2MCl_6 structure with five or six different metal ions occupying the M-site in nearly equimolar ratios, leading to complex vibrational and electronic structures. These structures exhibit unique energy transfer interactions between confined exciton states, which are crucial for photovoltaic applications such as solar cells and light-emitting diodes (LEDs). The study also explores the stability and optoelectronic properties of these materials, paving the way for their potential integration into electronic device architectures.

5.3.4 Conclusion

HEAs exhibit significant magnetic properties due to the presence of ferromagnetic elements (Fe, Co, Ni). These alloys can be designed as hard or soft magnetic materials. Recent developments in high-entropy soft magnetic materials involve microadditions of non-ferromagnetic elements like Cr, Cu, Si, Ga, Al, and Ti to enhance magnetic and mechanical properties.

1. Studies by Zhang et al. and Zuo et al. have shown that careful composition design can achieve high saturation magnetization, low coercivity, and good mechanical properties. For example, $FeCoNi(AlSi)_x$ HEAs demonstrate exceptional

magnetic and electrical properties with optimal performance at $x = 0.2$ [10]. Processing methods and heat treatment significantly affect the magnetic properties of HEAs. Cold rolling and heat treatment can enhance the strength of alloys like $FeCoNiAl_{0.2}Si_{0.2}$ HEAs [11].

2. HEMs, including HEA nitride films, have attracted attention for their unique microstructures and superior properties such as high hardness, wear resistance, and corrosion resistance. These materials show promise in solar energy absorption coatings due to their excellent optical properties and thermal stability. Research by Zhang et al. has shown that the optical properties of FeCrV $Ta_{0.4}W_{0.4}$ HEA nitride films can be tailored by adjusting the nitrogen content. HEFs in the pseudobinary system (Cr, Fe, V) – (Ta, W) exhibit high solar absorptivity, particularly at the transition from amorphous to BCC structure. High-entropy nitride films have been shown to enhance solar energy absorption and photothermal conversion efficiency, with applications in concentrating solar power systems [16].
3. HEAs and HEOs have been explored for their potential in optoelectronic applications, including photocatalysis and photoelectrocatalysis. HEA nanoparticles have demonstrated excellent stability and performance in photocatalytic CO_2 reduction, with significant enhancements when immobilized on TiO_2. HEO SNSs induced by POM clusters show strong light absorption and efficient charge carrier dynamics, making them promising for photoelectrocatalytic applications in Zn–air batteries. High-entropy halide chalcogenide single crystals synthesized at room or low temperature exhibit unique energy transfer interactions, crucial for photovoltaic applications like solar cells and LEDs [19].

In summary, HEMs offer significant potential in magnetic and optical applications, with ongoing research exploring their use in soft magnetic materials, solar energy absorption, and optoelectronic devices. These materials continue to push the boundaries of functional materials science.

5.4 Materials Design for Functional Applications

The design methodology for functional applications of HEMs requires the integration of traditional materials science and modern computational techniques to realize high-entropy functional materials with excellent comprehensive performance. The core concept is to optimize the mechanical and functional properties of materials through microstructural modulation, such as grain refinement, phase transition, and precipitation phase control. Computational materials science can also play a key role in accelerating the discovery of new materials by predicting material properties and guiding experimental design through first principles and MD simulations. Meanwhile, the application of ML and artificial intelligence makes it possible to extract valuable structure property and even functional property relationships from a large amount of literature and data, further accelerating the iterative process of high-entropy functional material design [22] (e.g., Figure 5.22). For rapid screening and testing of materials, high-throughput experimental methods can be used in

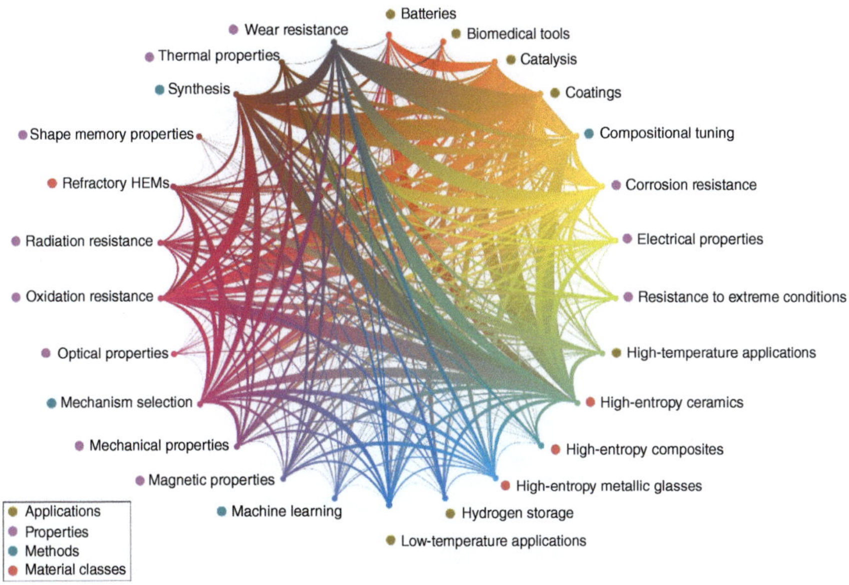

Figure 5.22 Functional features of HEMs and connections between them as identified by a natural language analysis. Overview of properties explored of functional HEMs, showing promising connections among different functional characteristics of HEMs, which reveal avenues for multifunctional material design. The width of the lines indicates stronger (coarser) and weaker (finer) connections among different characteristics. [22] / with permission of Springer Nature.

combination with computational methods to achieve multiobjective optimization of material design. In addition, the application of interface engineering and nonequilibrium state-processing techniques can provide new means to regulate the microstructure and properties of materials. In element selection, in addition to mixing entropy and solubility, electronic structure and thermodynamic stability are taken into account to achieve the desired physicochemical properties. Finally, sustainability is also an important consideration in the design process, requiring a balance between material performance, cost, environmental impact, and resource availability. This interdisciplinary design approach not only advances materials science but also provides new ideas for achieving comprehensive optimization of material properties.

5.4.1 Microstructure Regulation

Under certain conditions, the functional properties of HEMs can be regulated by altering their microstructures. Pei et al. [23] conducted research discussing the development of high-performance bulk thermoelectric materials capable of efficiently converting heat directly into electrical energy. They focused on enhancing the thermoelectric figure of merit (zT) by manipulating the electronic band structure of Pb $Te_{1-x}Se_x$ alloys. They achieved a remarkable zT value of 1.8 at 850 K by converging multiple valence bands through tuning doping and composition (e.g., Figure 5.23), which led to high valley degeneracy (e.g., Figure 5.24). The study highlights that

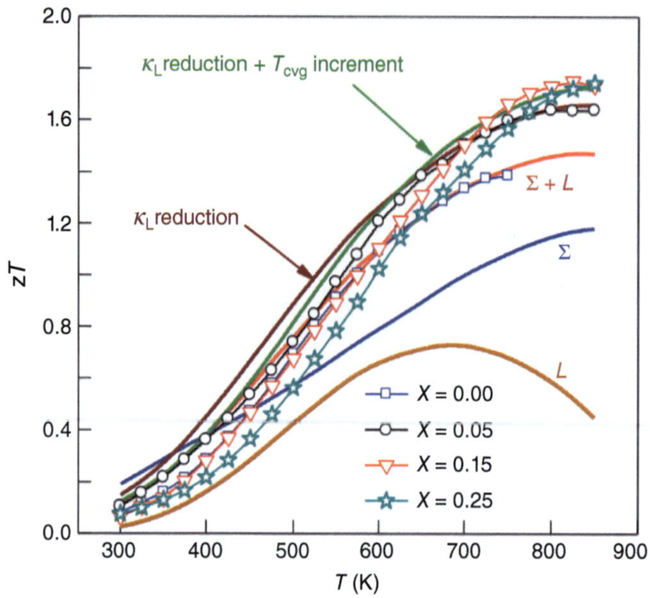

Figure 5.23 Temperature dependence of the zT of p-$PbTe_{1-x}Se_x$ materials doped with 2 atom% Na. Symbols are experimental data. Curves are calculated results, based on a three-band model with a total hole density of $2.5 \times 1020\ cm^{-3}$. [23] / with permission of Springer Nature.

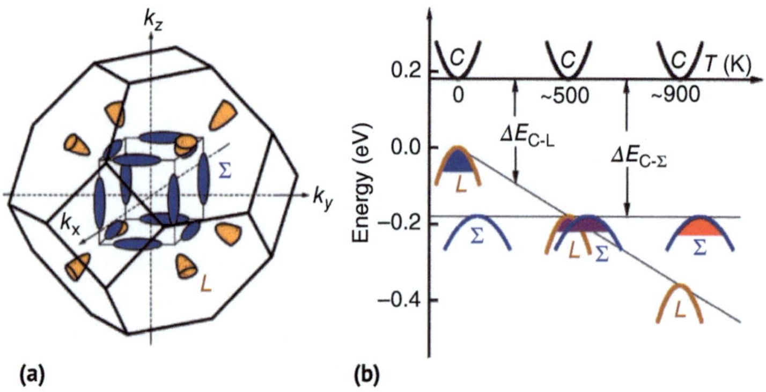

Figure 5.24 Valence band structure of $PbTe_{1-x}Se_x$. (a) Brillouin zone showing the low degeneracy hole pockets (orange) centered at the L point, and the high degeneracy hole pockets (blue) along the Σ line. The figure shows eight half-pockets at the L point so that the full number of valleys, N_v, is 4, while the valley degeneracy of the Σ band is $N_v = 12$. (b) Relative energy of the valence bands in $PbTe_{0.85}Se_{0.15}$. At ~500 K the two valence bands converge, resulting in transport contributions from both the L and Σ bands. C, conduction band; L, low degeneracy hole band; Σ, high degeneracy hole band. [23] / with permission of Springer Nature.

band engineering to increase valley degeneracy is a promising strategy for improving bulk thermoelectric materials, as it can simultaneously enhance the Seebeck coefficient and electrical conductivity. The researchers also developed a comprehensive three-band model to understand the transport properties and validate their findings. The results indicate that bulk materials with high valley degeneracy, like heavily doped p-$PbTe_{1-x}Se_x$, can significantly boost thermoelectric performance, especially when combined with strategies that reduce lattice thermal conductivity. This research opens up new avenues for developing advanced thermoelectric materials for waste heat recovery applications.

Currently, most strategies to enhance the stability of catalysts typically come at the expense of their activity and vice versa. This trade-off has limited the further development of water electrolysis technologies. Therefore, the development of oxygen evolution reaction (OER) catalysts that possess both high activity and high stability is of great significance for advancing water electrolysis technologies. To address this challenge, Liu et al. [24] from the University of Science and Technology Beijing have proposed a novel approach to overcome the inherent difficulty of balancing activity and stability in traditional catalysts. They have developed a dual-layer nanostructure on the surface of porous HEA catalysts, where each layer is designed to control activity and stability, respectively (e.g., Figure 5.25). The catalyst features an outer high-entropy amorphous oxide layer for high OER activity and an inner layer for dynamic replenishment, ensuring long-term stability in alkaline solutions. The NP-$(FeCoNi)_2Nb$ catalyst demonstrates remarkable electrocatalytic activity and stability, requiring only 305 mV overpotential for 100 mA cm^{-2} current density in 1.0 M KOH, outperforming other catalysts. It maintains constant potential for over 1600 hours at 500 mA cm^{-2}, showcasing exceptional durability. The catalyst's performance is attributed to the synergistic high-entropy effect of the outer layer and the dynamic replenishment from the inner NbO_x layer. The high-entropy effect

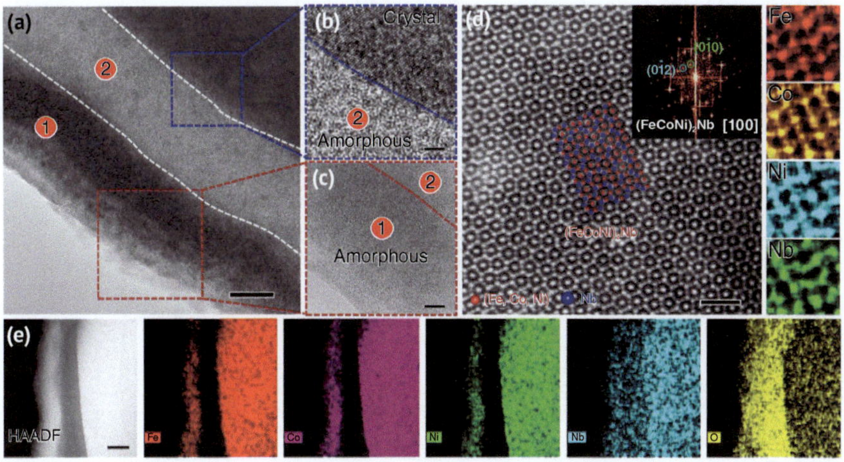

Figure 5.25 Structure characterization of the NP-$(FeCoNi)_2Nb$ catalysts. (a) TEM image of the surface of NP-$(FeCoNi)_2Nb$. Scale bar, 10 nm. The different contrast in layer 1 is due to the variation in the thickness (thin outside and thick inside). (b) High-resolution TEM image of the interface between layer 2 and the matrix. Scale bar, 2 nm. (c) High-resolution TEM image of the interface between layers 1 and 2. Scale bar, 5 nm. (d) STEM image taken from the <10 10> direction revealing the Fe_2Nb-type ordered structure of the $(FeCoNi)_2Nb$ alloys. Scale bar, 5 nm. The inset shows the fast Fourier transform patterns. And the corresponding atomic-resolution elemental maps (right). Scale bar, 0.25 nm. (e) HAADF-STEM images and the corresponding elemental mappings of the double-layered nanostructure of the NP-$(FeCoNi)_2Nb$. Scale bar, 20 nm. Reproduced with permission from [24] / Elsevier.

enhances electron transport and reaction kinetics, while the NbO_x layer compensates for material loss during OER, maintaining activity and stability. The study also highlights the catalyst's industrial potential, as it can be mass-produced and used in an anion-exchange-membrane water electrolyzer with stable operation at high current densities. This work presents a new paradigm for designing electrocatalysts that balance activity and stability, offering a promising approach for large-scale electrolysis applications.

Reasonably increasing the local distortions in HEMs can also achieve improvements in their functional properties. Hu et al. [25] investigated the unconventional isotropic zero thermal expansion (ZTE) characteristics found in cubic $Sc_{1-x}M_xF_3$ (M = Ga,Fe) compounds over a wide temperature range (300–900 K). This ZTE behavior, characterized by a linear coefficient of thermal expansion of $\alpha l = 2.34 \times 10^{-7}\ K^{-1}$, is rare due to the minimal change in lattice constant with temperature (e.g., Figure 5.26). The ZTE is attributed to local rhombohedral distortions introduced by the substitution of Ga^{3+} and Fe^{3+} for Sc^{3+} in the macroscopic cubic lattice, which reduces the contribution of transverse thermal vibration of fluorine atoms. Additionally, the ZTE compound $(Sc_{0.85}Ga_{0.05}Fe_{0.1})F_3$ exhibits high-temperature ferromagnetism ($T_c \approx 435$ K) and narrow-gap semiconductor characteristics. The study provides insights into designing ZTE materials with multifunctionality, such as magnetic-optical switching or magnetic semiconductor applications, by introducing local distortions through cation substitution. The findings could pave the way for developing advanced materials with tailored thermal expansion properties for various applications.

The multifaceted interfaces and defects within high-entropy perovskite materials give rise to polarization effects, bolstering their capacity to attenuate electromagnetic waves. Concurrently, severe lattice distortions engender ultra-high-density strains, offering a variety of conduction pathways for electron carriers and leading

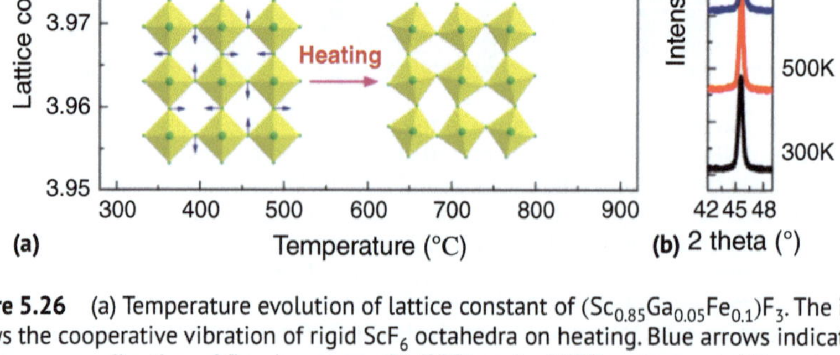

Figure 5.26 (a) Temperature evolution of lattice constant of $(Sc_{0.85}Ga_{0.05}Fe_{0.1})F_3$. The inset shows the cooperative vibration of rigid ScF_6 octahedra on heating. Blue arrows indicate the transverse vibration of fluorine atoms. (b) (200) peak of XRD patterns at different temperatures of $(Sc_{0.85}Ga_{0.05}Fe_{0.1})F_3$. [25] / with permission of American Chemical Society.

to dielectric dissipation. Electron holography has unveiled the accumulation of positive and negative charges at grain boundaries in high-entropy ceramics, along with strain concentration, which further augments dielectric polarization losses. The tunable magnetic properties of high-entropy ceramics enable transitions between paramagnetic and ferromagnetic states. Chen et al. [26] investigated the development of multiprincipal element M_2FeC (M = Ti, V, Nb, Ta, Zr) MAX phases as effective electromagnetic (EM) wave-absorbing materials. MAX phases are known for their high conductivity and nanolaminate structure, making them potential candidates for EM wave absorption. The study successfully synthesized two new magnetic MAX phases, $(Ti_{1/3}Nb_{1/3}Ta_{1/3})_2FeC$ and $(Ti_{0.2}V_{0.2}Nb_{0.2}Ta_{0.2}Zr_{0.2})_2FeC$, where Fe atoms replace Al atoms in the A sites. These phases were created through an isomorphous replacement reaction with Lewis acid salt ($FeCl_2$). Both phases exhibit ferromagnetic behavior with Curie temperatures of 302 and 235 K, respectively (e.g., Figure 5.27).

The study highlights a dual EM absorption mechanism involving both dielectric and magnetic loss in these multiprincipal MAX phases. The $(Ti_{1/3}Nb_{1/3}Ta_{1/3})_2$ FeC phase demonstrated particularly strong EM wave absorption, with a minimum reflection loss (RL) of −44.4 dB at 6.56 GHz and an effective bandwidth of

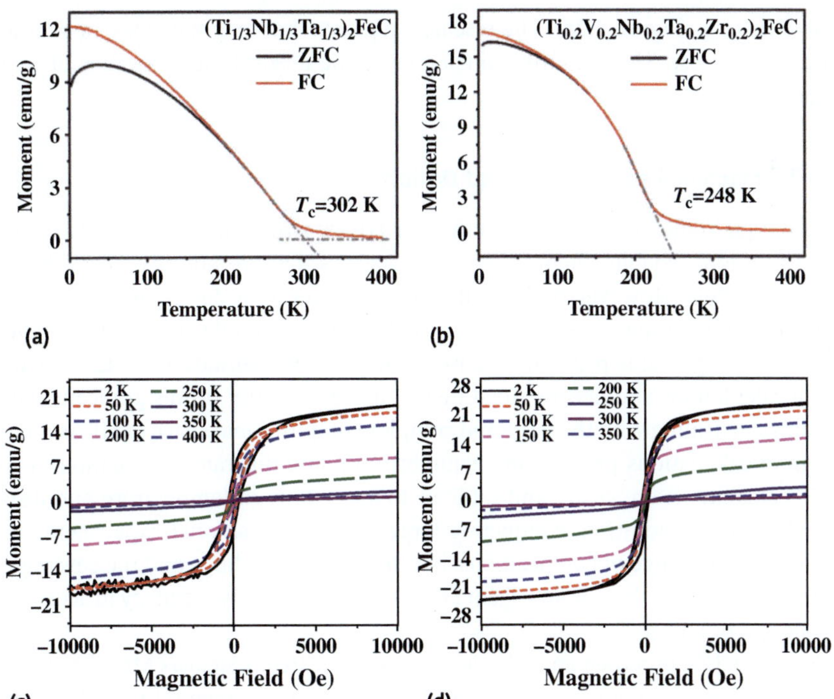

Figure 5.27 Magnetic behavior of new magnetic MAX phases. Zero-field-cooling curves (black lines) and field-cooled curves (red lines) of $(Ti_{1/3}Nb_{1/3}Ta_{1/3})_2FeC$ and $(Ti_{0.2}V_{0.2}Nb_{0.2}Ta_{0.2}Zr_{0.2})_2FeC$ are illustrated in (a) and (b). (c), (d) Field-dependent magnetization curves above the two MAX phases at different temperatures from −10,000 to 10,000 Oe. [26] / John Wiley & Sons / CC BY 4.0.

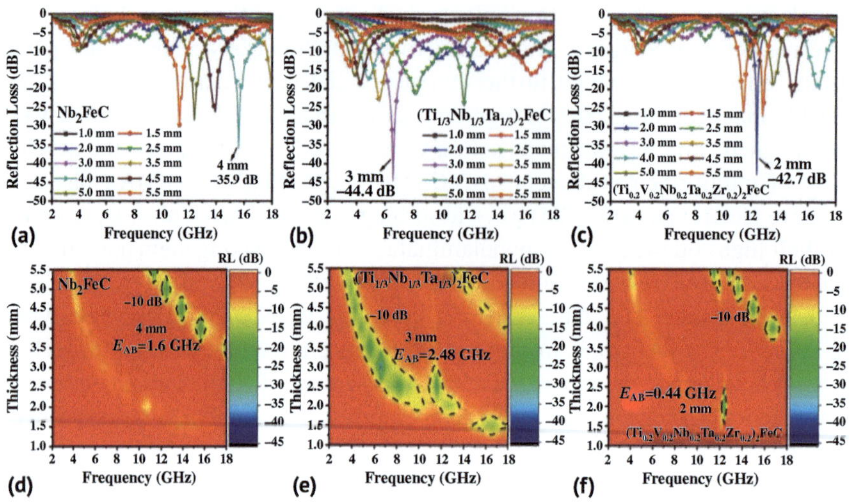

Figure 5.28 RL curves and its corresponding two-dimensional contour mappings of (a, d) Nb_2FeC. (b, e) $(Ti_{1/3}Nb_{1/3}Ta_{1/3})_2FeC$. (c, f) $(Ti_{0.2}V_{0.2}Nb_{0.2}Ta_{0.2}Zr_{0.2})_2FeC$. [26] / John Wiley & Sons / CC BY 4.0.

2.48 GHz (e.g., Figure 5.28). The research suggests that magnetic loss, in addition to dielectric loss, significantly contributes to the absorption properties of these materials.

5.4.2 Machine Learning-Assisted Design

The unique multielement configuration of HEAs, which are multimajor element solid solutions, enables them to realize properties that are not possible with conventional alloys. However, traditional thermodynamic alloy design principles fail in the multielement space, resulting in the inability to accurately and quickly uncover HEAs with high-value properties. Existing alloy design methods, such as phase diagram calculations and density flood theory, however, suffer from significant shortcomings in dealing with the large and complex compositional space of HEAs. Phase diagram calculations provide only equilibrium phase diagram information, while DFT calculations are costly and difficult to apply to high-temperature and disordered alloys. In addition, conducting large-scale compositional experiments can cover a certain range of compositions, but the workload is huge and the coverage is limited. Therefore, there is an urgent need for a more effective strategy to accelerate the design of HEAs.

Rao et al. [27] proposed an active learning strategy that integrates ML with density functional theory, thermodynamic calculations, and experiments to accelerate the design of high-entropy Invar alloys (e.g., Figure 5.29). These alloys have potential applications in industries requiring materials with low thermal expansion coefficients, such as in the transport of liquid hydrogen, ammonia, and natural gas. The researchers processed and characterized 17 new alloys, identifying two with exceptionally low thermal expansion coefficients around 2×10^{-6} per degree kelvin at 300 K (e.g., Figure 5.30). The

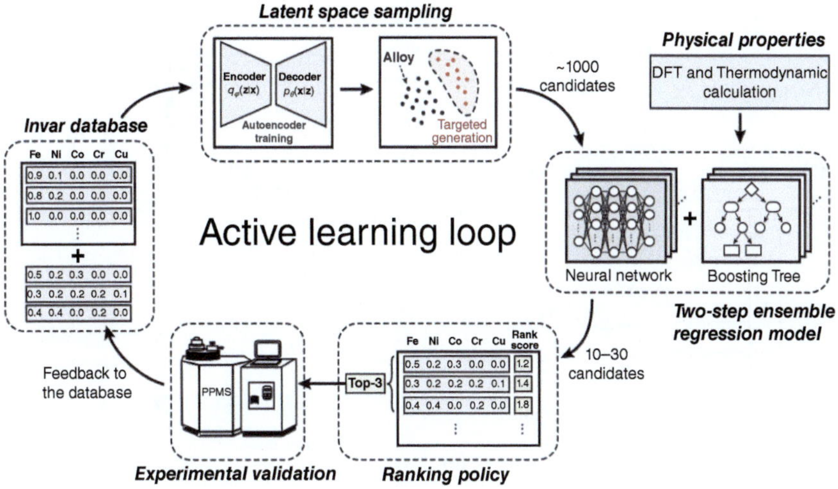

Figure 5.29 Approach overview. We developed an active learning framework for the targeted composition design and discovery of HEAs, which combines ML models, DFT calculations, thermodynamic simulations, and experimental feedback. [27] / with permission of The American Association for the Advancement of Science.

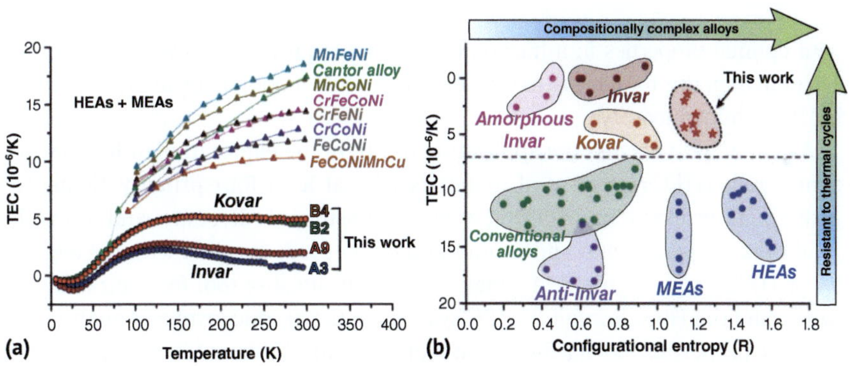

Figure 5.30 Summary of the properties of the ML-designed HEAs. (a) TEC (Thermal Expansion Coefficient) of the ML-designed HEAs as a function of the change in temperature. As a comparison, we plotted the thermal expansion curve of the HEAs and MEAs. A3 and A9 FeNiCoCr HEAs show extremely low TECs around 2×10^{-6}/K at 300 K, which can be used as Invar alloys. B2 and B4 FeNiCoCrCu HEAs show low TECs around 5×10^{-6}/K at 300 K, which qualifies them as Kovar alloys. (b) Configurational entropy plotted against the TEC values for various known alloys and alloys discovered in this work. ML enables this approach to efficiently discover new alloys with excellent properties (high resistance to thermal cycles) in an infinite phase spectrum (compositionally complex alloys). [27] / with permission of The American Association for the Advancement of Science.

active learning framework includes targeted composition generation, physics-informed screening, and experimental feedback. It utilizes a generative model to search for potential alloys and ensemble regression models for predicting properties. The study demonstrates a fivefold higher discovery rate compared to traditional trial-and-error methods. The authors conclude that their ML-driven approach can efficiently discover new alloys

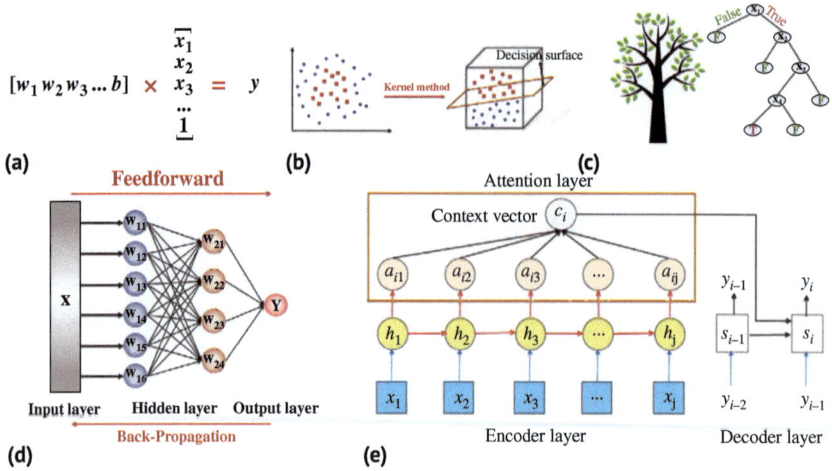

Figure 5.31 Schematic diagrams of (a) linear algorithms, (b) Kernel algorithms, (c) decision trees algorithms, (d) artificial neural network, and (e) attention mechanism in a transformer. In the encoder layer, x_i, h_i, a_{ij}, and c_i represent the input token, the hidden state of the encoder, the attention weight, and the context vector, respectively. For the decoder layer, y_i and s_i represent the output at the i step and the hidden state of the decoder, respectively. [28] / with permission of John Wiley & Sons.

with optimal properties in a fraction of the time required by conventional methods. They envision this framework as a generalizable tool for optimizing multiple properties in HEAs simultaneously.

Wan et al. [28] discuss the application of ML in the exploration of high-entropy compounds (HECs), a class of materials with at least four primary elements in near-equal proportions, offering vast compositional diversity and tunability. Traditional experimental and computational methods struggle with the complexity of HECs (e.g., Figure 5.31), making ML a promising tool for their study. ML can model the Hamiltonian of HEC systems at the atomic level and analyze macroscopic properties like hardness and melting point. Various ML algorithms, from traditional methods to deep neural networks, are applicable. Key steps include comprehensive data collection, feature engineering, and model training with cross-validation. ML aids in understanding phase structures and stability and facilitates the design of functional materials, though some areas like magnetic materials require further exploration. Wan et al. [28] outline ML's role in microscopic and macroscopic modeling of HECs, detailing representative algorithms and the importance of data quality. It emphasizes the potential of ML in predicting phase structures and stability, constructing simulation potentials, and guiding the design of functional materials. Despite challenges, ML is deemed indispensable for advancing HEC research, forming the basis of a new AI-assisted material exploration paradigm.

Meng et al. [29] established formation ability descriptors for high-entropy diborides (HEBs) using high-throughput experiments and ML (e.g., Figure 5.32). The authors synthesized 70 HEB samples and computed 22 formation ability descriptors. Through Pearson correlation coefficient analysis and ML methods, they identified an optimal

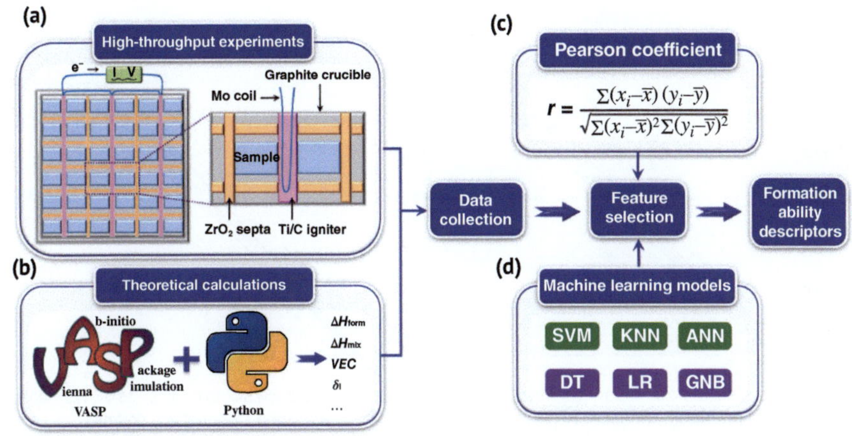

Figure 5.32 Flowchart for establishing the formation ability descriptors of HEBs. (a) High-throughput combustion synthesis process of 70 HEBs. (b) Theoretical calculations of 22 formation ability descriptors. (c) Removal of the strongly correlated features. (d) Selection of the optimal feature set based on eight ML models. [29] / with permission of Elsevier.

combination of four descriptors—Pauling electronegativity difference ($\delta\chi P$), average volume, average bulk modulus, and density difference—with a validation accuracy of 97.1%. Among these, $\delta\chi P$ was found to be the most significant, with smaller values indicating easier synthesis of HEBs. The study provides a theoretical guideline for discovering and designing HEBs with desirable properties, accelerating the development of these materials for applications in aerospace, cutting tools, and nuclear reactors.

Yadav et al. [30] utilize ML models to predict the thermal conductivity (κ) of HEAs for thermoelectric applications. The researchers compiled a dataset of experimentally measured κ values from the Materials Platform for Data Science and employed various ML algorithms to classify and predict κ.

Initially, classification models were used to categorize the alloys into low- or high-κ groups based on different cutoff values. Voting classifiers, which combine multiple models to reduce bias, achieved over 85% accuracy. Regression models were then built on the entire dataset and a subset containing only alloys with four or more elements. The latter showed significantly higher accuracy, with cross-validation R^2 (CV – R^2) values ranging from 0.80 to 0.81, compared to 0.65 for the entire dataset (e.g., Figure 5.33).

To account for experimental uncertainty, interval prediction models were developed using the gradient boost algorithm. These models predict a range of κ values, considering the inherent errors in experimental measurements. A novel confusion matrix was proposed to assess the performance of these interval prediction models. The best-performing model showed a nonzero overlap between experimental and predicted κ for 75% of the alloys, with a mean overlap size of 60–70%.

Yadav et al. [30] found that models built on targeted datasets, such as alloys with κ less than 2 W/mK, performed better in terms of accuracy, while models built on the entire dataset showed higher overfitting. Both regression and interval prediction models were validated on a test dataset containing unseen alloys, yielding

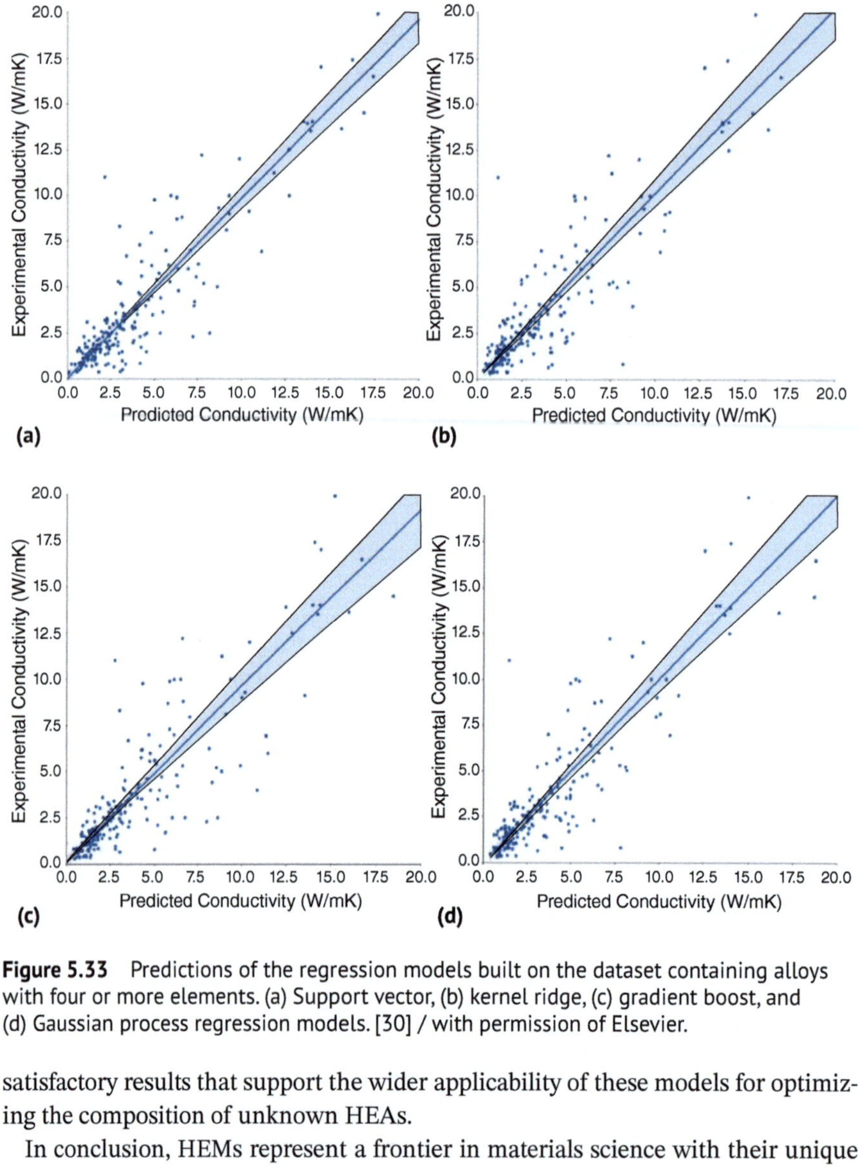

Figure 5.33 Predictions of the regression models built on the dataset containing alloys with four or more elements. (a) Support vector, (b) kernel ridge, (c) gradient boost, and (d) Gaussian process regression models. [30] / with permission of Elsevier.

satisfactory results that support the wider applicability of these models for optimizing the composition of unknown HEAs.

In conclusion, HEMs represent a frontier in materials science with their unique combination of properties that can be tailored for specific applications. Their potential in thermal, electrical, magnetic, and optical applications is significant, and ongoing research is expected to lead to the development of new materials with exceptional performance for various technological applications.

HEMs exhibit excellent thermal stability and low thermal conductivity, making them suitable for thermal barrier coatings and high-temperature structural materials. The high configurational entropy in HEMs leads to significant lattice distortion, which enhances phonon scattering and reduces lattice thermal conductivity. For instance, the addition of lanthanum (La) to PbSnTeSe HEAs has been shown to enhance the Seebeck coefficient and electrical conductivity while suppressing the bipolar effect, improving the thermoelectric figure of merit (zT)

to approximately 0.8 at 873 K [7]. This suggests that HEMs could be a promising material for thermoelectric applications if challenges related to the bipolar effect are addressed.

HEMs possess high electrical conductivity and thermoelectric properties, making them suitable for thermoelectric conversion and superconducting materials. The presence of multiple elements increases the conformational entropy, avoiding the formation of intermetallic compounds and facilitating the formation of HEMs with simple structures. ML techniques have been used to predict the lattice thermal conductivity of HEAs, offering a powerful tool for developing HEAs with tailored properties. For example, the Ta—Nb—Hf—Zr—Ti system, with a composition of $[TaNb]_{0.67}[HfZrTi]_{0.33}$, is the first reported HEA superconductor with a critical temperature (T_c) of 7.3 K, showing robust zero-resistance superconductivity up to 190 GPa [4].

HEAs containing ferromagnetic elements (Fe, Co, Ni) can be designed as hard or soft magnetic materials. Recent developments in high-entropy soft magnetic materials involve microadditions of non-ferromagnetic elements like Cr, Cu, Si, Ga, Al, and Ti to enhance magnetic and mechanical properties. For example, $FeCoNi(AlSi)_x$ HEAs demonstrate exceptional magnetic and electrical properties with optimal performance at $x = 0.2$ [10]. Processing methods and heat treatment significantly affect the magnetic properties of HEAs, with cold rolling and heat treatment enhancing the strength of alloys like $FeCoNiAl_{0.2}Si_{0.2}$ HEAs [11].

HEMs, including HEA nitride films, have attracted attention for their unique microstructures and superior properties such as high hardness, wear resistance, and corrosion resistance. These materials show promise in solar energy absorption coatings due to their excellent optical properties and thermal stability. Research has shown that the optical properties of $FeCrVTa_{0.4}W_{0.4}$ HEA nitride films can be tailored by adjusting the nitrogen content. HEFs in the pseudobinary system (Cr, Fe, V) – (Ta, W) exhibit high solar absorptivity, particularly at the transition from amorphous to BCC structure [16]. High-entropy nitride films have been shown to enhance solar energy absorption and photothermal conversion efficiency, with applications in concentrating solar power systems.

The unique combination of properties in HEMs offers significant potential for functional applications. Ongoing research is exploring their use in soft magnetic materials, solar energy absorption, and optoelectronic devices. The interdisciplinary design approach, combining traditional materials science with modern computational techniques, is expected to push the boundaries of functional materials science. ML and artificial intelligence are playing increasingly important roles in accelerating the discovery of new materials by predicting material properties and guiding experimental design. High-throughput experimental methods, combined with computational methods, are used for rapid screening and testing of materials, achieving multiobjective optimization of material design. The application of interface engineering and nonequilibrium state-processing techniques provides new means to regulate the microstructure and properties of materials. In element selection, in addition to mixing entropy and solubility, electronic structure and thermodynamic stability are considered to achieve the desired physicochemical properties. Finally, sustainability is an important consideration in the design process, requiring a balance between material performance, cost, environmental impact, and resource availability.

References

1 Lu J, Huang X A, Yue Y A. Estimating the lattice thermal conductivity of AlCoCrNiFe high-entropy alloy using machine learning. *Journal of Applied Physics*, 2024, 135(13): 8. https://doi.org/10.1063/5.0201042.

2 Singh P, Acemi C, Kuchibhotla A, et al. Alloying effects on the transport properties of refractory high-entropy alloys. *Acta Materialia*, 2024, 276: 14. https://doi.org/10.1016/j.actamat.2024.120032.

3 Yang J Y, Ren W J, Zhao X G, et al. Mictomagnetism and suppressed thermal conduction of the prototype high-entropy alloy CrMnFeCoNi. *Journal of Materials Science & Technology*, 2022, 99: 55–60. https://doi.org/10.1016/j.jmst.2021.04.077.

4 Guo J, Wang H H, Von Rohr F, et al. Robust zero resistance in a superconducting high-entropy alloy at pressures up to 190 GPa. *Proceedings of the National Academy of Sciences of the United States of America*, 2017, 114(50): 13144–13147. https://doi.org/10.1073/pnas.1716981114.

5 Stolze K, Tao J, Von Rohr F O, et al. Sc-Zr-Nb-Rh-Pd and Sc-Zr-Nb-Ta-Rh-Pd high-entropy alloy superconductors on a CsCI-type lattice. *Chemistry of Materials*, 2018, 30(3): 906–914. https://doi.org/10.1021/acs.chemmater.7b04578.

6 Marik S, Varghese M, Sajilesh K P, et al. Superconductivity in equimolar Nb-Re-Hf-Zr-Ti high entropy alloy. *Journal of Alloys and Compounds*, 2018, 769: 1059–1063. https://doi.org/10.1016/j.jallcom.2018.08.039.

7 Fan Z, Wang H, Wu Y, et al. Thermoelectric performance of PbSnTeSe high-entropy alloys. *Materials Research Letters*, 2017, 5(3): 187–194. https://doi.org/10.1080/21663831.2016.1244116.

8 Shi X, Song S W, Gao G H, et al. Global band convergence design for high-performance thermoelectric power generation in Zintls. *Science*, 2024, 384(6697): 757–762. https://doi.org/10.1126/science.adn7265.

9 王鹏将, 康慧君, 杨雄, 等. 熵调控抑制 ZrNiSn 基 half-Heusler 热电材料的晶格热导率. 无机材料学报, 2022, 37(7): 717–723. https://doi.org/10.15541/jim20210610.

10 Zhang Y, Zuo T T, Cheng Y Q, et al. High-entropy alloys with high saturation magnetization, electrical resistivity, and malleability. *Scientific Reports*, 2013, 3: 7. https://doi.org/10.1038/srep01455.

11 Zuo T, Ren S, Liaw P K, et al. Processing effects on the magnetic and mechanical properties of $FeCoNiAl_{0.2}Si_{0.2}$ high entropy alloy. *International Journal of Minerals, Metallurgy, and Materials*, 2013, 20: 549–555. https://doi.org/10.1007/s12613-013-0764-x.

12 Zuo T T, Li R B, Ren X J, et al. Effects of Al and Si addition on the structure and properties of CoFeNi equal atomic ratio alloy. *Journal of Magnetism and Magnetic Materials*, 2014, 371: 60–68. https://doi.org/10.1016/j.jmmm.2014.07.023

13 Zuo T T, Gao M C, Ouyang L Z, et al. Tailoring magnetic behavior of CoFeMnNiX (X = Al, Cr, Ga, and Sn) high entropy alloys by metal doping. *Acta Materialia*, 2017, 130: 10–18. https://doi.org/10.1016/j.actamat.2017.03.013.

14 Zhang Y, Zhang M, Li D, et al. Compositional design of soft magnetic high entropy alloys by minimizing magnetostriction coefficient in $(Fe_{0.3}Co_{0.5}Ni_{0.2})_{100-x}(Al_{1/3}Si_{2/3})_x$ system. *Metals*, 2019, 9(3): 382. https://doi.org/10.3390/met9030382.

15 Song X F, Liaw P K, Wei Z Y, et al. Evolution of the microstructures, magnetic and mechanical behaviors of $Co_{47.5}Fe_{28.5}Ni_{19}Si_{3.4}Al_{1.6}$ high-entropy alloy fabricated by laser powder bed fusion. *Additive Manufacturing*, 2023, 71: 13. https://doi.org/10.1016/j.addma.2023.103593.

16 王子鑫,张勇. FeCrVTa$_{0.4}$W$_{0.4}$ 高熵合金氮化物薄膜的微观结构与性能. 工程科学学报, 2021, 43(05): 684–692. https://doi.org/10.13374/j.issn2095-9389.2020.09.28.004.

17 Xing Q W, Ma J, Wang C, et al. High-Throughput Screening Solar-Thermal Conversion Films in a Pseudobinary (Cr, Fe, V)-(Ta, W) System. *ACS Combinatorial Science*, 2018, 20(11): 602–610. https://doi.org/10.1021/acscombsci.8b00055.

18 He C Y, Zhao P, Zhang H, et al. Efficient warming textile enhanced by a high-entropy spectrally selective nanofilm with high solar absorption. *Advanced Science*, 2023, 10(3): 10. https://doi.org/10.1002/advs.202204817.

19 Huang H, Zhao J, Guo H, et al. Noble-metal-free high-entropy alloy nanoparticles for efficient solar-driven photocatalytic CO_2 reduction. *Advanced Materials*, 2024, 36: 2313209. https://doi.org/10.1002/adma.202313209.

20 Ge H Y, Zheng L R, Yuan G B, et al. Polyoxometallate cluster induced high-entropy oxide Sub-1 nm nanosheets as photoelectrocatalysts for Zn-Air batteries. *Journal of the American Chemical Society*, 2024, 146(15): 10735–10744. https://doi.org/10.1021/jacs.4c00652.

21 Folgueras M C, Jiang Y X, Jin J B, et al. High-entropy halide perovskite single crystals stabilized by mild chemistry. *Nature*, 2023, 621(7978): 282. https://doi.org/10.1038/s41586-023-06396-8.

22 Han L L, Zhu S Y, Rao Z Y, et al. Multifunctional high-entropy materials. *Nature Reviews Materials*, 2024: 20. https://doi.org/10.1038/s41578-024-00720-y.

23 Pei Y Z, Shi X Y, Lalonde A, et al. Convergence of electronic bands for high performance bulk thermoelectrics. *Nature*, 2011, 473(7345): 66–69. https://doi.org/doi:10.1038/nature09996.

24 Li Z B, Wu R Y, Duan D B, et al. Empowering multicomponent alloys with unique nanostructure for exceptional oxygen evolution performance through self-replenishment. *Joule*, 2024, 8(10): 19. https://doi.org/10.1016/j.joule.2024.06.023.

25 Hu L, Chen J, Fan L, et al. Zero thermal expansion and ferromagnetism in cubic $Sc_{1-x}M_xF_3$ (M = Ga, Fe) over a wide temperature range. *Journal of the American Chemical Society*, 2014, 136(39): 13566-13569. https://doi.org/10.1021/ja5077487.

26 Chen L, Li Y, Zhao B, et al. Multiprincipal element M_2FeC (M=Ti, V, Nb, Ta, Zr) MAX phases with synergistic effect of dielectric and magnetic loss. *Advanced Science*, 2023, 10(10): 2206877. https://doi.org/10.1002/advs.202206877.

27 Rao Z Y, Tung P Y, Xie R W, et al. Machine learning-enabled high-entropy alloy discovery. *Science*, 2022, 378(6615): 78–84. https://doi.org/10.48550/arXiv.2202.13753.

28 Wan X H, Li Z Y, Yu W, et al. Machine learning paves the way for high entropy compounds exploration: challenges, progress, and outlook. *Advanced Materials*, 2023: 26. https://doi.org/10.1002/adma.202305192.

29 Meng H, Yu R W, Tang Z Y, et al. Formation ability descriptors for high-entropy diborides established through high-throughput experiments and machine learning. *Acta Materialia*, 2023, 256: 9. https://doi.org/10.1016/j.actamat.2023.119132.

30 Yadav N, Chakraborty N, Tewari A. Interval prediction machine learning models for predicting experimental thermal conductivity of high entropy alloys. *Computational Materials Science*, 2022, 214: 9. https://doi.org/10.1016/j.commatsci.2022.111754.

6

Nanoscale High-Entropy Materials

*Yifan Wan, Yuhua Wang**

College of Science, Wuhan University of Science and Technology, Wuhan, P.R. China

Corresponding email: wangyuhua@wust.edu.cn

6.1 Introduction to Evolution from Bulk to Nanoscale HEMs

The evolution of high-entropy materials (HEMs) from bulk to nanoscale structures marks a transformative advancement in materials science. Initially, bulk HEMs gained attention due to their exceptional properties stemming from their high configurational entropy, which imparts outstanding mechanical strength, corrosion resistance, and thermal stability. These bulk materials, which are made up of five or more principal elements in nearly equal atomic proportions, show a high level of structural complexity and phase stability. Therefore, they have been acclaimed as a significant advancement in the creation of advanced materials for challenging applications in the aerospace, energy, and environmental fields (Figure 6.1a).

However, as technological progress demands higher precision and enhanced functional performance, the limitations of bulk HEMs—particularly their reduced surface-to-volume ratios and lower reactivity—become apparent. The transition to nanoscale HEMs presents a profound shift in material design, one that amplifies the intrinsic properties of HEMs and introduces new capabilities. Nanoscaling these materials results in a significant increase in the surface-to-volume ratio, which not only boosts catalytic activity and reactivity but also offers unique opportunities for optimizing performance in energy conversion, storage, and environmental remediation [3–5].

This transition is not merely a miniaturization of bulk materials but rather a profound change in the behavior of the materials themselves. Nanoscale HEMs exhibit unique phenomena such as quantum effects, size-dependent phase stability, and

Functional High Entropy Alloys and Compounds, First Edition. Edited by Yong Zhang, Yonggang Yao and Haojie Liu.

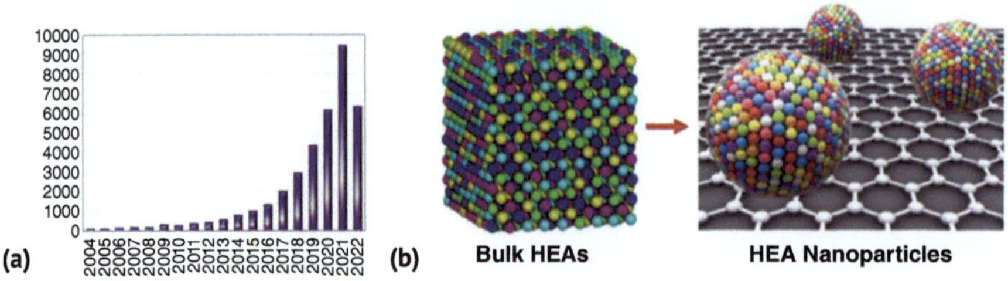

Figure 6.1 (a) Papers published concerning high-entropy alloy nanoparticles (HEA-NPs) by July 2022 [1]/with permission of Elsevier. (b) Schematic illustration of bulk HEAs into nanoscale HEAs [2]/John Wiley & Sons/CC BY 4.0.

enhanced atomic diffusivity, all of which contribute to their superior performance in various applications. The ability to fine-tune these materials at the nanoscale enables the design of systems with tailored properties, such as high-energy density, improved charge transport, or exceptional mechanical strength at reduced sizes (Figure 6.1b).

The significance of this evolution is underscored by the increasing demand for materials that not only function under extreme conditions but also minimize resource usage and environmental impact. Nanoscale HEMs open up new frontiers in catalysis, energy storage, and environmental applications, where performance often hinges on surface reactivity, nanoscale structure, and atomistic design. Moreover, the synthesis of nanoscale HEMs represents a complicated process that requires new fabrication methods and precise control over the material's composition, structure, and phase formation. Remarkable techniques like high-energy ball milling, chemical vapor deposition, and sol–gel processes have been devised to address these challenges, with each method affecting the material's properties in unique ways.

Therefore, the "metamorphosis" from bulk to nanoscale HEMs is not just a progression in material development but also a fundamental reconsideration of optimizing these materials for next-generation applications. This chapter will explore the crucial significance of this shift, the scientific theories supporting it, and the cutting-edge techniques employed to unleash the potential of nanoscale HEMs.

6.2 Synthesis and Advanced Characterization

The "Hume-Rothery criterion" is widely regarded as an essential conceptual foundation in physical metallurgy for predicting phase formation during intercomponent alloying [6, 7]. For conventional alloys, factors such as similar atomic size, crystal structure, valence electron concentration, and electronegativity are crucial for solid-solution formation. However, the multicomponent nature of HEMs introduces unique challenges in phase stability and immiscibility between the constituent elements. This is a key obstacle in the synthesis of these advanced materials.

When the family of elements involved in a system expands, the entropy of mixing rises, leading to a more unstable alloy or compound system. According to Gibbs free energy law, this increase in entropy can make the system more prone to phase separation and segregation. In addition, when refining HEMs to the nanoscale, the

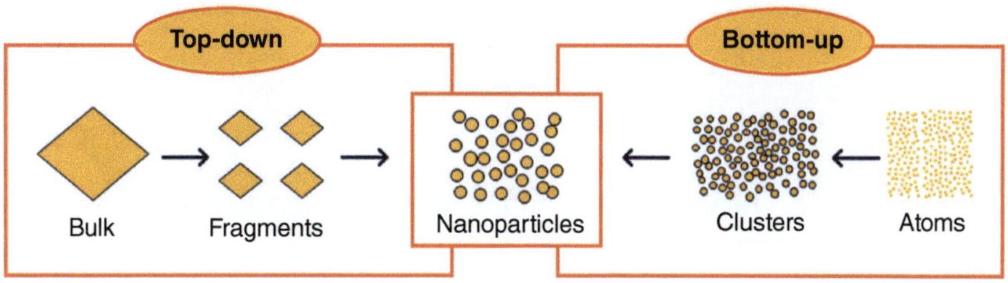

Figure 6.2 Scheme of two NP synthesis approaches.

extremely short diffusion length of nanoparticles (NPs) further exacerbates these issues, facilitating the formation of phase-segregated structures, which marks the failure of alloying and impedes the synthesis process [6, 8–11].

In response to these challenges, researchers have made significant strides in developing various methods for the preparation of nanoscale HEMs. In recent years, remarkable progress has been achieved in the preparation of HEM NPs under extreme conditions. These methods have made the successful synthesis of many branches of nanoscale HEMs, which are currently being investigated for their improved properties in various fields like catalysis, energy storage, and environmental remediation.

In general, the synthetic methods for nanoscale HEM preparation can be divided into two conceptually different approaches: top-down and bottom-up. As shown in Figure 6.2, the key difference between these two approaches lies in their start points. In top-down approaches, the process usually begins with a bulk form of the HEM as the starting material, which is then broken down into smaller particles or structures through certain processes. On the other hand, bottom-up approaches start with metal atoms or raw precursors, and nanoscale HEMs are assembled or synthesized via various chemical or physical processes.

Both approaches encompass a range of techniques that differ in how they break down or build up the materials. The top-down methods often involve mechanical grinding, ball milling, or laser ablation, while bottom-up methods include chemical vapor deposition, sol–gel synthesis, and atomic layer deposition. Each of these methods introduces specific challenges in terms of controlling phase formation, material stability, and ensuring the uniformity of the nanoscale HEMs produced. By examining the advantages and limitations of each method, it becomes clear that the synthesis of nanoscale HEMs requires a careful balance of material composition, structural integrity, and processing conditions. This chapter will explore the different synthesis strategies for high-entropy NPs, with a focus on overcoming the immiscibility challenges, controlling phase stability, and optimizing the material properties for advanced applications.

6.2.1 Top-Down Method

The pulverization of the bulk material is achieved through externally applied energy. Within the top-down approach, the various synthetic methods differ based on the type of energy used to break down the bulk material, which can include mechanical, chemical, thermal, or laser irradiation (Figure 6.3). Among the more commonly used

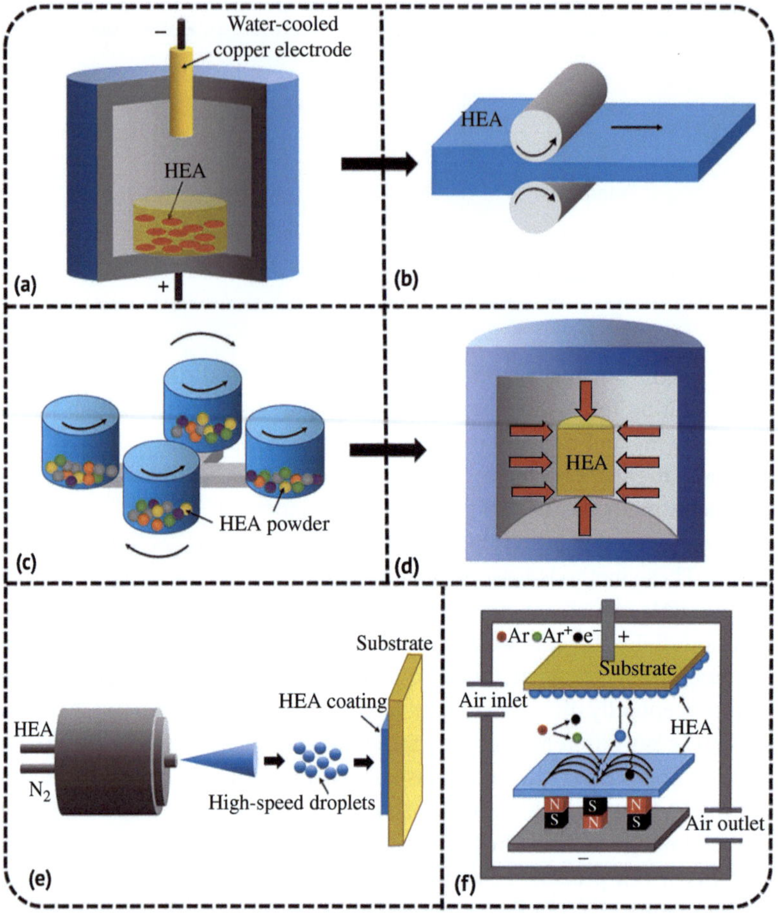

Figure 6.3 Schemes of the preparation method of HEAs: (a) arc melting, (b) rolling, (c) mechanical alloying, (d) sintering, (e) Sputter deposition method, and (f) magnetron sputtering [26]. [23] / De Gruyter / CC BY 4.0.

top-down methods are ball milling (mechanical energy), arc melting (thermal energy), dealloying (chemical energy), laser ablation (laser irradiation), and photolithography (electromagnetic radiation). For further details, the reader is directed to some excellent recent reviews [12–21]. However, almost all the top-down approaches have a limitation in the scale of particles (100 nm–1 μm). Laser ablation is an exception, which is fabricated on a 5 nm scale [18], with defective surface, unpredictable nature, or sometimes drastic changes in physiochemical properties [15, 17, 21, 22]. In a word, top-down approaches are usually considered simple, cheap, and scalable procedures.

Mechanical Alloying. Kumar et al. first stated that large-scale high-entropy alloy (HEA)-NPs with a size of less than 10 nm can be successfully prepared through cryo-grinding massive HEA ingots. The low-temperature milling technique they used offers many advantages, including low pollution, reduced oxidation, inhibition of recycling, and recrystallization. In the experiment, the main principle is to prepare alloy ingots, divide them into small pieces, and then mechanically cryo-grind them (cryo-grinding mainly consists of three steps: particle flattening, flake fracture, and steady-state NP formation). They also accurately assessed the daily obtainable quantity of HEA-NPs, demonstrating their suitability for potential mass manufacturing.

The mechanical alloying process is an experimentally straightforward and efficient way to prepare HEAs by physically crushing the required elements into powder. After calculating the composition of the HEA powder, pure metal powder precursors are mixed in a specific proportion and placed into a high-energy planetary ball mill along with the grinding medium. During the high-speed and high-pressure grinding process, cold welding occurs, leading to bonding between different metal surfaces. Throughout the milling process, metal particles continuously experience impact, collision, and fracture, with smaller particles being prone to lattice defects. Moreover, the multidirectional collisions generate significant heat, promoting defect diffusion, and facilitating the formation of chemical bonds between different metal atoms. From the previous description, it can be seen that the mechanical ball milling process is capable of preparing HEAs with stable structures. Varalakshmi et al. prepared the AlFeTiCrZnCu system using the mechanical alloying method [23]. This system is a BCC-structured nano equiatomic high-entropy solid solution composed of two to six elements. The resulting NPs exhibit good stability and can remain stable even after being annealed at high temperatures (800 °C) for 1 hour.

Magnetron Sputtering. Magnetron sputtering is physical vapor deposition (PVD). It ejects metal atoms from a metal target surface by bombarding it with high-energy gas ions in a vacuum environment. These metal atoms with high kinetic energy are transported to subsystem surfaces to form nuclei and then condense, eventually constructing thin films or metal NPs. Xiao et al. achieved equiatomic nanocrystals (NCs) and CrMnFeCoNi HEAs (7.2 nm) with extremely small grains through magnetron sputtering [24]. The deposited sample consists of layered nanoscale amorphous or crystalline biphasic structures. Due to the unique deformation mechanism of the interaction between amorphous phases and NCs, this layered nanostructure ultimately produces an ultrahigh strengthening effect and reaches the highest reported hardness of 13.76 GPa.

Sputter Deposition Method. Sputtering deposition is a PVD method that uses high-energy gas ions to bombard the surface of metal targets in a vacuum or oxygen-free environment, causing them to eject metal atoms. These metal atoms with high kinetic energy are transported to the surface scratch of the substrate, then nucleate and condense, ultimately forming thin films of gold NPs. The ion implantation method is an important part of the sputtering deposition method. Hao et al. successfully prepared CuCoNiFeZn high-entropy metallic glass (HEMG) NPs by improving the ion implantation method [25]. This consists of injecting a target ion beam into the matrix at room temperature with a metal vapor vacuum arc implanter, using an appropriate and selected accelerating voltage. The fundamental principle is that five groups of high-energy metal plasma ions beam into the matrix, causing cascading collisions among the metal ions. Between the metal ions and the glass matrix atoms, the latter ones will gradually lose their energy. The beams surpass the solid-solution limit of the matrix, enabling various single metal element ions to uniformly install within the substrate. Afterward, the implanted matrix will be tempered, allowing the individual metal element ions to aggregate, grow, and nucleate with each other, thereby forming smaller NPs with a more uniform distribution of elements.

6.2.2 Bottom-Up Method

The bottom-up approach constructs HEA-NPs with raw materials in the forms of clusters of atoms or molecules [27]. The distinction of these approaches depends on the interaction between atoms and molecules, which are triggered and regulated by chemical reactions per the specific method employed. This is similar to the concepts in conventional NP synthesis and has been boosted by the progress in wet-chemical synthesis, which has offered methods like hydrothermal and solvothermal synthesis [15, 17, 21, 22, 28, 29], ultrasonication [30], and colloidal synthesis [31, 32]. However, other methods such as carbothermal shock synthesis [8] and co-deposition of multiple precursors [33, 34] perform excellently as well as synthetic methods of HEA-NPs. Differing from the top-down approaches, the synthetic methods emphasized that all the methods mentioned earlier share the advantage of controllable parameters. This enables a more targeted fabrication of HEA-NPs regarding size (100–1 nm), composition (uniformity), and fewer surface defects. For more detailed descriptions of the synthetic methods, the reader is directed to these manuscripts [8, 13–22, 28–36].

Carbon Thermal Shock. The initial breakthrough in synthesizing HEA-NPs with various compositions and large elemental numbers (8 for maximum) was achieved through a high-temperature "thermal shock" process developed by the Hu group at the University of Maryland [8]. This method enables the formation of NPs from immiscible element combinations. A core factor in this process is the cooling rate, which controls the degree of structural ordering of the elements by adjusting the nonequilibrium, as illustrated by temperature–time-transformation diagrams in physical metallurgy. The resulting NP structures can be diverse, including metallic glass (disordered lattices), normal HEA-NPs (randomly mixed in crystalline lattices), intermetallic nanoparticles (ordered sublattices), and heterogeneous nanoparticles (phase separation) (Figure 6.4).

The rapid quenching and brief duration of thermal shock synthesis also promote the formation of small, uniform particles, which can be further refined through defect engineering and substrate modulation. Besides this "shock"-based method, a variety of other techniques have been developed for synthesizing high-entropy NPs, such as vapor phase spark discharge, rapid radiative heating, chemical reduction, hydrogen spillover, sputtering, electrosynthesis, and plasma, laser, or microwave heating. These methods, powered by kinetic forces, allow for the efficient production of nanocatalysts [37–41].

The Ellingham diagram is often imported to guide the thermochemical synthesis of high-entropy NPs, as it illustrates the oxidation potential of elements at

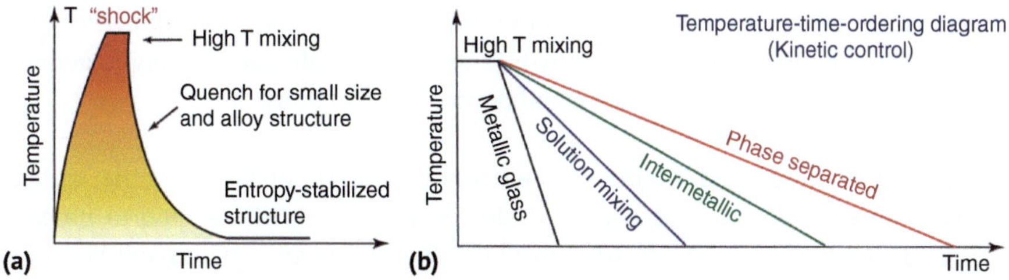

Figure 6.4 (a) Thermal shock synthesis of high-entropy NPs needs a high-temperature pulse that promotes element mixing, followed by rapid temperature quenching to preserve the high-entropy structure. (b) The temperature–time-transformation diagram, illustrating how cooling rates in high temperatures can be adjusted to produce NPs with varying degrees of structural and chemical ordering in kinetic ways [40]/The American Association for the Advancement of Science.

different temperatures. Initially made for bulk metallurgy, it has been found suitable for nanoscale, shock-driven reactions. Elements at the top of the diagram (e.g., noble metals, Fe, Co, Cu) possess lower oxidation potentials and can form alloy NPs via high-temperature synthesis. On the other hand, elements at the bottom (e.g., Zr, Ti, Hf, Nb) form high-entropy oxides. For elements with intermediate oxidation potentials (e.g., Mo, W, Mn), diverse synthesis strategies can switch between metallic and oxide states, thus widening the range of possible HEAs and oxides. High-entropy compounds, including sulfides and carbides, have also been synthesized in various sizes, shapes, and phases, further expanding the potential applications of these materials.

Fast-Moving Bed Pyrolysis (FMBP). Gao et al. proposed a novel FMBP method to synthesize HEA-NPs on various carriers. For example, CuPdSnPtAu HEA-NPs were synthesized using graphene oxide (GO) as the support. Metal chloride salts coordinated with 1,10-phenanthroline (Phen) were introduced onto GO through wet impregnation, followed by ultrasonic drying at 323 K. Subsequently, the salt-loaded GO was heated to 923 K in an Ar atmosphere with a flow rate of 20 cm/s, and this temperature was maintained for 120 minutes. Phen acted as a reducing agent, facilitating the simultaneous pyrolysis of the metal salts and the rapid nucleation of HEA-NPs (Figure 6.5).

In addition to GO, other supports such as γ-Al_2O_3, zeolite, and carbon black were employed for synthesizing various HEA-NPs. The FMBP method showed reliability and versatility, yielding HEA-NPs with compositions ranging from five-element

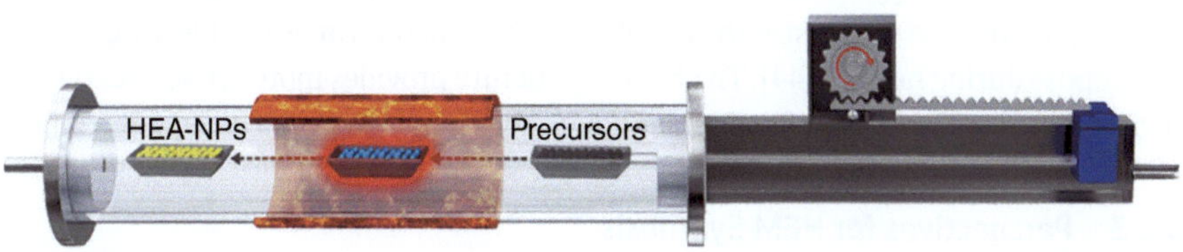

Figure 6.5 FMBP for the general synthesis of HEA nanocatalysts [4]/Springer Nature/CC BY 4.0.

alloys (e.g., CuPdSnPtAu) to ten-element alloys (e.g., MnCoNiCuRhPdSnIrPtAu) at 923 K. Notably, FeCoPdIrPt/GO HEA-NPs exhibited high electrocatalytic activity and stability for hydrogen evolution and desorption reactions.

Nanodroplet-Mediated Electrodeposition (NMED). Inspired by the carbon thermal shock method, Glasscott et al. developed the NMED method to prepare HEA-NPs with an amorphous structure [43]. By mixing an aqueous solution of a metal salt precursor $\left(MCl_xH_y,\ M = Pt, CO,\ Ni, etc.\right)$ with dichloroethane, a large number of "water-in-oil" droplets were formed in the microemulsion "microreactor." Under the action of an electric field, the dotted droplets collided with the surface of the highly oriented pyrolytic graphite (HOPG) cathode electrode, and the metal precursors in the water droplets were reduced. Similarly to the carbon thermal shock method, sufficiently high cathodic potential and short pulses allow different metal ions to be simultaneously reduced and precipitated on the support.

This method can control not only the composition of metal particles by adjusting the stoichiometric ratio of metal salts in nanodroplets but also the size of NPs by adjusting the roughness and defect distribution of the HOPG surface. The authors used this method to successfully synthesize low-entropy alloys (unary [Fe and Pt], binary [CuPt and CoNi], ternary [CoCrNi and CrCoGd], and quaternary [CrVNiCo] NPs). They also used the same method to synthesize different HEA samples such as eight-membered ones(CoCrCuGdInMnNiV) and other amorphous HEA-NPs, among which CoFeLaNiPt HEA-NPs showed excellent catalytic performance and catalytic activity in hydrogen evolution reaction (HER) and oxygen evolution reaction (OER).

Wet Chemistry. While methods such as carbon thermal shock and NMED can produce HEAs with uniform size and composition, they require specialized equipment and are time-consuming. In 2020, Wu et al. introduced the polycloud alcohol method, a simpler and faster approach for synthesizing Pt-group (PGM) HEA-NPs [3]. In this method, six metal precursors in an equimolar ratio are added to preheated triethylene glycol (TEG) at 230 °C, with poly(*N*-vinyl-2-pyrrolidone) as a protective agent. TEG serves as both a reducing agent and a solvent, enabling the simultaneous reduction of all metal precursors to form a solid-solution alloy. The resulting PtPdRuRhIrOs PGM-HEA-NPs are uniformly dispersed with a narrow size distribution (3.1 ± 0.6 nm) and exhibit superior electrocatalytic activity for ethanol oxidation, outperforming commercial Pd/C and Pt/C catalysts by 3.2 and 3.4 times, respectively. To give a direct impression, Figure 6.6 displays a typical procedure to synthesize HEA by wet chemistry method.

Additionally, the preparation of hollow-structured polymetallic NPs has emerged as a promising strategy for efficient catalysis. Wang et al. developed a continuous particle drip method to synthesize these hollow NPs, which form as gas-blowing agents decompose during heating [44]. The hollow structure provides more active sites on the surface, making these NPs effective candidates for energy and catalytic applications.

6.2.3 Perspectives for HEM Synthesis

HEMs' composition can be engineered through high-entropy mixing and element modulation, enabling the design of alloys with specific properties. Morphology control

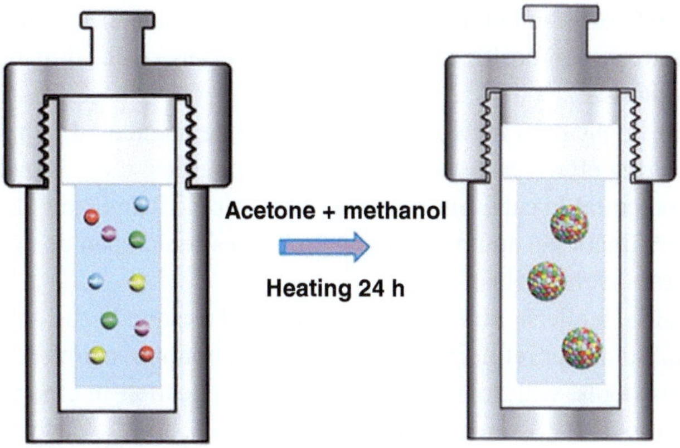

Figure 6.6 Scheme of the precise and controllable synthesis process of HEA-NPs by wet chemistry method [41]/American Chemical Society.

involves producing HEAs in various dimensional forms, such as NPs, nanowires, and nanosheets. Structure control enables the formation of solid solutions, intermetallic phases, amorphous structures, and quasicrystals. Surface/interface engineering, through modification and reconstruction techniques, allows the creation of multiphase and hybrid structures, including core-shell and hierarchical forms. These advanced strategies have enabled the fabrication of nanoscale HEM products with controlled atomic arrangements, paving the way for atomic-precision manipulation of their properties.

Despite these advances, several challenges remain in the controllable synthesis of HEMs. The inherent compositional and structural complexity of HEMs limits the current methods for precise fabrication. There is a need for better control over various properties such as composition, size uniformity, morphology, exposed facets, and phase transitions. For example, two-dimensional HEMs with atomic-scale thicknesses could offer unique physical properties suitable for quantum and optoelectronic applications. Moreover, the development of strategies to control short-range order, phase interfaces, and the distribution of individual elements is crucial for optimizing their structural performance.

To overcome these challenges, a deeper understanding of the growth mechanisms and dynamic evolution of HEMs under operational conditions is necessary. In situ characterization techniques, such as liquid or environmental electron microscopy, provide real-time insights into the growth and structural changes of HEMs. These techniques can also reveal how phase structure correlates with mechanical and catalytic properties, aiding in the fine-tuning of HEMs for specific applications. In situ technologies like X-ray absorption spectroscopy, Raman spectroscopy, and photoemission spectroscopy are key to exploring the chemical valence states and electronic structures, which are essential for understanding catalytic mechanisms.

Furthermore, the "tailoring" of HEMs has thus far been largely experimental, driven by trial and error. However, high-throughput computational methods, including machine learning techniques, can significantly streamline this process by predicting desirable compositions and properties. By combining computational predictions with experimental screening, it will be possible to establish reliable correlations between composition, structure, and performance, thus enabling more efficient synthesis.

Finally, the application fields of HEMs need to be broadened. While most structural engineering efforts focus on mechanical properties, there is increasing interest in their potential in catalysis, photocatalysis, biomedicine, and plasmonics. The integration of functional and mechanical properties could open up new avenues for HEMs in these diverse fields.

In conclusion, while significant progress has been made in the controllable synthesis of HEAs, further refinement of synthesis techniques, coupled with advanced computational tools, will drive their successful application in emerging technologies. The future of HEA research holds great promise for the development of versatile, high-performance materials with tailored properties for a wide range of applications.

6.3 Diverse Applications of Nanoscale HEMs

6.3.1 HEMs as Electrocatalysts

Since 2018, HEMs have been addressed as highly promising electrocatalysts for a wide range of energy-related applications, especially in energy conversion and storage [45–52]. Their incomparable physical and chemical properties distinguish themselves from conventional metal and metal-alloy-based electrocatalysts [53]. One key factor is their unique high-entropy mixing, which ensures a uniform distribution of elements without any phase segregation, forming a stable solid solution. Mixing elements with different electronegativity provides partially charged and continuous adsorption sites, resulting in a synergistic effect. The high lattice distortion and surface potential energy, caused by varying atomic sizes, also contribute to exceptional electrical and thermal conductivities. Furthermore, this distortion impedes atom diffusion within the crystal structure, thus enhancing the thermal stability of HEMs [54]. In addition, the broad composition possibilities of HEMs potentially eliminate dependence on single elements for their electrocatalytic performance [48].

In many energy-related applications, bulk and nanosized HEMs have found extensive use. However, highly porous structured nano-HEMs are known to enhance electrocatalytic performance [55]. As a result, significant efforts have been devoted to synthesizing nano-HEMs, in either sizes or structures, and they have often been achieved through top-down approaches starting from bulk materials. These nanosized HEMs can be efficiently synthesized, commonly through wet-chemical and sol–gel combustion techniques [56, 57]. Now, carbothermal shock synthesis has been established as a large-scale synthesis technology [8]. Among the existing basis of HEMs in energy applications, HEAs and high entropy oxides (HEOs) stand out. HEO synthesis is notably simpler than HEAs, thanks to the feasible formation of surface native oxides. Moreover, HEO electrocatalysts show a large number of oxygen vacancies, which contributes to their outstanding electrocatalytic performances in comparison to HEAs. As a result, the intrinsic activity of HEMs exceeds that of other electrocatalysts [58]. Besides, HEMs greatly cut down on the use of noble metals without affecting performance, promoting the development of cost-effective, large-scale electrochemical cells [59].

OER. The OER is an important step in water splitting and various electrochemical processes. This reaction, with a sluggish four-electron transfer, often requires expensive catalysts like IrO_2 and RuO_2 [60]. However, the search for cost-effective and diverse electrocatalyst materials has resulted in the discovery of alternatives that are able to overcome the inherent kinetic barriers of OER $(4OH^- \rightarrow O_2(g) + 2H_2O\ (1) + 4e^-$, alkaline; $2H_2O(1) \rightarrow O_2 + 4H^+ + 4e^-$, acidic$)$ [49]. Recently, different types of HEMs have appeared as potential candidates because their highly tunable catalytic properties can be achieved through entropy stabilization effects, which allow for the incorporation of more than five elements into a single phase [61–64]. The HEAs and HEOs have demonstrated significant promise as OER electrocatalysts. In 2018, Cui et al. reported FeNiMnCrCu HEA as an OER electrocatalyst in alkaline electrochemical water splitting, where the metal compositions of the HEA were tailored to improve the electrocatalytic OER performance by adjusting the d-vacancy number [65] (Figure 6.7).

HER. HER means the cathodic process $(4H_2O + 4e^- \rightarrow 2H_2(g) + 4OH^-$@alkaline; $4H^+ + 4e^- \rightarrow 2H_2(g) + 4OH^-$@acidic$)$ in ECWS and should occur at a cell potential of 0 V versus the reversible hydrogen electrode. Currently, platinum (Pt) is regarded as an advanced HER electrocatalyst because of its optimal M—H bond strength, which leads to excellent HER performance. Nevertheless, the limited availability and high cost of Pt metal greatly restrict its practical large-scale applications [67, 68]. Therefore, it is necessary to identify efficient electrocatalysts with MH bond strengths similar to Pt. The HEMs have been acknowledged as advanced HER electrocatalysts in comparison to other types because of their highly adaptable chemical composition and surface chemistry, which enable the optimization of MH bond strength [67, 69]. Moreover, their high corrosion resistance in harsh alkaline and acidic conditions, which is attributed to the high-entropy mixing effect and disordered atomic structures, further enhances their suitability for HER applications [69].

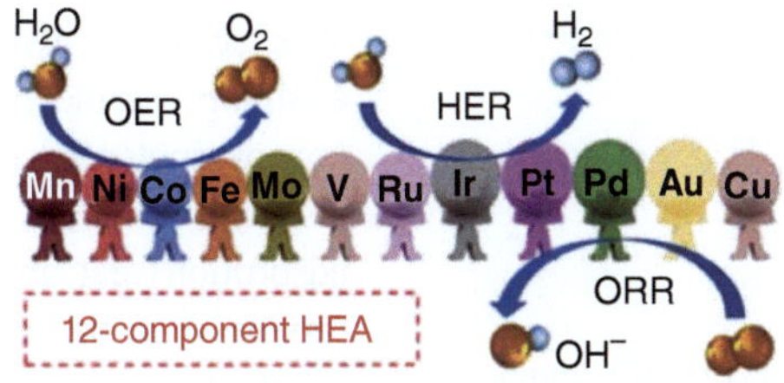

Figure 6.7 Illustration of trifunctional (OER, HER, and oxygen reduction reaction [ORR]) electrochemical performances of the HEMs [66]. [63] / American Chemical Society.

ORR. The ORR is a sluggish kinetic process involving complex multiple-electron transfers in various electrochemical applications. In acidic environments, oxygen is reduced into water or hydrogen peroxide through four and two electron–proton transfers, respectively $\left(O_2(g) + 4H^+ + 4e^- \rightarrow 2H_2O(1);\ O_2(g) + 2H^+ + 2e^- \rightarrow H_2O_2(1)\right)$. In alkaline media, the reaction gives out hydroxide ions (four electron–proton) or peroxide ions (two electron–proton), as follows: $O_2(g) + 2H_2O(1) + 4e^- \rightarrow 4OH^-$; $O_2(g) + H_2O(1) + 2e^- \rightarrow HO_2 + OH^-$ [70].

Traditionally, Pt-based materials have been regarded as superactive electrocatalysts for ORR, along with their well-established reaction mechanisms [71]. However, these noble materials still face obstacles like limited availability, structural degradation, and mutual dissolution in harsh electrolyte conditions. To fix these problems, non-noble metals or Pt-alloy-based ORR electrocatalysts have been explored [72].

Nitrogen Reduction Reaction (NRR). Electrochemical NRR provides an eco-friendly way to synthesize ammonia (NH_3) from nitrogen (N_2) and water (H_2O) as raw materials $\left(N_2(g) + 3H_2O(1) \rightarrow 2NH_3 + 3/2O_2(g)\right)$. However, activating and dissociating the N_2 molecule on the electrocatalyst surface are hard works, due to its inherent inertness. This results from the highly stable triple bond $N \equiv N$ and the lack of dipole moments [73, 74]. Also, electrons tend to catalyze hydrogen evolution (H_2) rather than promoting NRR on the catalyst surface. As a result, this idea provides the ability to achieve high NH_3 yield, minimize overpotential (activation energy), and ensure electrocatalyst stability, which are significant challenges in NRR [45, 73, 75]. Figure 6.8 shows NRR as a cathodic six electron–proton couple transfer reaction under external bias conditions [76].

Due to their distinctive catalytic properties and stability, the HEMs have attracted great attention in electrocatalytic NRR, providing a potential solution to existing challenges [45, 73, 74]. The schematic representation of an electrocatalytic NRR cell, with anodic and cathodic reactions taking place at the HEM electrodes, is illustrated in Figure 6.9 [77]. Dan et al. pioneered this field by introducing RuFeCoNiCu HEA-NPs as potential electrocatalysts for NRR [73]. Their research resulted in a high NH_3 yield and low overpotential, maintaining a remarkable Faradaic efficiency of 38.5% for 100 hours. According to density functional theory (DFT) calculations, the multimetal surroundings at the Fe site played a key role in N_2 adsorption and activation. Meanwhile, the Co—Cu and Ni—Ru sites acted as hydrogenation surfaces, enabling a rapid NRR process even at low overpotential [45, 73]. Figure 6.10 displays different chemical species' adsorption (reactants, intermediates, and products) on/between active sites at the HEM surface during NH_3 synthesis [77].

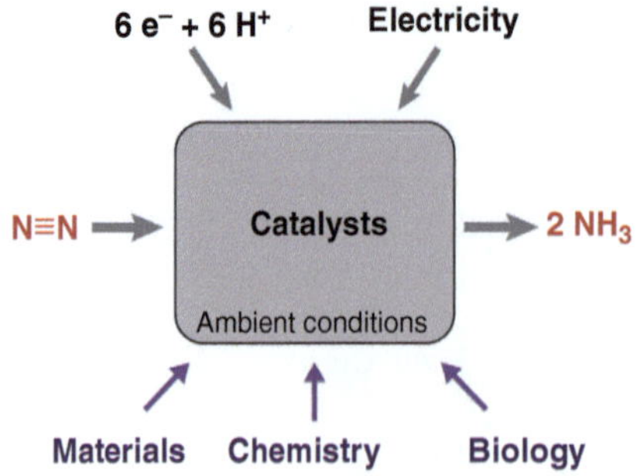

Figure 6.8 The process of NRR: ammonia synthesis—scheme of the flow cell for electrochemical synthesis of NH_3 via N_2 electrolysis [74] / with permission of Elsevier.

Furthermore, Rafael et al. clarified the role of local environments in N_2 adsorption on HEA surfaces based on

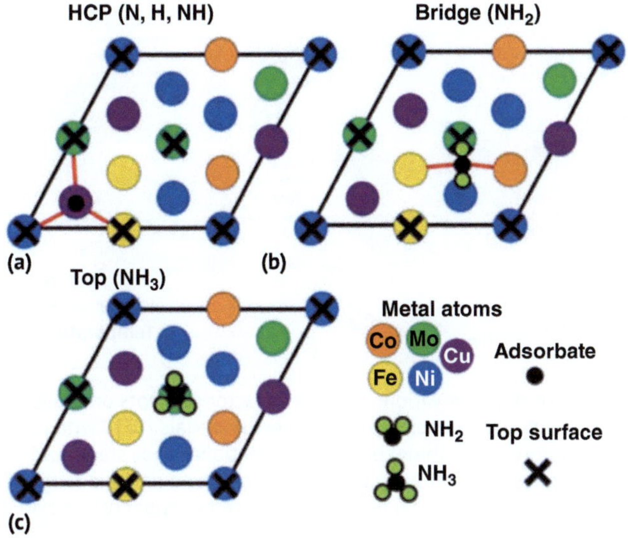

Figure 6.9 Various chemical species adsorption configurations, acting on the HEM surface (top view) [75] / with permission of Wissam A. Saidi/Researchgate/CC BY 4.0.

DFT results [75]. In HEA systems, although the d-band centers of adsorbing compounds initially adjusted the N_2 adsorption energy, these centers were further modulated by the local surrounding compounds (local environment) through electron donor–acceptor interactions. The interaction between direct active sites and other compounds on the HEAs' surface offered additional modulation capabilities. As a result, HEM electrocatalysts demonstrated stronger bond strength with N_2 molecules, effectively tackling N_2 fixation challenges [74].

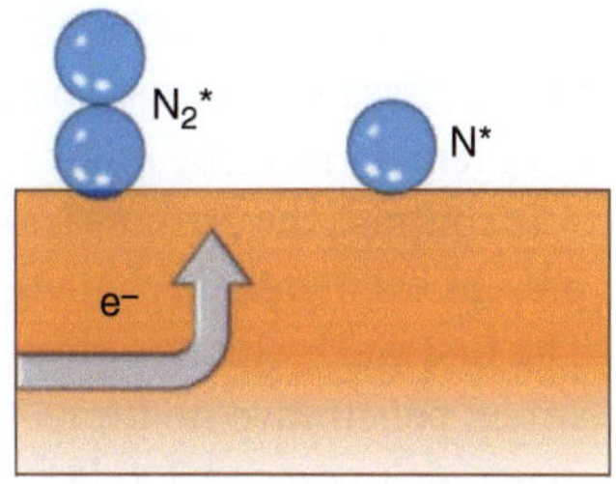

Figure 6.10 Descriptors (N* and N_2*) and their formation on the HEMs during the catalytic activity [76] / with permission of Elsevier.

Apart from carbon monoxide and dioxide reduction reactions, supercapacitors, Li-ion and Na-ion batteries, Li-sulfur batteries, CO/methanol/ethanol/CO_2 methanation, and ammonia decomposition reactions represent popular applications of HEMs in the field of electrocatalysis.

6.3.2 Nano-HEAs as an Irradiation-Resistant Structural Material

The main concerns regarding nuclear safety focus on the incompatibility between the operating environment inside reactors and structural materials. In comparison to other energy sources, the requirements for structural materials within nuclear reactors are extremely demanding.

As can be seen in Figure 6.11, most of the mainstream structural materials currently used in reactors, such as vanadium alloys, nickel-based superalloys, reduced

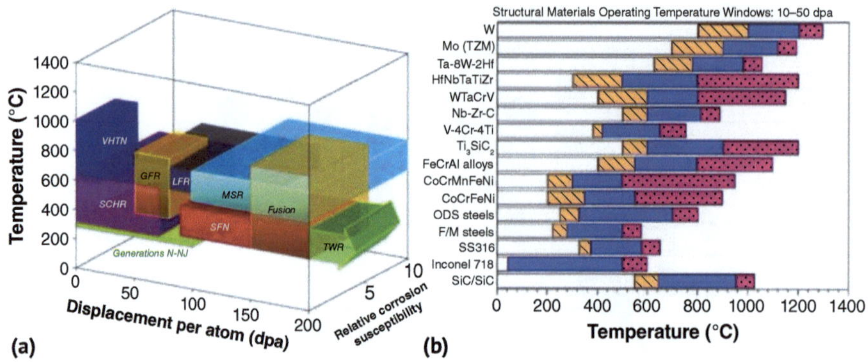

Figure 6.11 (a) Positioning of various advanced reactor concepts on scales depicting the ranges of environmental component values for core materials [77] / with permission of Elsevier. (b) Temperature range (blue) of some conventional alloy steels and new structural materials utilized in nuclear energy systems at an irradiation damage range of 10–50 dpa. The lower (orange) and higher (pink) temperature uncertainty bands are also displayed [78] / with permission of Elsevier.

activation ferritic/martensitic steels (RAFMs), oxide dispersion strengthened (ODS) steels, and SiC composites, encounter substantial challenges in meeting the service temperature and dose requirements of advanced, high-efficiency fourth-generation nuclear fission reactors. The performance limits of only a few materials come close to or remain near the minimum requirements.

In contrast, the extreme irradiation conditions and temperatures in fusion reactors present even greater challenges to the performance limits of structural materials. The first wall, which is directly exposed to the neutron radiation field, experiences strong penetrative neutrons that cause displacements per atom (dpa) in structural materials to exceed 200 [81, 82]. The poor compatibility of hydrogen and helium, produced by (n, p) and (n, α) transmutation reactions with the matrix, is a major concern. In terms of irradiation properties alone, such uncertainties that could threaten the safe operation of the reactor are unacceptable in the nuclear industry.

From the perspective of traditional alloying methods, the concept of HEAs as a "multicomponent" design approach that goes against prior experience is rather unconventional. However, over the past decades, HEAs have gradually been proven to possess extraordinary potential. Their compositional and structural complexity allows them to surpass the performance limits of existing conventional alloys, exhibiting excellent properties that are difficult to achieve in conventional alloys. Clearly, within the operating temperature window, HEAs show significant advantages over RAFMs and vanadium alloys. Regarding radiation damage levels, experimental evidence shows that, when subjected to ion irradiation at various doses, HEAs demonstrate overall superior radiation tolerance compared to conventional alloys. This superiority includes a smaller maximum lattice damage, a longer time to reach the damage peak, and a shallower radiation damage depth [83–85].

HEAs exhibit a lower rate of hardness increase [86–96] and reduced radiation swelling [96–103] under the same irradiation conditions. Regarding microstructure stability, it is worth noting that HEAs show less pronounced segregation phenomena under irradiation [104–106] and higher phase stability [106–108]. From the perspective of irradiation resistance, HEAs have demonstrated absolute potential to replace zirconium alloys [109], austenitic stainless steels [88, 110, 111], titanium alloys [112], and RAFMs [99, 113, 114].

Recent studies have put forward several radiation-resistance mechanisms in HEAs, dealing with defect formation and evolution at different scales—energy, quantum, atomic, and microstructural. At the energy scale, HEAs take advantage of a rough energy landscape, which provides higher defect formation and migration energies. At the quantum scale, shorter carrier means free paths, and strong electron–phonon coupling prolongs the thermal lifetime. At the atomic scale, increased atomic-level stress and 3D diffusion pathways are beneficial for defect annihilation. At the microstructural scale, the introduction of defect traps promotes the evolution and suppression of defects.

These mechanisms often coexist and are coupled within a single HEA. For example, in refractory HEAs, equilibrium vacancies supply sites for interstitial atoms, and the high melting point raises the defect formation energy. In eutectic HEAs, phase boundaries and precipitates restrain defect growth. High-density oxide particles in ODS HEAs block dislocation movement, trap radiation defects, and stabilize fine-grained microstructures, enhancing high-temperature mechanical properties and radiation resistance. Doping with interstitial atoms also refines the grain structure by suppressing diffusion. Rather than asserting that HEAs universally surpass traditional alloys in radiation resistance, it is more accurate to state that their complex chemical environments offer multiple ways to improve radiation resistance. This multitier-integrated design strategy is promising for future alloy development.

Nevertheless, challenges still exist. First, transforming computational simulations into practical design principles, especially concerning electronic interactions between elements for radiation-resistant HEAs, needs further research. Second, most simulations concentrate on simple Cantor HEAs, which may not be directly applicable to more complex, heterogeneous structures. More detailed modeling for HEAs is required. Lastly, attention should be paid to the comparability of radiation effects across studies with different irradiation conditions (particles, temperatures, doses) to standardize findings.

6.3.3 Nano-HEAs for Solid Hydrogen Storage Properties

The hydrogen storage (HS) properties of HEAs were first researched by Kao et al., who investigated the effect of Ti, V, and Zr content on the HS behavior of CoFeMnTiVZr-based HEAs. The samples were synthesized through vacuum arc melting and hydrogenated at 400 °C and 40 atm pressure. The HEAs exhibited high storage capacity with Ti content between 0.5 and 2 and Zr content between 0.4 and 2.3, although phase segregation occurred at higher Ti and Zr concentrations [115].

In another study, a C14 Laves phase ZrTiVCrFeNi HEA was synthesized using laser-engineered net shaping, with maximum HS capacities of 1.81 wt.% (as-cast) and 1.56 wt.% (annealed). The decrease in capacity after annealing was attributed to the reduced deformation of voids [116]. Similarly, TiZrNbMoV HEAs synthesized with different laser powers exhibited varied HS capacities, with the high-power HEA reaching 2.3 wt.% HS [117]. HEAs synthesized by ball milling, such as $MgZrTiFe_{0.5}Co_{0.5}Ni_{0.5}$, demonstrated phase transformations and absorbed up to 1.2 wt.% hydrogen [118]. Quinary MgVAlCrNi HEAs demonstrated low HS capacity as a result of poor hydride formation. In contrast, the AlCrFeMnNiW HEA displayed a capacity of 0.615 wt.% at room temperature and atmospheric pressure. CALPHAD (CALculation of PHAse Diagram) method, which describes the thermodynamic properties of the aiming system by constructing thermodynamic equilibria among all existing data, verified the formation of C14 Laves phases in TiZrNbFeNi HEAs, with the equiatomic alloy achieving an HS capacity of 1.64 wt.%. Further research revealed that equiatomic and non-equiatomic compositions possess different HS properties, with the former exhibiting better reversibility due to a higher valence electron concentration.

TiVZrNbHf HEAs demonstrated an HS capacity of 2.5 H/M ratio, equivalent to 2.7 wt.%, and released hydrogen at temperatures at 200 °C. Strozi et al. investigated Ti—V—Nb—M HEAs (M = Cr, Mn, Fe, Co, Ni) and found that increasing Cr content led to decreased lattice parameters and a higher HS capacity, with the $(TiVNb)_{85}Cr_{15}$ alloy reaching 3 wt.% storage [119].

6.3.4 Nano-HEAs for Biomedical Fields

Nano-HEMs have emerged as promising candidates in the biomedical field due to their unique structural and functional properties [120]. The combination of multiple elements in a single material matrix enhances the versatility of nano-HEMs, allowing for a wide range of properties, such as improved biocompatibility, enhanced stability, and tailored interactions with biological systems. These materials, particularly in the nanoscale form, exhibit high surface-to-volume ratios, which are critical for applications such as drug delivery, biosensing, and imaging.

In drug delivery, nano-HEMs can be designed to optimize the loading and release of therapeutic agents, taking advantage of their high surface area and tunable surface properties. For example, HEMs can be engineered with specific functional groups or incorporated into composite structures to enhance the affinity for target cells or tissues. The ability to adjust the composition and structure at the atomic level enables the fine-tuning of their release profiles, improving therapeutic efficacy and minimizing side effects.

In biosensing, nano-HEMs provide excellent platforms for the development of sensitive and selective sensors. Their diverse elemental composition allows for enhanced electrical, optical, and magnetic properties, which can be harnessed for the detection of biomarkers or pathogens. Furthermore, their stability and adaptability under different physiological conditions make them ideal candidates for in vivo applications, where they can be used for real-time monitoring of biological processes.

Moreover, the tunability of nano-HEMs' mechanical and thermal properties makes them suitable for tissue engineering and regenerative medicine. Their ability to mimic the mechanical properties of native tissues, combined with their biocompatibility, provides a foundation for creating scaffolds for cell growth and tissue repair. Additionally, their potential for radiation resistance is beneficial in therapeutic applications that require precise targeting and minimal damage to surrounding tissues. In the context of biomedical applications, nano-HEMs offer several advantages over traditional materials. Their high surface area and tunable composition allow for enhanced catalytic activity, which can be leveraged for biosensing and therapeutic purposes. For instance, nano-HEMs with peroxidase-like (POD-like) activity can catalyze the conversion of endogenous hydrogen peroxide (H_2O_2) into highly toxic hydroxyl radicals (•OH), which can be used to kill cancer cells effectively. Additionally, nano-HEMs exhibit excellent photothermal conversion efficiency, enabling their use in photothermal therapy (PTT) for tumor ablation (Figure 6.12). The synergistic effect of enzyme-like catalysis and photothermal action has been demonstrated to enhance the therapeutic efficacy of nano-HEMs in both in vitro and in vivo experiments [121].

Overall, nano-HEMs offer significant advantages in the biomedical field, with their multifunctional properties making them versatile materials for various applications, including drug delivery, biosensing, tissue engineering, and therapeutic interventions. Continued research into their synthesis, structure–property relationships, and biological interactions will further unlock their potential in advancing medical technologies.

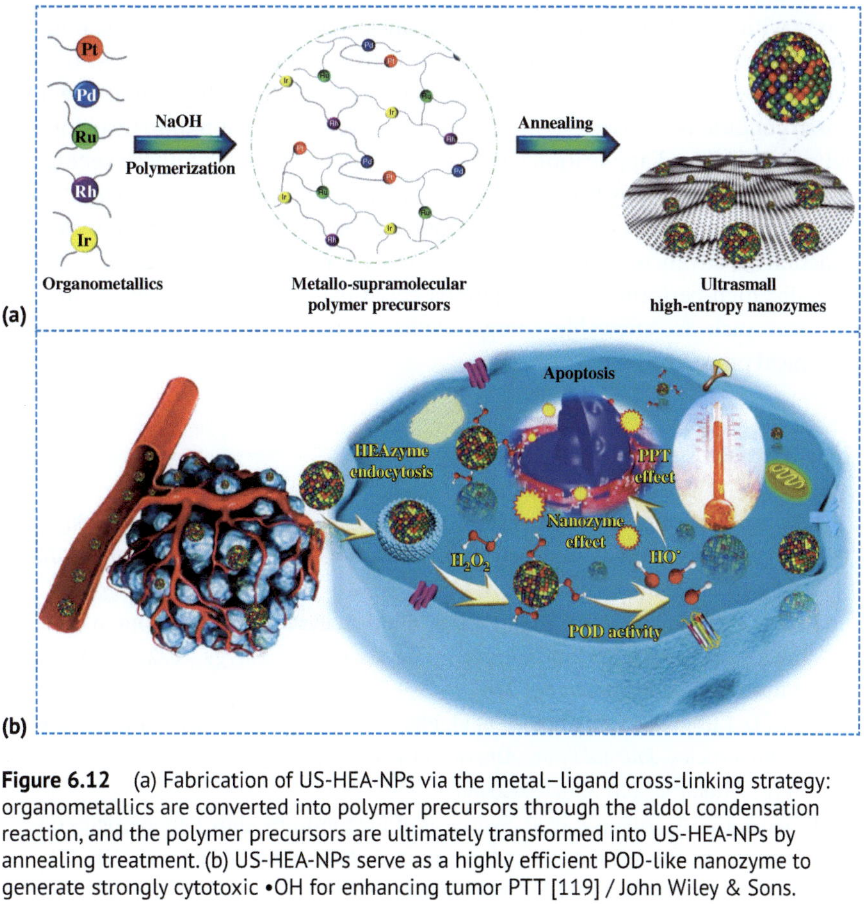

Figure 6.12 (a) Fabrication of US-HEA-NPs via the metal–ligand cross-linking strategy: organometallics are converted into polymer precursors through the aldol condensation reaction, and the polymer precursors are ultimately transformed into US-HEA-NPs by annealing treatment. (b) US-HEA-NPs serve as a highly efficient POD-like nanozyme to generate strongly cytotoxic •OH for enhancing tumor PTT [119] / John Wiley & Sons.

6.4 Emerging Nano-HEAs

Nano-HEAs offer significant advantages over their conventional counterparts due to their unique microstructure at the nanoscale. These materials combine the multielement nature of HEAs with nanoscale features, resulting in enhanced mechanical properties and particularly multifunctionalities. They are emerging as highly promising materials for catalytic and advanced applications due to their unique combination of high entropy, nanoscale microstructure, and excellent chemical stability. The high number of elements in HEAs, when combined with nanoscale features, leads to the formation of complex, highly active sites at the atomic level, enhancing the catalytic properties of the material, with a combination of superior catalytic activity, stability, and selectivity compared to traditional single-phase materials. The nanoscale size of the grains in nano-HEAs increases the surface area available for catalytic reactions, while the diverse elemental composition promotes synergistic effects between the constituent elements, further improving catalytic efficiency and reducing the susceptibility to poisoning. In advanced applications, nano-HEAs are also being explored for energy conversion and storage systems, such as in lithium-ion and sodium-ion batteries, supercapacitors, and fuel cells. The enhanced electrochemical performance of nano-HEAs, along with their high thermal and mechanical stability, make them ideal for use in high-performance, high-stress environments. Additionally, their resistance to corrosion and wear is critical for ensuring long-term durability and efficiency in harsh conditions. Looking to the future, the development of nano-HEAs for catalytic and advanced applications will focus on optimizing the synthesis techniques to control their composition, grain size, and phase structure at the nanoscale. This will allow for the precise tailoring of material properties to suit specific catalytic reactions or advanced technologies. The ability to design and engineer these materials for high-performance applications could lead to significant breakthroughs in areas such as renewable energy, environmental protection, and sustainable chemical processes.

References

1 Wang Y T, Wang Y H. High-entropy alloys in catalyses and supercapacitors: progress, prospects. *Nano Energy*, 2022, 104: 107958. https://doi.org/10.1016/j.nanoen.2022.107958

2 Yu L, Zeng K, Li C, et al. High-entropy alloy catalysts: from bulk to nano toward highly efficient carbon and nitrogen catalysis. *Carbon Energy*, 2022, 4: 731–761. https://doi.org/10.1002/cey2.228

3 Wu D, Kusada K, Yamamoto T, et al. Platinum-group-metal high-entropy-alloy nanoparticles. *Journal of the American Chemical Society*, 2020, 142(32): 13833. https://doi.org/10.1021/jacs.0c04807

4 Gao S, Hao S, Huang Z, et al. Synthesis of high-entropy alloy nanoparticles on supports by the fast moving bed pyrolysis. *Nature Communications*, 2020, 11: 2016. https://doi.org/10.1038/s41467-020-15934-1

5 Li H, Han Y, Zhao H, et al. Fast site-to-site electron transfer of high-entropy alloy nanocatalyst driving redox electrocatalysis. *Nature Communications*, 2020, 11: 5437. https://doi.org/10.1038/s41467-020-19277-9

6 Yao Y G, Huang Z, Hughes L A, et al. Extreme mixing in nanoscale transition metal alloys. *Matter-US*, 2021, 4: 2340–2353. https://doi.org/10.1016/j.matt.2021.04.014

7 F. C, T. The Structure of Metals and Alloys. *Nature*, 1936, 138: 7–8. https://doi.org/10.1038/138007a0

8 Yao Y G, Huang Z, Xie P, et al. Carbothermal shock synthesis of high-entropy-alloy nanoparticles. *Science*, 2018, 359: 1489–1494. https://doi.org/10.1126/science.aan5412

9 Kwon S G, Krylova G, Phillips P J, et al. Heterogeneous nucleation and shape transformation of multicomponent metallic nanostructures. *Nature Materials*, 2015, 14: 215–223. https://doi.org/10.1038/Nmat4115

10 Fenton J L, Steimle B C, Schaak R E. Tunable intraparticle frameworks for creating complex heterostructured nanoparticle libraries. *Science*, 2018, 360: 513–517. https://doi.org/10.1126/science.aar5597

11 Chen P C, Liu X, Hedrick J L, et al. Polyelemental nanoparticle libraries. *Science*, 2016, 352: 1565–1569. https://doi.org/10.1126/science.aaf8402

12 Al Zoubi W, Putri R A K, Abukhadra M R, et al. Recent experimental and theoretical advances in the design and science of high-entropy alloy nanoparticles. *Nano Energy*, 2023, 110: 108362. https://doi.org/10.1016/j.nanoen.2023.108362

13 Chen X T, Si C, Gao Y, et al. Multi-component nanoporous platinum-ruthenium-copper-osmium-iridium alloy with enhanced electrocatalytic activity towards methanol oxidation and oxygen reduction. *Journal of Power Sources*, 2015, 273: 324–332. https://doi.org/10.1016/j.jpowsour.2014.09.076

14 Ito T, Okazaki S. Pushing the limits of lithography. *Nature*, 2000, 406: 1027–1031. https://doi.org/10.1038/35023233

15 Li H N, Zhu H, Zhang S, et al. Nano high-entropy materials: synthesis strategies and catalytic applications. *Small Structures*, 2020, 1: 2000033. https://doi.org/10.1002/sstr.202000033

16 Li S Y, Tang X, Jia H, et al. Nanoporous high-entropy alloys with low Pt loadings for high-performance electrochemical oxygen reduction. *Journal of Catalysis*, 2020, 383: 164–171. https://doi.org/10.1016/j.jcat.2020.01.024

17 Tomboc G M, Kwon T, Joo J, et al. High entropy alloy electrocatalysts: a critical assessment of fabrication and performance. *Journal of Materials Chemistry A*, 2020, 8: 14844–14862. https://doi.org/10.1039/d0ta05176d

18 Waag F, Li Y, Ziefuß A R, et al. Kinetically-controlled laser-synthesis of colloidal high-entropy alloy nanoparticles. *RSC Advances*, 2019, 9: 18547–18558. https://doi.org/10.1039/c9ra03254a

19 Wallraff G M, Hinsberg W D. Lithographic imaging techniques for the formation of nanoscopic features. *Chemical Reviews*, 1999, 99: 1801–1822. https://doi.org/10.1021/cr980003i

20 Wu Q, Wang Z, He F, et al. High entropy alloys: from bulk metallic materials to nanoparticles. *Metallurgical and Materials Transactions A*, 2018, 49: 4986–4990. https://doi.org/10.1007/s11661-018-4802-1

21 Xin Y, Li S, Qian Y, et al. High-entropy alloys as a platform for catalysis: progress, challenges, and opportunities. *ACS Catalysis*, 2020, 10: 11280–11306. https://doi.org/10.1021/acscatal.0c03617

22 Hao M Q, Wang Y, Chen D, et al. Ion implantation synthesis of long-term stable high-entropy metallic glass nanoparticles. *Journal of Alloys and Compounds*, 2022, 906: 164303.

23 Jamkhande P G, Ghule N W, Bamer A H, et al. Metal nanoparticles synthesis: an overview on methods of preparation, advantages and disadvantages, and applications. *Journal of Drug Delivery Science and Technology*, 2019, 53: 101174. https://doi.org/10.1016/j.jddst.2019.101174

24 Varalakshmi S, Kamaraj M, Murty B S. Synthesis and characterization of nanocrystalline AlFeTiCrZnCu high entropy solid solution by mechanical alloying. *Journal of Alloys and Compounds*, 2008, 460(1): 253–257.

25 Xiao L L, Zheng Z, Guo S, et al. Ultra-strong nanostructured CrMnFeCoNi high entropy alloys. *Materials & Design*, 2020, 194: 108895.

26 Chen J, Jiang X, Sun H, et al. Phase transformation and strengthening mechanisms of nanostructured high-entropy alloys. *Nanotechnology Reviews*, 2021, 10: 1116–1139. https://doi.org/doi:10.1515/ntrev-2021-0071

27 Nugroho F A A, Iandolo B, Wagner J B, et al. Bottom-up nanofabrication of supported noble metal alloy nanoparticle arrays for plasmonics. *ACS Nano*, 2016, 10: 2871–2879. https://doi.org/10.1021/acsnano.5b08057

28 Broge N L N, Bertelsen A D, Søndergaard-Pedersen F, et al. Facile solvothermal synthesis of Pt–Ir–Pd–Rh–Ru–Cu–Ni–Co high-entropy alloy nanoparticles. *Chemistry of Materials*, 2023, 35: 144–153. https://doi.org/10.1021/acs.chemmater.2c02842

29 Bondesgaard M, Broge N L N, Mamakhel A, et al. General solvothermal synthesis method for complete solubility range bimetallic and high-entropy alloy nanocatalysts. *Advanced Functional Materials*, 2019, 29: 1905933. https://doi.org/10.1002/adfm.201905933

30 Liu M M. Zhang Z, Okejiri F, et al. Entropy-maximized synthesis of multimetallic nanoparticle catalysts via a ultrasonication-assisted wet chemistry method under ambient conditions. *Advanced Materials Interfaces*, 2019, 6: 1900015. https://doi.org/10.1002/admi.201900015

31 Singh M P, Srivastava C. Synthesis and electron microscopy of high entropy alloy nanoparticles. *Materials Letters*, 2015, 160: 419–422. https://doi.org/10.1016/j.matlet.2015.08.032

32 Quinson J. Colloidal surfactant-free syntheses of precious metal nanoparticles for electrocatalysis. *Current Opinion in Electrochemistry*, 2022, 34: 100977. https://doi.org/10.1016/j.coelec.2022.100977

33 Clausen C M, Krysiak O A, Banko L, et al. A Flexible theory for catalysis: learning alkaline oxygen reduction on complex solid solutions within the Ag-Pd-Pt-Ru composition space. *Angewandte Chemie International Edition*, 2023, 62: e202307187. https://doi.org/10.1002/anie.202307187

34 Batchelor T A A, Löffler T, Xiao B, et al. Complex-solid-solution electrocatalyst discovery by computational prediction and high-throughput experimentation.

Angewandte Chemie International Edition, 2021, 60: 6932–6937. https://doi.org/10.1002/anie.202014374

35 Wang B, Yao Y, Yu X, et al. Understanding the enhanced catalytic activity of high entropy alloys: from theory to experiment. *Journal of Materials Chemistry A*, 2021, 9: 19410–19438. https://doi.org/10.1039/d1ta02718b

36 Yao Y G, Liu Z, Xie P, et al. Computationally aided, entropy-driven synthesis of highly efficient and durable multi-elemental alloy catalysts. *Science Advances*, 2020, 6: eaaz0510. https://doi.org/10.1126/sciadv.aaz0510

37 Liu M, Zhang Z, Okejiri F, et al. Entropy-maximized synthesis of multimetallic nanoparticle catalysts via a ultrasonication-assisted wet chemistry method under ambient conditions. *Advanced Materials Interfaces*, 2019, 6: 1900015. https://doi.org/10.1002/admi.201900015

38 Wang B, Wang C, Yu X, et al. General synthesis of high-entropy alloy and ceramic nanoparticles in nanoseconds. *Nature Synthesis*, 2022, 1: 138–146. https://doi.org/10.1038/s44160-021-00004-1

39 Löffler T, Meyer H, Savan A, et al. Discovery of a multinary noble metal-free oxygen reduction catalyst. *Advanced Energy Materials*, 2018, 8: 1802269. https://doi.org/10.1002/aenm.201802269

40 Hao J, Zhuang Z, Cao K, et al. Unraveling the electronegativity-dominated intermediate adsorption on high-entropy alloy electrocatalysts. *Nature Communications*, 2022, 13: 2662. https://doi.org/10.1038/s41467-022-30379-4

41 Feng J, Chen D, Pikhitsa P.V, et al. Unconventional alloys confined in nanoparticles: building blocks for new matter. *Matter*, 2020, 3: 1646–1663. https://doi.org/10.1016/j.matt.2020.07.027

42 Yao Y. Dong Q, Brozena A, et al. High-entropy nanoparticles: synthesis-structure-property relationships and data-driven discovery. *Science*, 2022, 376: eabn3103. https://doi.org/10.1126/science.abn3103

43 Glasscott M W, Pendergast A D, Goines S, et al. Electrosynthesis of high-entropy metallic glass nanoparticles for designer, multi-functional electrocatalysis. *Nature Communications*, 2019, 10(1): 2650.

44 Wang X Z, Dong Q, Qiao H, et al. Continuous Synthesis of Hollow High-Entropy Nanoparticles for Energy and Catalysis Applications. *Advanced Materials*, 2020, 32(46): 2002853.

45 Zhang D, Shi Y, Zhao H, et al. The facile oil-phase synthesis of a multi-site synergistic high-entropy alloy to promote the alkaline hydrogen evolution reaction. *Journal of Materials Chemistry A*, 2021, 9: 889–893. https://doi.org/10.1039/d0ta10574k

46 Xie P F, Yao Y, Huang Z, et al. Highly efficient decomposition of ammonia using high-entropy alloy catalysts. *Nature Communications*, 2019, 10: 4011. https://doi.org/10.1038/s41467-019-11848-9

47 Sarkar A, Velasco L, Wang D I, et al. High entropy oxides for reversible energy storage. *Nature Communications*, 2018, 9: 3400. https://doi.org/10.1038/s41467-018-05774-5

48 Liu X F, Li X, Li Y, et al. High-entropy oxide: a future anode contender for lithium-ion battery. *Ecomat*, 2022, 4: e12261. https://doi.org/10.1002/eom2.12261

49 Liu F M, Yu M, Chen X, et al. Defective high-entropy rocksalt oxide with enhanced metal-oxygen covalency for electrocatalytic oxygen evolution.

Chinese Journal of Catalysis, 2022, 43: 122–129. https://doi.org/10.1016/S1872-2067(21)63794-4

50 Li H J, Huang H, Chen Y, et al. High-entropy alloy aerogels: a new platform for carbon dioxide reduction. *Advanced Materials*, 2023, 35: 2209242. https://doi.org/10.1002/adma.202209242

51 Hussain I, Lamiel C, Ahmad M, et al. High entropy alloys as electrode material for supercapacitors: a review. *The Journal of Energy Storage*, 2021, 44: 103405. https://doi.org/10.1016/j.est.2021.103405

52 Batchelor T A A, Pedersen J K, Winther S H, et al. High-entropy alloys as a discovery platform for electrocatalysis. *Joule*, 2019, 3: 834–845. https://doi.org/10.1016/j.joule.2018.12.015

53 Ye Y F, Wang Q, Lu J, et al. High-entropy alloy: challenges and prospects. *Materials Today*, 2016, 19: 349–362. https://doi.org/10.1016/j.mattod.2015.11.026

54 Ruiz Esquius J, Liu L. High entropy materials as emerging electrocatalysts for hydrogen production through low-temperature water electrolysis. *Materials Futures*, 2023, 2: 022102. https://doi.org/10.1088/2752-5724/accbd8

55 Zheng H, Luo G Y, Zhang A J, et al. The synthesis and catalytic applications of nanosized high-entropy alloys. *ChemCatChem*, 2021, 13: 806–817. https://doi.org/10.1002/cctc.202001163

56 Qiao H Y, Saray M T, Wang X, et al. Scalable synthesis of high entropy alloy nanoparticles by microwave heating. *Acs Nano*, 2021, 15: 14928–14937. https://doi.org/10.1021/acsnano.1c05113

57 Niu B, Zhang F, Ping H, et al. Sol-gel autocombustion synthesis of nanocrystalline high-entropy alloys. *Scientific Reports—UK*, 2017, 7: 3421. https://doi.org/10.1038/s41598-017-03644-6

58 Anandkumar M, Trofimov E. Synthesis, properties, and applications of high-entropy oxide ceramics: current progress and future perspectives. *Journal of Alloys and Compounds*, 2023, 960: 170690. https://doi.org/10.1016/j.jallcom.2023.170690

59 Cheng Y S, Sun Y, Deng X, et al. High-performance high-entropy quinary-alloys as anode catalysts for direct ethylene glycol fuel cells. *International Journal of Hydrogen Energy*, 2023, 48: 8156–8164. https://doi.org/10.1016/j.ijhydene.2022.11.214

60 Abdelhafiz A, Wang B M, Harutyunyan A R, et al. Carbothermal shock synthesis of high entropy oxide catalysts: dynamic structural and chemical reconstruction boosting the catalytic activity and stability toward oxygen evolution reaction. *Advanced Energy Materials*, 2022, 12: 2200742. https://doi.org/10.1002/aenm.202200742

61 Talluri B, Yoo K, Kim J. High entropy spinel metal oxide $(CoCrFeMnNi)_3O_4$ nanoparticles as novel efficient electrocatalyst for methanol oxidation and oxygen evolution reactions. *Journal of Environmental Chemical Engineering*, 2022, 10: 106932. https://doi.org/10.1016/j.jece.2021.106932

62 Qiao H Y, Wang X, Dong Q, et al. A high-entropy phosphate catalyst for oxygen evolution reaction. *Nano Energy*, 2021, 86: 106029. https://doi.org/10.1016/j.nanoen.2021.106029

63 Li F, Sun S K, Chen Y, et al. Bottom-up synthesis of 2D layered high-entropy transition metal hydroxides. *Nanoscale Advances*, 2022, 4: 2468–2478. https://doi.org/10.1039/D1NA00871D

64 Hegde C, Lim C H J, Teng T H, et al. In situ synthesis and microfabrication of high entropy alloy and oxide compounds by femtosecond laser direct writing under ambient conditions. *Small*, 2022, 18: 2203126. https://doi.org/10.1002/smll.202203126

65 Cui X D, Zhang B, Zeng C, et al. Electrocatalytic activity of high-entropy alloys toward oxygen evolution reaction. *MRS Communications*, 2018, 8: 1230–1235 https://doi.org/10.1557/mrc.2018.111

66 Yu T, Zhang Y, Hu Y, et al. Twelve-component free-standing nanoporous high-entropy alloys for multifunctional electrocatalysis. *ACS Materials Letter*, 2022, 4: 181–189. https://doi.org/10.1021/acsmaterialslett.1c00762

67 Zhang G L, Ming K, Kang J, et al. High entropy alloy as a highly active and stable electrocatalyst for hydrogen evolution reaction. *Electrochim Acta*, 2018, 279: 19–23. https://doi.org/10.1016/j.electacta.2018.05.035

68 Ma P Y, Zhao M, Zhang L, et al. Self-supported high-entropy alloy electrocatalyst for highly efficient H evolution in acid condition. *Journal of Materiomics*, 2020, 6: 736–742. https://doi.org/10.1016/j.jmat.2020.06.001

69 Yang J, Fan Y, Liu Y, et al. Self-supporting porous high-entropy MAX electrode for highly active electrocatalyst H_2 evolution in alkali solution. *Journal of Porous Materials*, 2022, 29: 693–704. https://doi.org/10.1007/s10934-022-01205-5

70 Huo X R, Yu H S, Xing, B, et al. Review of high entropy alloys electrocatalysts for hydrogen evolution, oxygen evolution, and oxygen reduction reaction. *The Chemical Record*, 2022, 22: e202200175. https://doi.org/10.1002/tcr.202200175

71 Jin Z, Lyu J, Hu K, et al. Eight-component nanoporous high-entropy oxides with low Ru contents as high-performance bifunctional catalysts in Zn-air batteries. *Small*, 2022, 18: 2107207. https://doi.org/10.1002/smll.202107207

72 Chun H, Lee E, Nam K, et al. First-principle-data-integrated machine-learning approach for high-throughput searching of ternary electrocatalyst toward oxygen reduction reaction. *Chem Catalysis*, 2021, 1: 855–869. https://doi.org/10.1016/j.checat.2021.06.001

73 Zhang D, Zhao H, Wu X, et al. Multi-site electrocatalysts boost pH-universal nitrogen reduction by high-entropy alloys. *Advanced Functional Materials*, 2021, 31: 2006939. https://doi.org/10.1002/adfm.202006939

74 Sun Y T, Yu L, Xu S, et al. Battery-driven N_2 electrolysis enabled by high-entropy catalysts: from theoretical prediction to prototype model. *Small*, 2022, 18: 2106358. https://doi.org/10.1002/smll.202106358

75 Araujo R B, Edvinsson T N. Adsorption on high-entropy alloy surfaces: unveiling the role of local environments. *Journal of Materials Chemistry A*, 2023, 11: 12973–12983. https://doi.org/10.1039/d2ta09348k

76 Deng J, Iñiguez J A, Liu C. Electrocatalytic nitrogen reduction at low temperature. *Joule*, 2018, 2: 846–856. https://doi.org/10.1016/j.joule.2018.04.014

77 Saidi W A, Shadid W, Veser G. Optimization of high-entropy alloy catalyst for ammonia decomposition and ammonia synthesis. *The Journal of Physical Chemistry Letters*, 2021, 12: 5185–5192. https://doi.org/10.1021/acs.jpclett.1c01242

78 Yasin G, et al. High-Entropy Alloys: Design, Manufacturing, and Emerging Applications. Elsevier, 2024.

79 Was G S, Petti D, Ukai S, et al. Materials for future nuclear energy systems. *Journal of Nuclear Materials*, 2019, 527: 151837. https://doi.org/10.1016/j.jnucmat.2019.151837

80 Cheng Z, Sun, J., Gao, X, et al. Irradiation effects in high-entropy alloys and their applications. *Journal of Alloys and Compounds*, 2023, 930: 166768. https://doi.org/10.1016/j.jallcom.2022.166768

81 Zinkle S J. & Was G S. Materials challenges in nuclear energy. *Acta Materialia*, 2013, 61: 735–758. https://doi.org/10.1016/j.actamat.2012.11.004

82 Bloom E E. The challenge of developing structural materials for fusion power systems. *Journal of Nuclear Materials*, 1998, 258–263: 7–17. https://doi.org/10.1016/S0022-3115(98)00352-3

83 Zhang Y, Jin K, Xue H, et al. Influence of chemical disorder on energy dissipation and defect evolution in advanced alloys. *Journal of Materials Research*, 2016, 31: 2363–2375. https://doi.org/10.1557/jmr.2016.269

84 Lear C R, Jin K, Xue H, et al. Effects of helium cavity size and morphology on the strength of pure titanium. *Scripta Materialia*, 2022, 212: 114531. https://doi.org/10.1016/j.scriptamat.2022.114531

85 Chen Y, Liao X, Qiu R, et al. Primary radiation damage in tungsten-based high-entropy alloy: interatomic potential and collision cascade simulations. *Journal of Nuclear Materials*, 2023, 585: 154646. https://doi.org/10.1016/j.jnucmat.2023.154646

86 Zhang H, Zhu Z, Huang H, et al. Microstructures, mechanical properties, and irradiation tolerance of the Ti–Zr–Nb–V–Mo refractory high-entropy alloys. *Intermetallics*, 2023, 157: 107873. https://doi.org/10.1016/j.intermet.2023.107873

87 Yang J, Shi K, Chen Q, et al. Effect of Au-ion irradiation on the surface morphology, microstructure and mechanical properties of amorphous AlCrFeMoTi HEA coating. *Surface and Coatings Technology*, 2021, 418: 127252. https://doi.org/10.1016/j.surfcoat.2021.127252

88 Yang L, Ge H, Zhang J, et al. High He-ion irradiation resistance of CrMnFeCoNi high-entropy alloy revealed by comparison study with Ni and 304SS. *Journal of Materials Science & Technology*, 2019, 35: 300–305. https://doi.org/10.1016/j.jmst.2018.09.050

89 Waseem O A, Woller K B, Sweidan F B, et al. Effects of F^{3+} ion implantation on the properties of W and $W_{0.5}(TaTiVCr)_{0.5}$ for depth marker-based plasma erosion analysis. *Nuclear Materials and Energy*, 2020, 25: 100806. https://doi.org/10.1016/j.nme.2020.100806

90 Shen S, Hao L, Liu X, et al. The design of Pd-containing high-entropy alloys and their hardening behavior under He ion irradiation. *Acta Materialia*, 2023, 261: 119404. https://doi.org/10.1016/j.actamat.2023.119404

91 Sadeghilaridjani M, Ayyagari A, Muskeri S, et al. Ion irradiation response and mechanical behavior of reduced activity high entropy alloy. *Journal of Nuclear Materials*, 2020, 529: 151955. https://doi.org/10.1016/j.jnucmat.2019.151955

92 Moorehead M, Nelaturu P, Elbakhshwan M, et al. High-throughput ion irradiation of additively manufactured compositionally complex alloys. *Journal of Nuclear Materials*, 2021, 547: 152782. https://doi.org/10.1016/j.jnucmat.2021.152782

93 Lu Y, Huang H, Gao X, et al. A promising new class of irradiation tolerant materials: $Ti_2ZrHfV_{0.5}Mo_{0.2}$ high-entropy alloy. *Journal of Materials Science & Technology*, 2019, 35: 369–373. https://doi.org/10.1016/j.jmst.2018.09.034

94 Kareer A, Waite J C, Li B, et al. Short communication: 'low activation, refractory, high entropy alloys for nuclear applications'. *Journal of Nuclear Materials*, 2019, 526: 151744. https://doi.org/10.1016/j.jnucmat.2019.151744

95 Chen W-Y, Kirk M A, Hashimoto N, et al. Irradiation effects on $Al_{0.3}CoCrFeNi$ and CoCrMnFeNi high-entropy alloys, and 316 H stainless steel at 500 °C. *Journal of Nuclear Materials*, 2020, 539: 152324. https://doi.org/10.1016/j.jnucmat.2020.152324

96 Chang S, Tseng K K, Yang T Y, et al. Irradiation-induced swelling and hardening in HfNbTaTiZr refractory high-entropy alloy. *Materials Letters*, 2020, 272: 127832. https://doi.org/10.1016/j.matlet.2020.127832

97 Cheng T, Wei G, Jiang S, et al. Enhanced resistance to helium irradiations through unusual interaction between high-entropy-alloy and helium. *Acta Materialia*, 2023, 248: 118765. https://doi.org/10.1016/j.actamat.2023.118765

98 Huang S S, Guan H Q, Zhong Z H, et al. Effect of He on the irradiation resistance of equiatomic CoCrFeMnNi high-entropy alloy. *Journal of Nuclear Materials*, 2022, 561: 153525. https://doi.org/10.1016/j.jnucmat.2022.153525

99 Parkin C, Moorehead M, Elbakhshwan M, et al. Phase stability, mechanical properties, and ion irradiation effects in face-centered cubic CrFeMnNi compositionally complex solid-solution alloys at high temperatures. *Journal of Nuclear Materials*, 2022, 565: 153733. https://doi.org/10.1016/j.jnucmat.2022.153733

100 Xia S Q, Yang X, Yang T F, et al. Irradiation resistance in $Al_xCoCrFeNi$ high entropy alloys. *JOM*, 2015, 67: 2340–2344. https://doi.org/10.1007/s11837-015-1568-4

101 Su Z, Shi T, Yang J, et al. The effect of interstitial carbon atoms on defect evolution in high entropy alloys under helium irradiation. *Acta Materialia*, 2022, 233: 117955. https://doi.org/10.1016/j.actamat.2022.117955

102 Ivanov I A, Ryskulov A, Kurakhmedov A, et al. Radiation swelling and hardness of high-entropy alloys based on the TiTaNbV system irradiated with krypton ions. *Journal of Materials Science: Materials in Electronics*, 2021, 32: 27260–27267. https://doi.org/10.1007/s10854-021-07095-8

103 Zhao Y F, Chen, H H, Zhang D D, et al. Unusual He-ion irradiation strengthening and inverse layer thickness-dependent strain rate sensitivity in transformable high-entropy alloy/metal nanolaminates: a comparison of $Fe_{50}Mn_{30}Co_{10}Cr_{10}$/Cu vs $Fe_{50}Mn_{30}Co_{10}Ni_{10}$/Cu. *Journal of Materials Science & Technology*, 2022, 116: 199–213. https://doi.org/10.1016/j.jmst.2021.10.036

104 Patel D, Richardson M D, Jim B, et al. Radiation damage tolerance of a novel metastable refractory high entropy alloy $V_{2.5}Cr_{1.2}WMoCo_{0.04}$. *Journal of Nuclear Materials*, 2020, 531: 152005. https://doi.org/10.1016/j.jnucmat.2020.152005

105 Lin W T, Yeli G M, Wang G, et al. He-enhanced heterogeneity of radiation-induced segregation in FeNiCoCr high-entropy alloy. *Journal of Materials Science & Technology*, 2022, 101: 226–233. https://doi.org/10.1016/j.jmst.2021.05.053

106 El-Atwani O, Li N, Li M, et al. Outstanding radiation resistance of tungsten-based high-entropy alloys. *Science Advances*, 2019, 5: eaav2002. https://doi.org/10.1126/sciadv.aav2002

107 Yang T, Guo W, Poplawsky J D, et al. Structural damage and phase stability of $Al_{0.3}CoCrFeNi$ high entropy alloy under high temperature ion irradiation. *Acta Materialia*, 2020, 188: 1–15. https://doi.org/10.1016/j.actamat.2020.01.060

108 Gwalani B, Soni V, Choudhuri D, et al. Stability of ordered L12 and B2 precipitates in face centered cubic based high entropy alloys—$Al_{0.3}CoFeCrNi$ and $Al_{0.3}CuFeCrNi_2$. *Scripta Materialia*, 2016, 123: 130–134. https://doi.org/10.1016/j.scriptamat.2016.06.019

109 Dong Q, Qin H, Yao Z, et al. Irradiation damage and hardening in pure Zr and Zr-Nb alloys at 573 K from self-ion irradiation. *Materials & Design*, 2019, 161: 147–159. https://doi.org/10.1016/j.matdes.2018.11.017

110 Xiu P, Osetsky Y N, Jiang L, et al. Dislocation loop evolution and radiation hardening in nickel-based concentrated solid solution alloys. *Journal of Nuclear Materials*, 2020, 538: 152247. https://doi.org/10.1016/j.jnucmat.2020.152247

111 Jiang L, Song M, Yang L, et al. A comparison study of void swelling in additively manufactured and cold-worked 316 L stainless steels under ion irradiation. *Journal of Nuclear Materials*, 2021, 551: 152946. https://doi.org/10.1016/j.jnucmat.2021.152946

112 Dey S, Dutta A, Mukherjee P, et al. Characterization of ion induced damage as a function of depth in proton irradiated pure Ti and Ti–6Al–4V. *Journal of Alloys and Compounds*, 2020, 821: 153441. https://doi.org/10.1016/j.jallcom.2019.153441

113 Saikumaran A, Mythili R, Magudapathy P, et al. Comparison of irradiation-induced hardening behavior of P91 ferritic martensitic steel and CrFeMoV high-entropy alloy. *Journal of Materials Engineering and Performance*, 2021, 30: 3547–3555. https://doi.org/10.1007/s11665-021-05661-z

114 Kim S, Heo J, Kim H, et al. The effect of tempering temperatures on the surface swelling of helium-ion-implanted Ti/Ta-added reduced-activation ferritic/martensitic steel. *Fusion Engineering and Design*, 2020, 154: 111509. https://doi.org/10.1016/j.fusengdes.2020.111509

115 Kao Y.-F, Chen S K, Sheu J H, et al. Hydrogen storage properties of multi-principal-component CoFeMnTixVyZrz alloys. *International Journal of Hydrogen Energy*, 2010, 35: 9046–9059. https://doi.org/10.1016/j.ijhydene.2010.06.012

116 Kunce I, Polanski M, Bystrzycki J. Structure and hydrogen storage properties of a high entropy ZrTiVCrFeNi alloy synthesized using laser engineered net shaping (LENS). *International Journal of Hydrogen Energy*, 2013, 38: 12180–12189. https://doi.org/10.1016/j.ijhydene.2013.05.071

117 Kunce I, Polanski M, Bystrzycki J. Microstructure and hydrogen storage properties of a TiZrNbMoV high entropy alloy synthesized using laser engineered net shaping (LENS). *International Journal of Hydrogen Energy*, 2014, 39: 9904–9910. https://doi.org/10.1016/j.ijhydene.2014.02.067

118 Floriano R, Zepon G, Edalati K, et al. Hydrogen storage in TiZrNbFeNi high entropy alloys, designed by thermodynamic calculations. *International Journal of Hydrogen Energy*, 2020, 45: 33759–33770. https://doi.org/10.1016/j.ijhydene.2020.09.047

119 Strozi R B, Leiva D R, Zepon G, et al. Effects of the chromium content in $(TiVNb)_{100-x}Cr_x$ body-centered cubic high entropy alloys designed for hydrogen storage applications. *Energies*, 2021, 14: 3068.

120 Chang L, Jing H, Liu C, et al. High-entropy materials for prospective biomedical applications: challenges and opportunities. *Advanced Science*, 2024, 11: 2406521. https://doi.org/10.1002/advs.202406521

121 Ai Y, He M Q, Sun H, et al. Ultra-small high-entropy alloy nanoparticles: efficient nanozyme for enhancing tumor photothermal therapy. *Advanced Materials*, 2023, 35: 2302335. https://doi.org/10.1002/adma.2020233

7

High-Entropy Materials for Thermal-Catalytic Applications

*Huibin Wang, Pengfei Xie**

College of Chemical and Biological Engineering, Zhejiang University, Hangzhou, China

*Corresponding email: pfxie@zju.edu.cn

7.1 Introduction to Thermal Catalysis and Important Concepts

7.1.1 Definition and Background

Thermal catalysis accelerates chemical reactions by using a catalyst that lowers the activation energy, allowing reactions to proceed more quickly at elevated temperatures. The catalyst provides an alternative reaction pathway and is not consumed in the process, enabling repeated use (Figure 7.1) [1]. Catalysis has significantly impacted science and industry. Early examples include using yeast for fermentation in bread and alcohol production. The Haber–Bosch process, developed in the early 20th century, used iron catalysts to produce ammonia from nitrogen and hydrogen, revolutionizing synthetic fertilizer production. Zeolites, discovered mid-20th century, transformed the petroleum refining industry by aiding gasoline cracking. Catalytic technologies, like those in car converters, have also helped reduce environmental pollution [2–4].

Today, catalysis plays a key role in green chemistry, with a focus on developing catalysts that reduce harmful by-products, lower energy use, and utilize renewable feedstocks. Research into CO_2 conversion catalysis is a prime example. As traditional catalysts face performance and sustainability challenges, the introduction of high-entropy materials (HEMs) marks a significant advancement in catalyst design, offering enhanced stability and functionality [5–10].

The introduction of HEMs in catalysis heralds a new era in catalyst design. Comprising five or more elements in nearly equiatomic proportions, HEMs present a novel paradigm owing to their exceptional properties, including high entropy,

Functional High Entropy Alloys and Compounds, First Edition. Edited by Yong Zhang, Yonggang Yao and Haojie Liu.

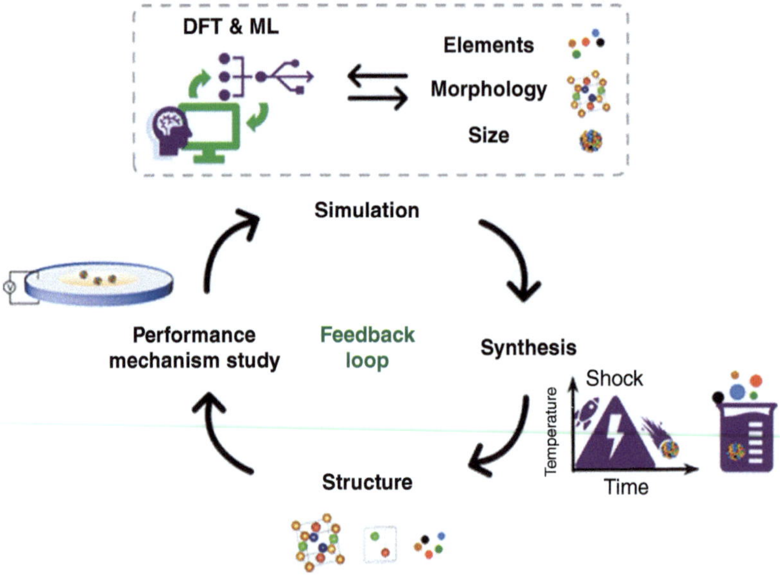

Figure 7.1 A strategy for the design of HEAs in catalysis. [1] / with permission of the American Chemical Society.

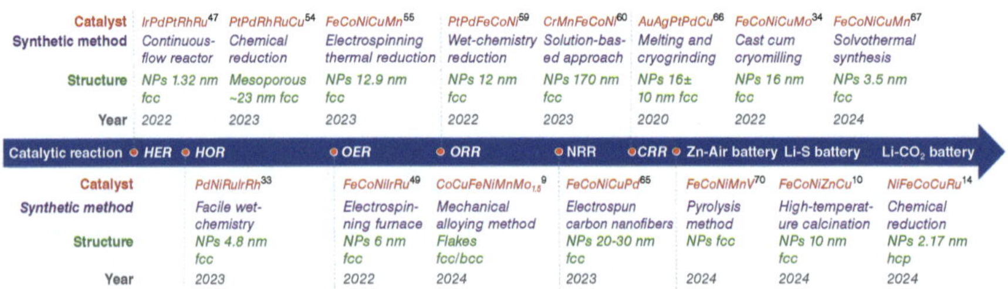

Figure 7.2 State-of-the-art HEAs as catalysts in various applications. [1] / with permission of the American Chemical Society.

intricate structures, and tunable electronic characteristics. They exhibit immense potential across diverse applications, ranging from energy conversion to environmental catalysis [11–19] (Figure 7.2).

7.1.2 The Role of Catalysts in Thermal Reactions

Catalysts play a critical role in industrial chemical processes by accelerating reactions without being consumed, offering economic and environmental benefits. They enable reactions under milder conditions—lower temperatures and pressures—reducing energy consumption and infrastructure costs. Their selectivity minimizes by-product formation, improving product quality and reducing waste, contributing to sustainable processes in line with green chemistry principles. Additionally, catalysts can safely handle hazardous materials, enhancing the safety of industrial

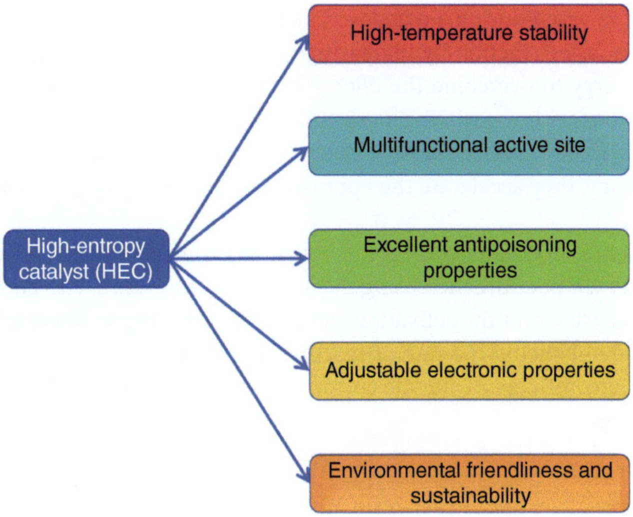

Figure 7.3 Advantages of high-entropy catalyst in thermal reaction.

processes. In thermal reactions, catalysts are crucial, though conventional ones face challenges like sintering and deactivation at high temperatures. HEMs offer promising solutions, outperforming traditional catalysts due to their unique structure and diverse composition. Key advantages of HEMs in thermal reactions include the following (Figure 7.3).

High-temperature stability: HEMs maintain stability at elevated temperatures due to their multimetal composition and high-entropy effect, preventing sintering and deactivation, thus ensuring long-term catalytic activity [20–23].

Multifunctional active sites: HEMs feature a wide range of active sites, enabling them to catalyze multiple reaction steps simultaneously, improving efficiency in processes like thermal cracking and reforming [24, 25].

Antipoisoning properties: The multicomponent nature of HEMs makes them resistant to poisoning, with unaffected components retaining catalytic activity, thus prolonging catalyst life and maintaining performance in the presence of impurities or toxic components.

Adjustable electronic properties: By modifying their composition, HEMs allow precise control over electronic structures, optimizing reaction performance and improving selectivity in thermal reactions.

Environmental and economic sustainability: HEMs reduce reliance on rare or precious metals, lowering costs and minimizing the environmental impact of resource extraction. Their stability and longevity further reduce catalyst replacement frequency, decreasing environmental burdens.

7.1.3 Thermodynamics and Kinetics in Catalysis

In catalysis, both enthalpy (ΔH) and entropy (ΔS) changes play vital roles in determining the rate and equilibrium of reactions. Enthalpy changes reflect the heat absorbed or released during a reaction. Catalysts lower the activation energy (ΔH), facilitating a faster reaction by increasing the fraction of reactant molecules with sufficient energy to overcome the energy barrier. Entropy, a measure of disorder, favors reactions with a positive change ($\Delta S > 0$) as it indicates increased randomness, which is thermodynamically favorable [26]. While catalysts don't directly change entropy, they accelerate the approach to equilibrium by increasing reaction rates.

The Arrhenius equation describes how temperature affects the rate constant (k), with a higher temperature increasing the number of molecules with enough energy to react. Catalysts lower the activation energy (E_a), providing an alternative reaction pathway and enhancing the rate of reaction. In HEMs, both thermodynamic and dynamic factors influence catalytic performance due to their complex multicomponent structure.

$$k = A \cdot e^{-\frac{E_a}{RT}}$$

where

k is the rate constant of the reaction.
A is the pre-exponential factor.
E_a is the activation energy of the reaction.
R is the universal gas constant.
T is the temperature in Kelvin.

High-entropy alloys (HEAs) have unique catalytic properties due to their diverse compositions, which affect both enthalpy and entropy changes. HEAs often exhibit low enthalpy changes, reducing activation energy and increasing reaction rates. The multicomponent nature of HEAs provides multiple active sites, further lowering activation energy. Additionally, their high configuration entropy (ΔS) stabilizes the material, improving resistance to poisoning and sintering, which helps maintain active sites. HEAs' complex surfaces also promote molecular adsorption and dissociation, increasing disorder in the reaction system and improving entropy. According to the Arrhenius equation, temperature increases the rate constant, and HEAs accelerate reactions by reducing E_a and offering multiple reaction pathways, selecting the lowest-energy path for the reaction. This multipathway feature boosts catalytic efficiency.

In summary, the multicomponent properties of HEMs enable faster reactions by reducing activation energy and increasing entropy. These materials excel in improving reaction stability and antipoisoning properties, making them highly effective catalysts. While catalysts don't alter the fundamental thermodynamics (ΔH and ΔS) of a reaction, they create an alternative route with reduced activation energy, significantly enhancing the reaction rate (Figure 7.4).

7.1.4 Catalyst Activation and Deactivation

Catalyst activation refers to the preparation of a catalyst to optimize its performance, typically through reduction, calcination, and the addition of additives [27, 28].

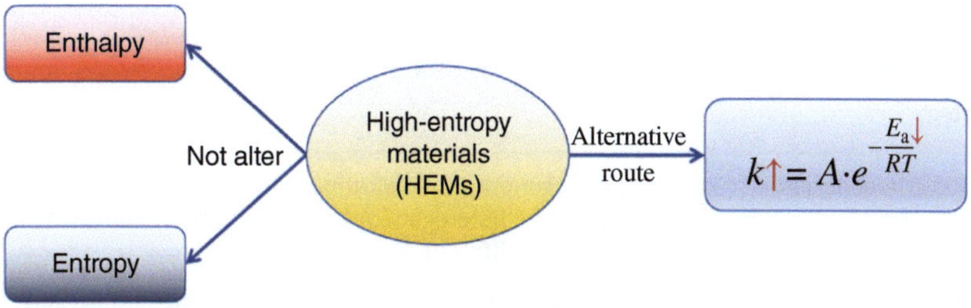

Figure 7.4 Effects of HEMs on entropy and enthalpy and on reaction thermodynamics and kinetics.

Reduction is the process of converting metal oxides into their metallic form using a reducing agent, such as hydrogen. This is crucial for many metal catalysts, as the metallic state often has the desired active sites. In HEA catalysts, reduction exposes more active sites and may lead to the formation of multiple active metal sites, enhancing catalytic performance.

Calcination involves heating a solid material at high temperatures, usually in the air, to remove volatile impurities and form the desired catalytic phase. For HEA catalysts, calcination not only removes impurities but also optimizes the surface structure and pore distribution, improving surface area and stability. The complex multicomponent nature of HEAs helps form a stable multiphase structure during calcination, enhancing thermal stability.

Additives are small substances added to catalysts to improve activity, selectivity, or stability. They can modify the surface properties of the catalyst, prevent the sintering of active sites, provide additional active sites or form new catalytic phases, and alter the electronic properties to adjust reactant or intermediate adsorption.

In HEA catalysts, additives play a key role in stabilizing polymetallic components, improving overall performance. They can also prevent the separation or deactivation of certain components at high temperatures. Catalyst deactivation refers to the loss of catalytic activity, which can be temporary or permanent. Key deactivation mechanisms include the following:

Deactivation mechanism	
Poisoning	Occurs when substances like sulfur or lead irreversibly adsorb to the active site. In HEAs, the diversity of components helps mitigate poisoning by allowing certain metals to tolerate poisons or preferentially adsorb them, protecting other sites
Sintering	The merging of catalyst particles at high temperatures, reducing active sites. HEAs are resistant to sintering due to their high configuration entropy and polyphase structure, maintaining activity at high temperatures
Carbonization	The deposition of carbon on the catalyst surface, which blocks active sites. HEA catalysts resist carbon deposition because their complex surfaces hinder carbon formation, and their diverse active sites help disperse the effects of carbon buildup

Strategies to slow deactivation include the following:

Strategy	
Periodic Regeneration	Regular regeneration to eliminate poisons or reverse sintering. The heterogeneous nature of HEA catalysts aids in regenerating different active sites
Enhanced Thermal Stability	Improve thermal stability to reduce sintering and carbon accumulation
Cleaner Raw Materials	Minimize poisoning risks by trapping impurities
Adjust Reaction Conditions	Minimize carbon accumulation and sintering by optimizing reaction parameters
Use of Inhibitors	Prevent carbon buildup during the reaction

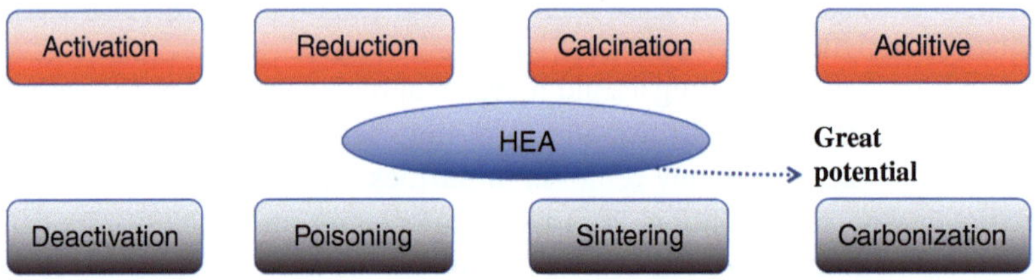

Figure 7.5 Potential of HEA catalysts in catalyst activation and deactivation.

Understanding catalyst activation and deactivation mechanisms allows chemists to develop strategies to maintain catalyst performance and extend its life. HEAs, with their multicomponent, high-entropy, and multiphase properties, show significant potential in improving catalyst stability and resistance to deactivation (Figure 7.5).

7.1.5 Selectivity and Activity in Thermal Catalysis

In the context of catalysis, activity, selectivity, and stability are key performance indicators that are used to evaluate and compare the effectiveness of catalysts. These parameters are critical in the design and optimization of catalysts for various chemical reactions (Figure 7.6).

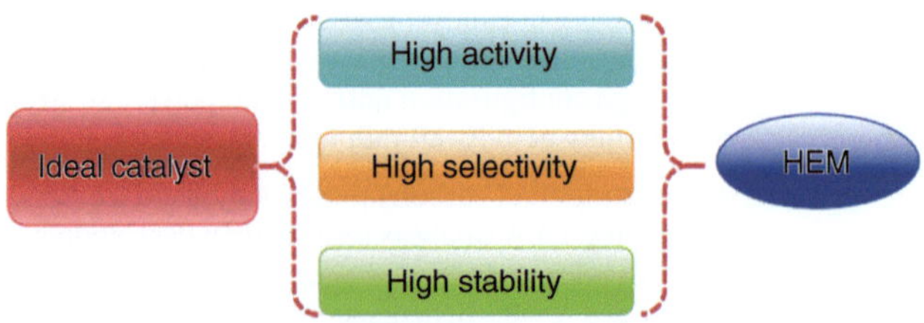

Figure 7.6 Advantages of HEMs in terms of activity, selectivity, and stability.

Activity refers to the rate at which a catalyst promotes a chemical reaction. A highly active catalyst can significantly increase the reaction rate, thereby speeding up the conversion of reactants to products. The activity of the catalyst is usually measured by determining the initial rate of the reaction under standard conditions, which can be achieved by various analytical techniques, such as gas chromatography, mass spectrometry, or spectroscopic methods.

HEM catalysts often provide more active sites due to their complex polymetallic composition. These diverse active sites can effectively promote the adsorption and activation of reactants, thereby improving the activity of the reaction. In addition, there is a synergistic effect between the components of the HEM, which can optimize the activity of the catalyst under different reaction conditions and provide multiple possible paths for the reaction to reduce the overall energy barrier.

Selectivity is a measure of the ability of a catalyst to direct the conversion of a reactant to a particular product or set of products. In reactions with multiple possible products, selectivity is critical because it quantifies the catalyst's preference for a particular reaction path. The polymetallic components of HEMs give their surfaces complex structures, and these different active sites and interactions between metals can selectively promote some reaction paths and inhibit others, significantly improving the selectivity of specific products. For example, by adjusting the metal composition in a HEM, its adsorption capacity for a specific reactant can be optimized to achieve high selectivity for the target product.

Stability refers to the length of time that a catalyst can maintain its activity and selectivity under a given reaction condition. Stable catalysts do not deactivate or lose their performance characteristics over a longer period of time. HEMs have a unique advantage in this regard, and their high configuration entropy makes them have better structural stability in high-temperature and complex reaction environments, thus reducing the possibility of sintering and poisoning. This allows HEMs to maintain high activity and selectivity at high temperatures and harsh reaction conditions, reducing the risk of catalyst deactivation.

The design of the catalyst is a very important step, catalysts are designed to optimize these parameters to meet the specific needs of chemical processes:

1. HEMs have unique advantages in catalyst design; through complex combinations of multiple components, they demonstrate a high degree of activity and flexibility under different reaction conditions. The synergistic effect of different metal elements in HEMs can effectively improve the reaction rate, thereby increasing production efficiency, reducing energy consumption, and accelerating reaction time.
2. High selectivity is essential to reduce the generation of by-products, which helps reduce the need for subsequent treatment and purification steps, thereby reducing the cost and environmental impact of the overall process. The diverse components in a HEM can optimize selectivity for the target reaction path through different interactions, thereby reducing unwanted by-products.
3. The high stability of HEMs comes from the interaction between their multiple components and the high configuration entropy, which allows them to maintain good catalytic performance even under high temperatures or harsh

conditions, thereby reducing the frequency of catalyst replacement and operating costs and maintaining consistent product quality.

In practical application, the ideal catalyst should have high activity, high selectivity, and high stability. However, there are often trade-offs between these three factors. For example, a catalyst may be highly active and selective but unstable. Or it may be stable and highly selective but less active. Therefore, the design of the catalyst requires a careful balance between these properties to meet the needs of the specific application. For HEMs, these characteristics can be optimized to a great extent by adjusting the proportion and structure of the polymetallic components, and the ideal catalytic performance can be achieved. In addition, measuring these parameters can provide important insights into reaction mechanisms and guide the modification of catalysts to improve their performance.

7.2 Catalysts Evolution to HEMs

7.2.1 Transition from Single-Component to Multicomponent Catalysts

The evolution toward multicomponent catalysts has been a significant trend in the field of catalysis, driven by the demand for improved performance, enhanced selectivity, and the development of new functionalities. This shift is a result of the continuous quest to optimize chemical reactions for various applications, from industrial production to environmental management. The activity of a catalyst is crucial for its commercial viability, as it determines the rate at which a chemical reaction proceeds. Early catalysts were often based on single metals or simple metal oxides. However, it became apparent that combining multiple metals or elements could lead to synergistic effects that enhanced the overall activity. For instance, bimetallic catalysts often exhibit higher activity than their monometallic counterparts due to electronic and geometric effects that alter the reactivity of the surface. Selectivity refers to the ability of a catalyst to favor the formation of a specific product from a set of possible reactions. Multicomponent catalysts can provide enhanced selectivity by stabilizing certain reaction intermediates or transition states over others, leading to a preferred reaction pathway. This is particularly important in the production of fine chemicals and pharmaceuticals, where the desired product must be synthesized with high purity [29, 30].

As new chemical transformations and processes are developed, there is a need for catalysts with novel functionalities to enable these reactions. Multicomponent catalysts can provide these functionalities by combining the properties of different metals or elements. For example, the integration of a redox-active component with a metal can create a catalyst capable of facilitating multielectron processes, such as the oxygen reduction reaction (ORR) in fuel cells [31]. Some catalysts are susceptible to poisoning, where specific adsorbates strongly bind to the catalyst surface and inhibit its activity. Multicomponent catalysts can be designed to resist poisoning by distributing the active sites among different types of metal atoms, making it more

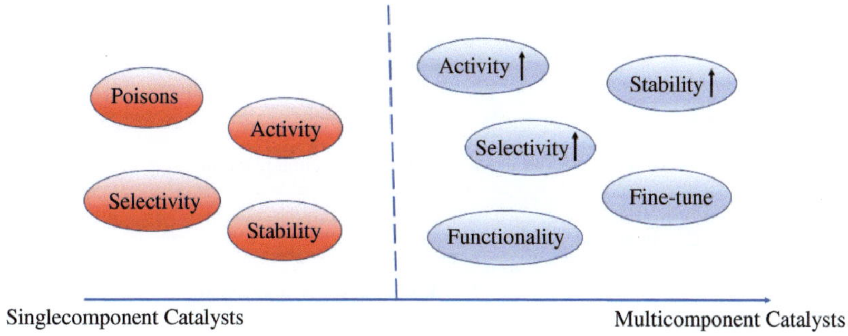

Figure 7.7 Evolution of single-component catalyst to multicomponent catalyst.

difficult for poisons to occupy all the active sites. Catalyst stability is essential for long-term operation, and multicomponent catalysts often exhibit enhanced stability due to their compositional diversity. For example, alloys or intermetallic compounds formed from multiple metals can be more resistant to sintering and other forms of degradation.

One of the key advantages of multicomponent catalysts is the ability to fine-tune their properties. By adjusting the ratio of different metals or elements or by introducing promoters and stabilizers, the electronic, geometric, and chemical properties of the catalyst can be modulated to achieve the desired performance. With the advent of high-throughput screening, computational-aided design has helped a lot; the growth of computational chemistry and materials science has enabled the prediction and simulation of catalysts' behavior before they are synthesized. This has facilitated the rational design of multicomponent catalysts with tailored properties. The environment and sustainability are also important drivers in the catalyst development process. The need for sustainable and environmentally friendly processes has driven the development of multicomponent catalysts that can operate under milder conditions, utilize renewable feedstocks, and minimize waste (Figure 7.7).

Based on the potential advantages of HEAs in catalysis, HEAs have shown promise in a variety of catalytic applications, including hydrogen production in both acidic and alkaline environments; ORR in fuel cells; carbon dioxide (CO_2) conversion to valuable chemicals, such as methanol and formate; ammonia decomposition into nitrogen and hydrogen, a process that could provide a source of clean hydrogen fuel; and so on (Figure 7.8). The study and application of HEAs in catalysis is a rapidly growing field, with researchers continually discovering new compositions and strategies to harness their unique properties for a wide range of chemical transformations [13, 32–36].

7.2.2 Strategies to Stabilize High-Entropy Catalysts

Stabilizing HEAs is crucial for their practical application, especially in demanding environments such as catalysis. Several strategies have been developed to enhance the stability of HEAs. Incorporating late transition metals into the alloy composition can improve the stability of HEAs. Late transition metals, such as platinum, palladium, and

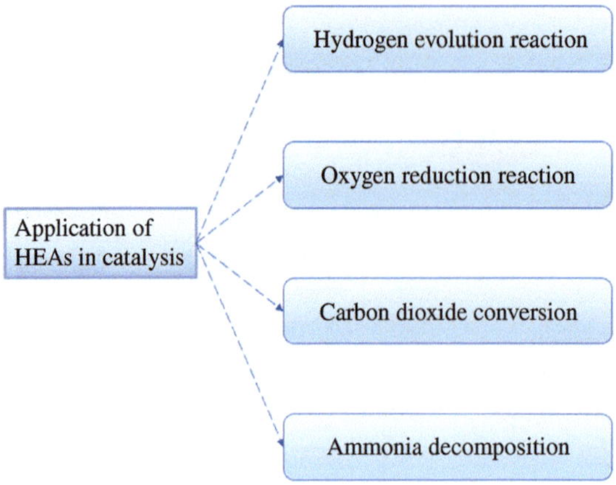

Figure 7.8 Application of HEAs in catalysis.

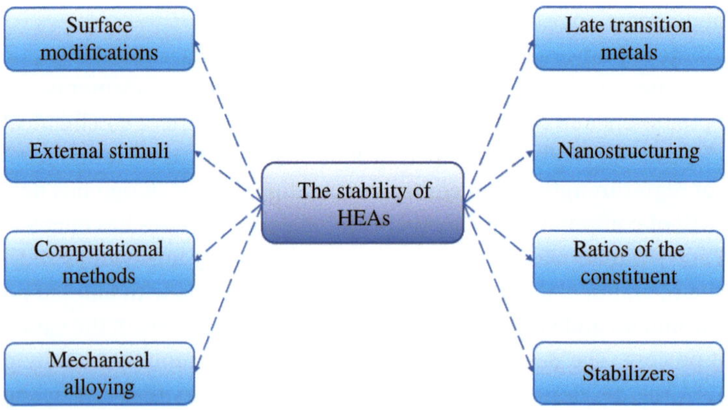

Figure 7.9 Several strategies improve the stability of HEAs.

iridium, often have a high resistance to corrosion and sintering, which can be beneficial for the stability of the alloy. These metals can alter the electronic structure of the HEA, leading to stronger metal–metal and metal–ligand bonds that resist phase segregation. Nanostructuring HEAs can enhance their thermal stability and mechanical properties. Nanoscale grains can provide additional barriers to phase separation by increasing the energy required for diffusion processes that lead to segregation. Furthermore, nanostructures can improve the catalytic performance of HEAs by increasing the number of active sites and enhancing mass transport properties (Figure 7.9).

The synthesis conditions play a vital role in determining the stability of HEAs. Parameters such as temperature, pressure, cooling rate, and the presence of specific elements during synthesis can significantly affect the final microstructure and stability of the alloy. For example, rapid solidification processes can "freeze in" the high-entropy state of the alloy, preventing phase separation. Adjusting the concentration ratios of the constituent elements can help stabilize certain HEAs.

By fine-tuning the concentrations, it is possible to optimize the balance between entropy and enthalpy, leading to a more stable solid solution. The addition of certain elements, known as stabilizers or modifiers, can help stabilize the HEA structure. These elements can alter the crystallography of the alloy, promote the formation of specific phases, or enhance the alloy's resistance to environmental degradation.

Understanding the microstructure and phase distribution of HEAs is essential for their stabilization. Advanced microscopy and spectroscopy techniques can provide insights into the alloy's structure at the atomic level, guiding the development of stabilization strategies. Surface modifications, such as the introduction of specific adsorbates or coatings, can enhance the stability of HEAs by passivating surface sites that might otherwise be prone to degradation or sintering. The application of external stimuli such as electric or magnetic fields can influence the stability of HEAs. These fields can affect the movement and arrangement of atoms within the alloy, potentially stabilizing certain phases or configurations. Computational methods, including density functional theory (DFT) and molecular dynamics (MDs) simulations, can predict the stability of different HEA compositions and guide how to design more stable alloys. Mechanical alloying, a method that involves grinding and mixing the elemental powders in a high-energy ball mill, can lead to the formation of amorphous or nanostructured HEAs that exhibit improved stability.

These strategies can be applied individually or in combination to develop HEAs with the desired stability for specific applications. The stabilization of HEAs is an active area of research, and ongoing work aims to uncover new methods and improve existing approaches to enhance their performance and durability.

7.3 Typical Reactions and Their High-Entropy Catalysts

7.3.1 Hydrotreatment Reactions

Semihydrogenation of Alkynes. HEMs have emerged as promising catalysts for the semihydrogenation of alkynes, a reaction crucial for producing alkenes with high selectivity. The unique composition and structure of HEMs enable enhanced catalytic performance through synergistic effects among multiple metal components. Liu et al. [37]. proposed that PdFeCoNiCu (PdM) HEM catalysts exhibited remarkable performance variations depending on their local chemical order. The highly ordered PdM catalysts demonstrated exceptional selectivity for semihydrogenation of nonalkyl-substituted alkynes, achieving nearly full conversion with 93–98% selectivity. This superior performance was attributed to the increased electron density of surface Pd atoms, which enhanced the adsorption strength of alkynes without leading to overhydrogenation. However, for alkyl-substituted alkynes, the activity of the highly ordered PdM catalysts decreased significantly due to the steric hindrance imposed by the alkyl groups, which weakened the adsorption of alkynes on the catalyst surface. In contrast, less-ordered PdM catalysts showed higher overall activity but lower selectivity for nonalkyl-substituted alkynes, often resulting in

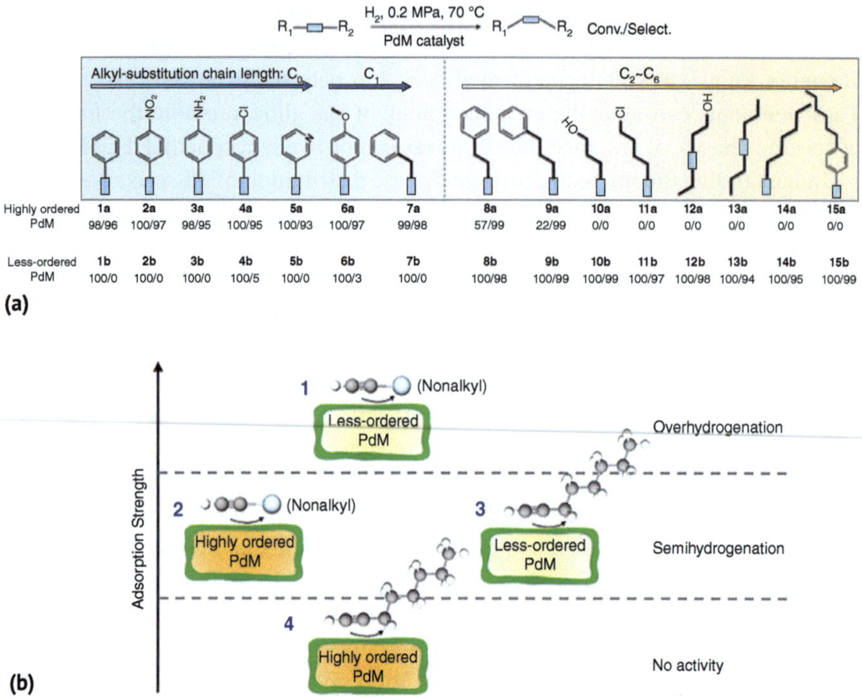

Figure 7.10 (a) Semihydrogenation of alkynes. The chain lengths of substituted alkyls: C_0–C_1 (1–7) and C_2–C_6 (8–15). Conversion/selectivity values are shown. (b) Schematic showing the relative adsorption strength of alkyl- and nonalkyl-substituted alkynes on core–shell high-entropy intermetallic (HEI) PdM with different chemical orders. [37] / with permission of American Chemical Society.

overhydrogenation. Interestingly, these less-ordered catalysts exhibited high selectivity for alkyl-substituted alkynes, suggesting that the disordered surface provided a more favorable environment for selective hydrogenation despite the presence of bulky alkyl groups. This study highlights the delicate balance between surface electronic properties and steric effects in determining the catalytic performance of HEM catalysts [38] (Figure 7.10).

Furthermore, Liu et al [37]. emphasized the importance of composition in tuning the near-surface structure and electronic properties in Pd_xFeCoNiCu (Pd_x-HEA) catalysts. The performance of Pdx-HEA catalysts showed a "volcano" plot with respect to Pd content, with Pd_{21}-HEA emerging as the optimal catalyst. This catalyst exhibited high resistance to overhydrogenation and exceptional activity for a broad range of alkynes. The superior performance of Pd_{21}-HEA was attributed to two key factors: (1) the appropriate surface lattice expansion induced by the rational substitution of Pd atoms, which optimized the adsorption energy of alkynes. (2) the unique electronic environment around Pd sites, characterized by the maximum electron transfer from Co/Ni to Cu/Pd. This electron transfer resulted in a more negatively charged surface Pd in Pd_{21}-HEA, which weakened the adsorption of intermediates and thus inhibited overhydrogenation. The study also demonstrated the high durability of the Pd_{21}-HEA catalyst, maintaining its performance over multiple reaction cycles. This robustness is essential for practical applications, where catalyst stability is a critical factor [36].

Overall, these studies collectively demonstrate the potential of HEM catalysts to achieve high selectivity and activity in alkyne semihydrogenation. By manipulating the local chemical order, surface lattice strain, and electronic properties, these catalysts can be fine-tuned to meet specific industrial needs. The findings not only provide a deeper understanding of the structure–performance relationship in HEM catalysts but also offer practical guidelines for designing next-generation catalysts with enhanced efficiency and selectivity. Future research in this area could focus on exploring other high-entropy compositions and further optimizing synthesis methods to achieve even better catalytic performance.

CO_2 Hydrogenation Reaction. Many studies on the CO_2 hydrogenation reaction using high HEA-NPs as catalysts have demonstrated remarkable performance improvements. For instance, the CoNiCuRuPd HEA-NPs synthesized on a TiO_2 support exhibited exceptional catalytic activity and stability, outperforming monometallic Pd/TiO_2 and other supports like MgO and Al_2O_3. During the CO_2 hydrogenation reaction conducted at atmospheric pressure and temperatures ranging from 300 to 400 °C, the HEA-NP catalyst showed a significantly higher yield of hydrogenated products, with methane selectivity reaching up to 68.3% at 400 °C. This superior performance was attributed to the unique electronic properties of the HEA surface, which promoted stronger interactions with CO intermediates, leading to higher methane selectivity. The apparent activation energy for CO_2 hydrogenation over CoNiCuRuPd/TiO_2 was determined to be 37.7 kJ/mol, lower than that for Pd/TiO_2 (44.2 kJ/mol), indicating a more favorable reaction pathway. Additionally, the HEA-NPs demonstrated remarkable structural robustness, with minimal particle growth and maintained elemental distribution even after prolonged use, highlighting their potential for practical applications in sustainable energy conversion processes.

The outstanding performance of CoNiCuRuPd HEA-NPs in the CO_2 hydrogenation reaction can be attributed to several key mechanisms that enhance both activity and stability. First, the hydrogen spillover effect on TiO_2 plays a crucial role in the formation and performance of the HEA-NPs. The process begins with the partial reduction of Pd^{2+} precursors to form nuclei, followed by the dissociation of hydrogen molecules on these Pd nuclei. The resulting active hydrogen atoms migrate over the TiO_2 surface via a coupled proton–electron transfer mechanism, facilitating the simultaneous reduction of other metal precursors (Co, Ni, Cu, Ru). This leads to the formation of HEA-NPs with a uniform distribution of all five elements, ensuring a high surface area and enhanced catalytic activity. The hydrogen spillover effect not only enables the low-temperature synthesis of HEA-NPs but also ensures a more homogeneous and rapid reduction process compared to non-reducible supports like Al_2O_3.

Second, the unique electronic properties of the HEA surface significantly impact the catalytic performance. The CoNiCuRuPd HEA-NPs exhibit stronger interactions with CO intermediates compared to monometallic Pd/TiO_2. This is evidenced by the lower desorption temperature of CO on CoNiCuRuPd/TiO_2, which promotes further hydrogenation to form methane. The stronger binding strength of CO on the HEA surface delays its desorption, allowing more opportunities for hydrogenation reactions to occur. DFT calculations support this observation, showing that the adsorption energy of CO on the HEA surface is higher than that on pure Pd, leading to higher methane selectivity.

Third, the structural robustness of HEA-NPs contributes to their exceptional stability during the CO_2 hydrogenation reaction. The sluggish diffusion effect in HEA-NPs, originating from the mixing of multiple elements and lattice distortion, suppresses particle growth and agglomeration. This is critical for maintaining high catalytic activity over prolonged periods. Transmission electron microscopy observations under electron beam irradiation revealed minimal structural changes in CoNiCuRuPd/TiO_2 compared to Pd/TiO_2, highlighting the superior resistance of HEA-NPs against radiation damage. DFT and MD simulations further confirmed that the HEA-NPs have higher cohesive energy and lower diffusion coefficients compared to monometallic clusters, reinforcing their durability.

In summary, the combination of hydrogen spillover-driven synthesis, unique electronic properties, and structural robustness of CoNiCuRuPd HEA-NPs on TiO_2 provides a comprehensive explanation for their enhanced performance in the CO_2 hydrogenation reaction. These mechanisms not only enable high activity and selectivity but also ensure long-term stability, making HEA-NPs a promising candidate for sustainable CO_2 conversion technologies (Figure 7.11).

7.3.2 Oxidation and Ammoxidation

HEAs have attracted significant interest for their potential in selective oxidation reactions, crucial for producing fine chemicals and intermediates in industries such as pharmaceuticals, agrochemicals, and fragrances. Their multimetal structure allows for the fine-tuning of electronic and geometric properties to enhance catalytic selectivity, enabling more efficient control over reaction pathways. The synergistic effects between different metals, where each metal plays a role in activating either the oxidizing agent or the substrate, significantly improve overall efficiency and product selectivity. Additionally, HEAs exhibit excellent thermal stability and resistance to sintering, which ensures sustained catalytic performance under extreme reaction conditions. Examples include vanadium-based HEAs for oxidative dehydrogenation; iron-, cobalt-, nickel-, and chromium-based HEAs for alcohol oxidation to aldehydes or ketones; and platinum-based HEAs for ammoxidation reactions, such as acrylonitrile production from propane. These attributes make HEAs a versatile and robust platform for selective oxidation, offering high efficiency, stability, and selectivity (Figure 7.12).

Methane Combustion Reaction. Conventional PdO_x catalysts used in methane (CH_4) combustion often face deactivation issues due to nanoparticle sintering and the transformation of PdO_x into metallic Pd during the reaction. To enhance the energy efficiency of the catalytic CH_4 combustion process, developing Pd-based catalysts with improved performance and stability is essential. Li et al. introduced a rapid, non-equilibrium method to synthesize multielement oxide (MEO) nanoparticles with a single-phase structure for catalytic CH_4 combustion. By incorporating

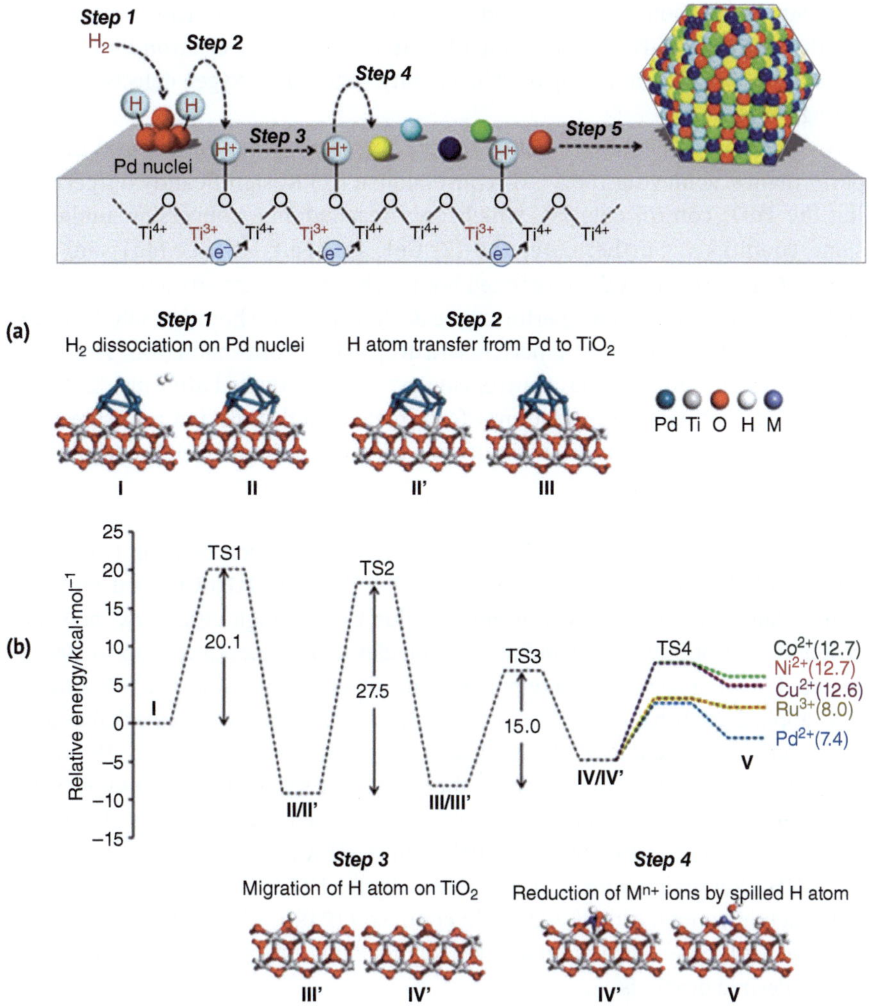

Figure 7.11 (a) Schematic illustration of the elementary steps. (b) The potential energy profile of processes on the $TiO_2(101)$ as obtained from DFT calculations. The values in parentheses are the calculated energy barriers for each M^{n+} cation in Step 4. [32] / Springer Nature / CC BY 4.0.

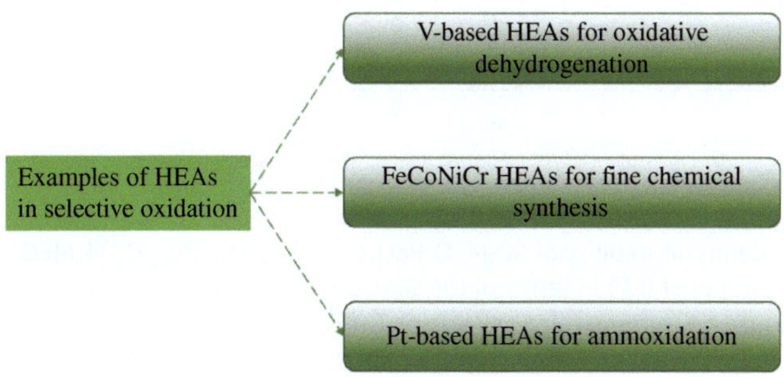

Figure 7.12 Several examples of HEAs in selective oxidation.

elements from various groups, including alkali metals, 3d–5d transition metals, and Pd, they created catalysts with distinct functions: promoting electron transfer during redox processes, improving redox capability, generating oxygen defects, and activating CH_4 [40].

Among the MEO catalysts, $(Zr,Ce)_{0.6}Mg_{0.3}Pd_{0.1}O_x$ (4-MEO-Pd) showed the best performance, achieving 100% CH_4 conversion at 823 K, significantly outperforming the PdO_x control catalyst, which only reached 74% conversion under the same conditions. Furthermore, the $(Zr,Ce)_{0.6}(Mg,La,Y,Hf,Ti,Cr,Mn)_{0.3}Pd_{0.1}O_{2-x}$ (10-MEO-PdO) catalyst demonstrated both high catalytic activity and exceptional stability, with no noticeable performance decline after 100 hours at 648 K. In contrast, the 4-MEO-Pd catalyst experienced a drop in CH_4 conversion from 31% to 20% after 26 hours, and the PdO_x control catalyst was deactivated after just 22 hours. This rapid synthesis method allows for targeted catalyst design and screening, and high-entropy MEOs show significant promise as robust, high-performance catalysts (Figure 7.13).

NH_3 Oxidation Reaction. In the ammonia (NH_3) oxidation reaction (AOR), the undesirable by-products are N_2O and N_2 [13, 35]. While PtPdRh-based multimetallic catalysts are widely used in industry due to their high efficiency, their high cost and susceptibility to degradation at the elevated temperatures (up to 800 °C) required to promote NO_x yield remain significant challenges. Yao et al. [41] successfully applied PtPdRhRuCe HEA nanoparticles in AOR, achieving remarkable activity, selectivity, and stability. The individual contributions of Pt, Pd, and Rh provide excellent catalytic performance, while the addition of Ru and Ce reduces the overall cost and enhances the catalyst's performance through the formation of a highly homogeneous alloy structure. The PtPdRhRuCe HEA nanoparticles demonstrated 100% NH_3 conversion with over 99% selectivity for $NO_x(NO + NO_2)$ at a relatively low reaction temperature of 700 °C. In contrast, PtPdRhRuCe multimetallic nanoparticles with phase separation showed significantly lower activity and selectivity for the desired products.

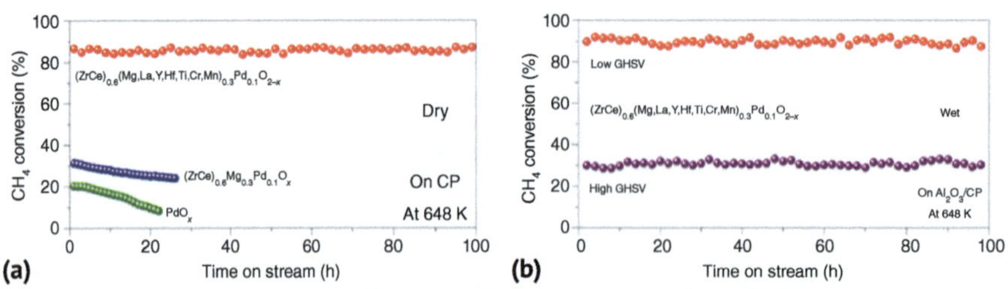

Figure 7.13 (a) Catalytic stability of 10-MEO-PdO, $(Zr,Ce)_{0.6}Mg_{0.3}Pd_{0.1}O_x$ (4-MEO-Pd) and PdO_x on the carbon paper (CP) in terms of the samples' CH_4 conversion at 648 K [38] under dry conditions using a Gas Hourly Space Velocity (GHSV) of 10,800 $l\,g_{Pd}^{-1}\,h^{-1}$. (b) Catalytic stability of 10-MEO-PdO on the Al_2O_3-coated CP performed under wet conditions at 648 K using different GHSVs of 10,800 $l\,l_{Pd}^{-1}\,h^{-1}$ and 108,000 $L\,g_{Pd}^{-1}\,h^{-1}$. [38] / with permission of Springer Nature.

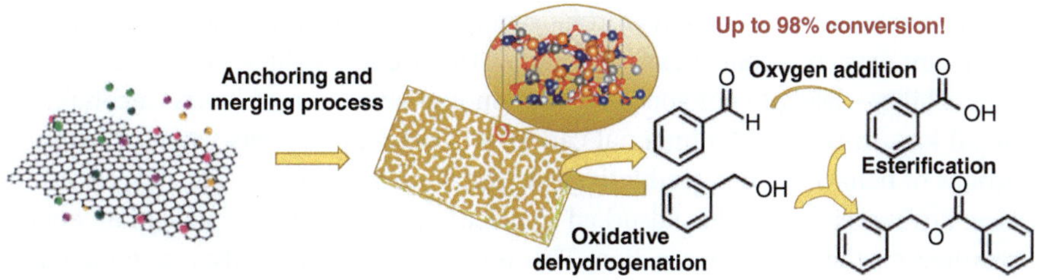

Figure 7.14 Schematic on the oxidation of benzyl alcohol by HEOs. [18] / with permission of Elsevier.

Benzyl Alcohol Oxidation. The high-entropy metal oxide (HEMO) catalyst, specifically the holey lamellar high-entropy oxide (HEO) material with the composition $Co_{0.2}Ni_{0.2}Cu_{0.2}Mg_{0.2}Zn_{0.2}O$, has demonstrated remarkable performance in the solvent-free aerobic oxidation of benzyl alcohol. Achieving an unprecedented conversion rate of up to 98% within just 2 hours under atmospheric conditions, this catalyst sets a new benchmark for benzyl alcohol oxidation efficiency. The exceptional performance of the HEMO catalyst can be attributed to several key factors that work synergistically to enhance its catalytic activity and selectivity (Figure 7.14).

The synergistic effect of multiple metal ions (Co, Ni, Cu, Mg, Zn) in near-equimolar ratios creates a unique catalytic environment. The random distribution of these metal ions on the atomic scale provides a diverse array of active sites that can interact with benzyl alcohol molecules in various ways, facilitating multiple reaction pathways and significantly boosting overall catalytic efficiency. This diversity in active sites allows for the simultaneous participation of each metal ion in the catalytic process, contributing to the high conversion rates observed.

The HEMO material is characterized by a high density of oxygen vacancies. These vacancies play a crucial role by providing additional active sites for the adsorption and activation of benzyl alcohol molecules. The presence of oxygen vacancies increases the surface reactivity of the catalyst, enabling stronger interactions with the substrate and facilitating the oxidation process. Moreover, these vacancies enhance the delivery of oxygen to the reaction sites, which is essential for the oxidation of benzyl alcohol to products like benzoic acid and benzaldehyde. The enhanced oxygen delivery capacity due to the presence of oxygen vacancies significantly lowers the activation energy required for the oxidation reaction, making the process more efficient.

The HEMO material has a porous, holey lamellar structure composed of interconnected nanoparticles. This unique structure significantly increases the surface area of the catalyst, providing a greater number of active sites for the reaction. The porous framework allows for better accessibility of benzyl alcohol molecules to the active sites, enhancing the overall reaction rate. The interconnected nanoparticles also facilitate the diffusion of reactants and products, ensuring efficient catalysis and contributing to the high activity observed. The high surface area and numerous active sites provided by the porous structure are critical for achieving the ultrahigh catalytic performance of the HEMO catalyst.

The combination of these features—synergistic effects from multiple metal ions, abundant oxygen vacancies, and a porous structure that enhances surface area and accessibility—leads to enhanced adsorption and activation of benzyl alcohol. The metal ions in the HEMO material can form strong interactions with the hydroxyl groups of benzyl alcohol, facilitating the initial adsorption step. The oxygen vacancies further stabilize the adsorbed molecules, lowering the activation energy required for the oxidation reaction. This results in efficient catalysis and high conversion rates.

Additionally, the HEMO catalyst offers the versatility of selective product formation. By adjusting the composition of the catalyst and the reaction conditions, the main product of the oxidation can be selectively optimized. For instance, under certain conditions, benzoic acid can be the dominant product, while, under others, benzaldehyde can be preferentially formed. This tunability is due to the diverse active sites and reaction pathways available in the HEMO material, allowing for precise control over the reaction outcome and making the catalyst highly adaptable for various industrial applications. Furthermore, the HEMO catalyst exhibits excellent stability and reusability. It maintains its performance over multiple reaction cycles without significant loss of activity, making it a practical choice for industrial applications. The robustness of the HEMO material under reaction conditions ensures that it can be used repeatedly without the need for frequent replacement, contributing to its cost-effectiveness and sustainability. In summary, the HEMO catalyst works effectively in the oxidation of benzyl alcohol through a combination of synergistic effects from multiple metal ions, the presence of abundant oxygen vacancies, and a porous structure that enhances surface area and accessibility. These features collectively contribute to its ultrahigh catalytic activity and selectivity, making it a promising candidate for sustainable and efficient oxidation processes. The exceptional performance, versatility, and stability of the HEMO catalyst highlight its significant potential for industrial applications in selective oxidation reactions, offering a sustainable and efficient catalytic solution.

7.3.3 Dehydrogenation and Decomposition Reaction

Propane dehydrogenation (PDH) is a crucial process for propylene production, which is in high demand globally. However, achieving high stability for Pt-based catalysts at temperatures ≥600 °C has been challenging due to issues such as coke formation and catalyst sintering. A novel class of catalysts based on HEIs has been developed, demonstrating exceptional performance in PDH. Specifically, the (PtCoCu)(GeGaSn)/Ca–SiO_2 catalyst, referred to as HEI (0.25), exhibited outstanding catalytic stability at 600 °C, with a mean catalyst life (τ) of 4146 hours (173 days), and almost no deactivation was observed over a period of 2 months. This remarkable stability is attributed to the unique structure of the HEI, which completely dilutes Pt–Pt ensembles using inert metals, thereby suppressing side reactions and coke formation (Figure 7.15) [38].

The catalytic mechanism of the HEI (0.25) catalyst is rooted in its multimetallic structure and site-isolation effect. The Pt atoms in the HEI are sufficiently isolated by Co and Cu, preventing undesired propylene decomposition and coke formation.

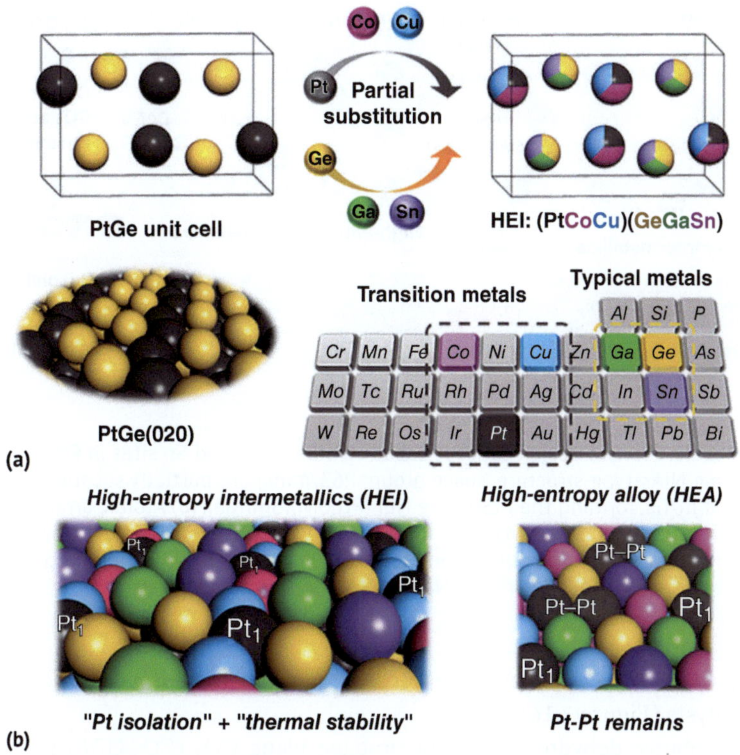

Figure 7.15 Catalyst design concept for thermally stable isolated Pt sites using the HEI. (a) Pt and Ge sites in PtGe are partially substituted by Co (Cu) and Ga (Sn), forming the PtGe-type HEI (PtCoCu)(GeGaSn). (b) Illustrations of the (020) surface of the (PtCoCu)(GeGaSn) HEI (left) and the (111) surface of the Pt-based senary HEA (right). [39] / with permission of American Chemical Society.

The incorporation of Ga and Sn further enhances the thermal stability of the catalyst due to the entropy effect. Detailed experimental studies and DFT calculations revealed that the isolated Pt sites in the HEI structure promote the desorption of propylene, inhibiting side reactions and maintaining high selectivity. The DFT calculations showed that the energy barrier for propylene desorption (ΔE) on the HEI surface is significantly higher than that on other bimetallic surfaces, indicating exceptional selectivity and stability. The Pt atoms act as the main active sites for C—H activation, while the presence of Co and Cu, along with the inert metals Ga and Sn, provides a stable environment that prevents Pt aggregation and enhances overall catalytic performance.

Moreover, the HEI (0.25) catalyst demonstrated excellent reusability and regenerability. After repeated regeneration processes involving O_2 and H_2 treatments, the catalyst showed no signs of deactivation, highlighting its durability. The long-term stability test in the presence of co-fed H_2 further confirmed the exceptional performance of the HEI (0.25) catalyst, with little deactivation observed over a period of 2 months. This outstanding stability, combined with high selectivity and activity, makes the HEI (0.25) catalyst a promising candidate

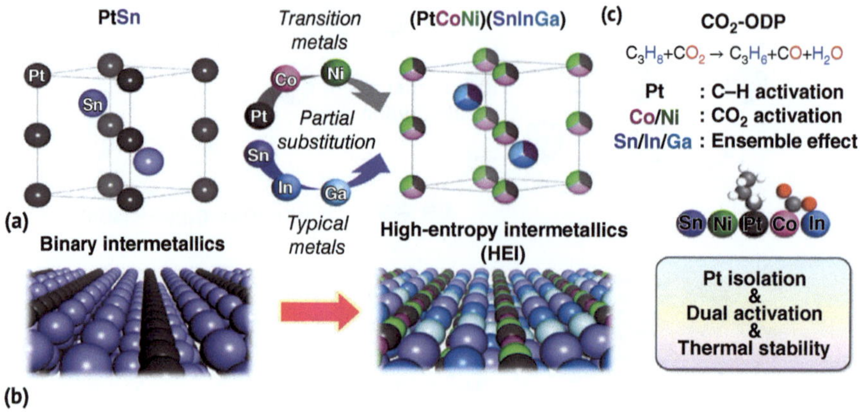

Figure 7.16 Catalyst design concept based on HEI. (a) Pt and Sn sites in PtSn (hexagonal NiAs-type structure, space group: P63/mmc) are partially substituted by Co/Ni and In/Ga, forming the PtSn-type HEI (PtCoNi)(SnInGa). (b) Atomic arrangement of the most stable (110) surfaces of PtSn (left) and (PtCoNi)(SnInGa) HEI (right). (c) Role of each metal and the effect of multimetallization on CO_2 oxidation and deoxygenation catalysis (CO_2-ODP). [40] / Springer Nature / CC BY 4.0.

for industrial PDH applications, offering a significant improvement over existing catalysts (Figure 7.16).

The oxidative dehydrogenation of propane using CO_2 (CO_2-ODP) is a highly promising process for the production of propylene while simultaneously utilizing CO_2, contributing to a more sustainable and carbon-neutral chemical industry. The development of the HEIs catalyst supported on ceria (CeO_2) has shown remarkable performance in this reaction. Specifically, the (PtCoNi)(SnInGa)/CeO_2 catalyst exhibits exceptional catalytic activity, with a propane conversion rate of approximately 30% and a propylene selectivity of up to 90% at 600 °C, significantly outperforming previously reported catalysts. This outstanding performance can be attributed to the unique structure and composition of the HEI catalyst, which effectively isolates Pt atoms to prevent propylene decomposition and coke formation, while enhancing CO_2 activation through the incorporation of Ni and Co [43].

The catalytic mechanism of the HEI/CeO_2 catalyst is deeply rooted in its multimetallic structure and the synergistic effects of the different metals involved. The apparent activation energy (E_a) for PDH over HEI/CeO_2 was found to be 128.0 kJ·mol^{-1}, significantly lower than that of PtSn/CeO_2 (143.9 kJ·mol^{-1}) and Pt—Co—In/CeO_2 (138.3 kJ·mol^{-1}). Similarly, the E_A for CO_2 reduction over HEI/CeO_2 was 102.5 kJ·mol^{-1}, much lower than that of Pt—Co—In/CeO_2 (130.3 kJ·mol^{-1}) and PtSn/CeO_2 (141.5 kJ·mol^{-1}). This indicates that both propane activation and CO_2 activation are kinetically promoted in the HEI structure. The Pt atoms play a crucial role in the activation of propane's C—H bonds, while the presence of Co and Ni significantly enhances the catalyst's ability to activate CO_2. This dual functionality allows for the efficient conversion of propane to propylene while simultaneously reducing CO_2 to CO, driving the reaction forward and maintaining high selectivity.

Moreover, the ceria support plays a vital role in the overall catalytic process. CeO_2 not only provides a basic environment that facilitates CO_2 capture and coke combustion but also releases lattice oxygen, which is essential for the continuous combustion of coke accumulated on the catalyst surface. The HEI/CeO_2 catalyst demonstrated a coke selectivity of only 0.001%, highlighting its outstanding coke resistance. The mean catalyst life (τ = kd − 1) of HEI/CeO_2 was 13 times that of the PtSn/CeO_2 catalyst and twice that of the Pt—Co—In/CeO_2 catalyst. The combination of these factors results in a catalyst with exceptional coke resistance, thermal stability, and regenerability, making it a highly attractive option for industrial applications in PDH.

NH_3 Decomposition Reaction. NH_3 decomposition is a critical reaction for hydrogen production, with potential applications in fuel cells and sustainable energy systems. However, the large-scale application of NH_3 decomposition has been limited by the need for precious metal ruthenium (Ru) as a catalyst. In this study, we report the development of highly efficient NH_3 decomposition catalysts using novel HEA nanoparticles composed of earth-abundant elements (Co, Mo, Fe, Ni, and Cu). These HEA nanoparticles demonstrate significantly enhanced catalytic activity and stability, with improvement factors exceeding 20 times compared to conventional Ru catalysts [33, 34, 41].

The catalytic performance of the HEA nanoparticles is robustly tunable by varying the Co/Mo atomic ratio, allowing for optimization of surface properties to maximize reactivity under different reaction conditions. The most active HEA catalyst ($Co_{25}Mo_{45}Fe_{10}Ni_{10}Cu_{10}$) achieved an NH_3 conversion rate of 84% at 500 °C, with a mass-specific reaction rate of 22.1 g_{NH3} g_{metals}^{-1} h^{-1}, representing a significant enhancement over both Ru and conventional Co—Mo catalysts. The superior performance of the HEA catalysts can be attributed to the robust tuning of surface adsorption properties enabled by the continuous variation of alloy composition. Nitrogen Temperature Programmed Desorption (TPD) studies revealed that the binding strength of nitrogen on the HEA catalysts can be continuously tuned by varying the Co/Mo ratio, with the most active catalyst having a nitrogen desorption energy (ΔE_N) of ~79 kJ/mol, close to the optimal value predicted for NH_3 decomposition.

The catalytic mechanism of the HEA catalysts is closely related to the mixed surface site mechanism derived from atomistic modeling. The HEA nanoparticles form a single solid-solution phase with well-mixed Co–Mo sites on the surface, consistent with the bulk compositions. This surface site-mixing mechanism allows for the continuous tuning of surface adsorption properties, which in turn governs the catalytic activity. The robust tunability of the alloy composition and surface adsorption properties demonstrated by the HEA catalysts indicates great potential for their implementation in practical reactors, where the catalytic performance can be optimized under various reaction conditions (Figure 7.17).

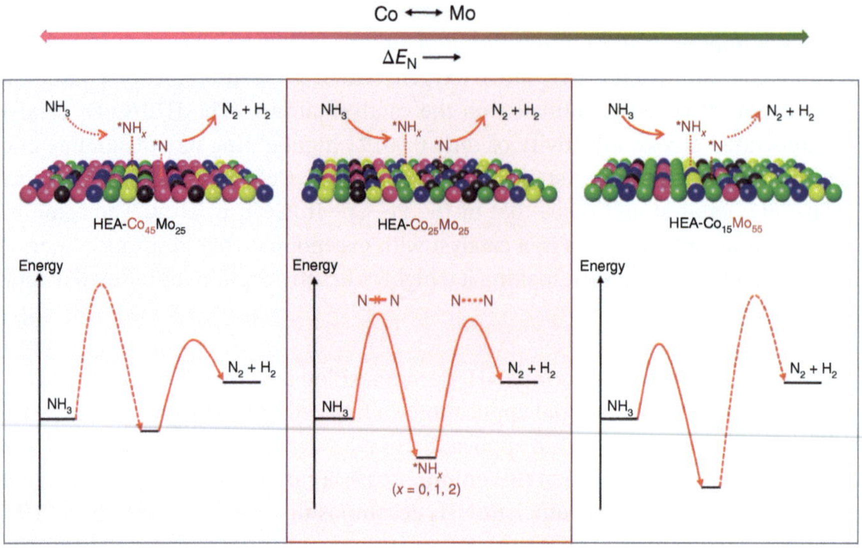

Figure 7.17 Schematic illustration of the rate-limiting factors in NH_3 decomposition. The rate-limiting factors are shown in the lower panel, with dashed lines indicating the key steps. On the Co-rich surface (left), the rate is limited by the activation or dehydrogenation of NH_3. On the Mo-rich surface, the rate is limited by the recombinative desorption of *N. The balance between these two steps is achieved on an intermediate composition, specifically HEA-$Co_{25}Mo_{45}$, where both steps are optimized for efficient NH_3 decomposition. [34] / Springer Nature / CC BY 4.0.

7.3.4 Catalyst Performance Comparison

HEAs have been gaining increasing attention in the field of catalysis due to their potential to outperform traditional catalysts in various thermal reactions. This section compares the performance of HEAs with traditional catalysts, highlighting the distinctive advantages of HEAs.

Catalytic Activity. Traditional Catalysts: Traditional catalysts typically consist of a single or a limited number of elements, resulting in fewer available active sites. While they are well-established in industrial applications, their simpler structures can limit their ability to achieve high reaction rates. The relatively straightforward composition often restricts the number and nature of active sites, potentially reducing catalytic efficiency.

HEAs: In contrast, HEAs, composed of multiple principal elements in near-equiatomic ratios, provide a vast array of potential active sites. This diversity results in a high density of active sites for reactant adsorption and enhanced reaction kinetics. The presence of multiple metals creates a varied landscape of active sites, each contributing uniquely to the catalytic process, often resulting in higher activity compared to traditional catalysts.

Catalytic Selectivity. Traditional Catalysts: Traditional catalysts can be highly selective for specific reactions but are often limited by their fixed composition and structure. Additionally, prolonged use under industrial conditions can lead to deactivation mechanisms, such as coking or poisoning, that diminish their selectivity over time.

HEAs: The tunable electronic properties of HEAs allow for improved selectivity. The multielement composition leads to unique synergistic effects, where interactions between different metals can stabilize particular reaction intermediates, favoring the formation of desired products. By fine-tuning the composition, HEAs can be optimized to exhibit enhanced selectivity and minimize unwanted side reactions, offering a significant advantage in complex chemical processes.

Catalytic Stability. Traditional Catalysts: Stability is a major concern for traditional catalysts, particularly under harsh reaction conditions. They can suffer from issues like sintering, poisoning, and corrosion, which lead to a shortened lifespan and reduced efficiency.

HEAs: HEAs are known for their exceptional thermal and chemical stability, making them well-suited for demanding reaction environments. Their multicomponent structure provides resistance to sintering, while the high-entropy configuration contributes to resilience against poisoning. This results in catalysts with longer operational lifespans and more consistent performance under extreme conditions.

Catalytic Resistance to Deactivation. Traditional Catalysts: Noble metal catalysts, such as platinum or palladium, can be highly effective but are often prone to deactivation due to impurities in the feedstock or coke formation. Such deactivation significantly impacts the performance and cost efficiency of catalytic processes.

HEAs: The multielement nature of HEAs makes them more resistant to these deactivation pathways. The variety of elements in HEAs can mitigate the effects of poisoning and coke buildup, leading to a more robust performance over extended periods. This resistance to deactivation is crucial for maintaining catalytic efficiency, especially when processing complex or impure feedstocks.

Catalytic Flexibility and Customization. Traditional Catalysts: The properties of traditional catalysts are often well-defined, but their scope for customization is limited without altering the fundamental composition, which can be both challenging and costly.

HEAs: HEAs offer a wide compositional space, enabling the fine-tuning of properties to meet the specific requirements of different reactions. This flexibility allows researchers to develop catalysts with optimized performance characteristics for each particular application, enhancing both activity and selectivity in targeted processes.

7.4 Understanding of High-Entropy Catalysts

7.4.1 Structural Characteristics Influencing Catalysis

The structural characteristics of HEMs are fundamental to their catalytic performance. These characteristics include the formation of solid solutions, the presence of lattice defects, and the degree of nanoporosity. Each of these structural features contributes uniquely to the catalytic properties of HEMs, making them versatile and efficient catalysts in various chemical processes.

The formation of a solid solution in HEMs is crucial for their catalytic performance as it leads to a homogeneous distribution of elements at the atomic level [40, 43–48]. This homogeneity provides several benefits that enhance catalytic performance: A solid solution ensures a uniform distribution of active sites, leading to consistent catalytic activity across the catalyst surface. This uniformity can result in higher activity and selectivity for the desired reactions, as reactants encounter a similar catalytic environment regardless of where they adsorb. The random distribution of multiple principal elements in a solid solution can generate synergistic effects that significantly enhance catalytic performance. These synergistic effects often involve electronic interactions between different elements, which can modify the surface reactivity, improve the adsorption and activation of reactants, and ultimately lead to more efficient catalysis. The high configurational entropy of HEMs discourages phase segregation, maintaining a stable and homogeneous catalytic surface over prolonged usage. This stability is critical for consistent catalytic performance, particularly under harsh reaction conditions where traditional catalysts may degrade or lose activity due to phase separation.

Lattice defects, such as vacancies, interstitials, and dislocations, also play a significant role in determining the catalytic performance of HEAs: Lattice defects create under-coordinated atoms on the catalyst surface, which are often more reactive than atoms in a perfect lattice. These under-coordinated sites serve as highly active sites for reactant adsorption and activation, thereby enhancing the overall catalytic activity. Defects can provide specific binding sites that stabilize reaction intermediates, facilitating their transformation into products. This stabilization can lower the activation energy of reactions, making the catalytic process more efficient. By controlling the type and concentration of lattice defects, it is possible to influence the selectivity of the catalyst. Certain defects may preferentially stabilize specific reaction intermediates or transition states, thereby favoring certain reaction pathways over others and leading to improved product selectivity. This control over reaction selectivity is particularly valuable in complex chemical processes where precise outcomes are desired [5, 29–33].

Nanoporous structures provide an extensive surface area, which increases the number of available active sites for catalytic reactions. This high surface area is crucial for enhancing catalytic activity, especially in reactions where surface interactions are the rate-limiting step. The porosity of HEAs influences the diffusivity of reactants, products, and even heat within the catalyst material. Optimizing the porosity can facilitate efficient mass and heat transfer, improving the overall efficiency of catalytic processes and preventing the formation of hot spots that could degrade the catalyst or lead to unwanted side reactions. Nanopores create a confined environment for chemical reactions, which can give rise to unique confinement effects. These effects can alter reaction mechanisms, favoring certain pathways and enhancing the selectivity for specific products. Confinement can also help stabilize reactive intermediates, further contributing to the overall selectivity and efficiency of the catalyst. The presence of nanoporosity can introduce internal stresses within the HEA, altering the electronic structure of surface atoms. These stresses can modify the reactivity of active sites, enhancing the catalytic performance in specific reactions. This aspect of stress engineering provides an additional layer of tunability, allowing for the optimization of HEAs for different catalytic applications.

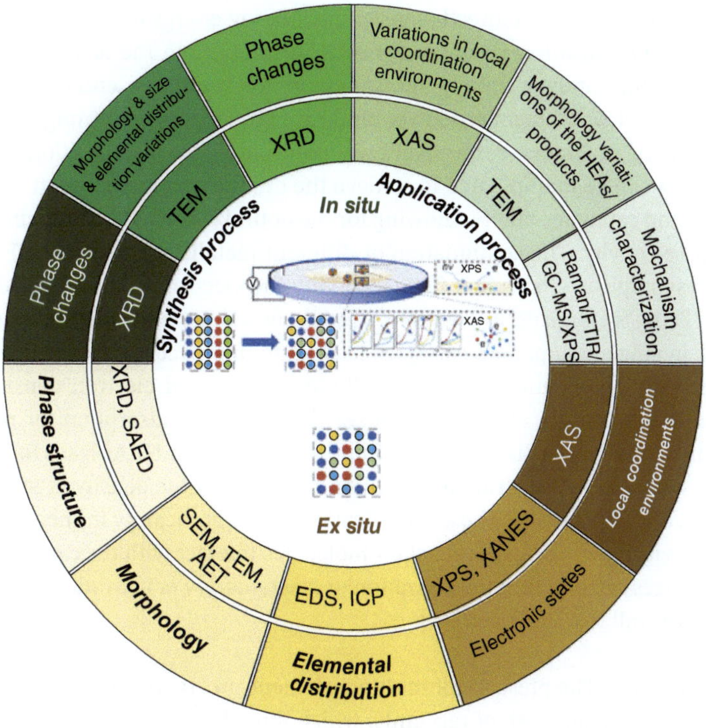

Figure 7.18 Characterization of HEAs. [1] / with permission of the American Chemical Society.

In sum, the structural characteristics of HEAs, including solid-solution formation, lattice defects, and nanoporosity, play a critical role in determining their catalytic performance. By tailoring these structural features, HEAs can be optimized for a wide range of catalytic applications, offering advantages such as enhanced activity, selectivity, and stability. However, a deeper understanding of these structures and their influence on catalysis is needed, particularly due to the complex and often unpredictable nature of HEAs. Future research should focus on unraveling the intricate relationships between HEA structure and catalytic behavior to fully harness their potential in industrial catalysis (Figure 7.18).

7.4.2 Electronic Effects and Ligand Effects

The multicomponent nature of HEAs gives rise to unique electronic and ligand effects that significantly influence their catalytic performance. These effects are crucial in determining the reactivity, selectivity, and stability of HEAs in various catalytic applications. Understanding these effects allows for the rational design and optimization of HEAs for specific catalytic processes.

Electronic Effects. The d-band center of a catalyst plays a critical role in its ability to bind adsorbates, influencing both the strength and nature of these interactions. In HEAs, the presence of multiple metallic elements leads to a variation in the d-band center, which directly affects the binding energy of reactants and intermediates. This variation provides a means to finely tune the adsorption properties, enhancing catalytic activity and selectivity for specific reactions. Shifts in the d-band center toward higher or lower binding energies result in stronger or weaker adsorption of reactants, respectively. By controlling the composition of HEAs, these shifts can be manipulated to achieve the desired balance between adsorption strength and catalytic activity, allowing for the optimization of reaction kinetics and product selectivity. The combination of different metallic elements in HEAs often leads to synergistic effects that further modify the electronic structure and d-band center. These synergistic interactions can create unique electronic properties not present in single-component catalysts, leading to improved catalytic performance. For instance, one element might contribute to enhanced electron density at the surface, while another element optimizes the energy of adsorption for particular reactants. The random distribution of multiple elements in HEAs leads to modifications in their electronic structure, including changes in the density of states, work function, and Fermi level. These modifications can significantly influence the catalytic behavior by altering how reactant molecules interact with the catalyst surface. Such changes can enhance the activation of reactants, lower activation barriers, and increase overall reaction rates [49].

Ligand Effects. The presence of multiple elements in HEAs can lead to alterations in the adsorption strength of reactants, which directly affects reaction kinetics and selectivity. This effect can be leveraged to create catalysts that favor specific reaction pathways, thereby improving the efficiency and selectivity of the catalytic process [20–23, 26, 50, 51].

Certain elements within HEAs can act as "ligands," assisting in the adsorption and activation of reactants. These ligand effects can lead to stronger adsorption and more favorable reaction pathways for particular reactants, thereby enhancing catalytic activity. For example, one metal might facilitate the initial adsorption of a reactant, while another metal supports the subsequent transformation, creating a cooperative effect that improves overall efficiency. The multielement composition of HEAs can also lead to competitive adsorption, where different elements play distinct roles in the catalytic process. One element might preferentially adsorb a specific reactant, while another element provides active sites for subsequent reaction steps. This division of roles can create a more selective catalytic process, minimizing side reactions and maximizing yields of desired products. The multicomponent nature of HEAs can give rise to steric effects, where the spatial arrangement of atoms on the catalyst surface influences the adsorption and reaction of molecules. These steric effects can ensure that only specific reactant molecules or intermediates can adsorb on the catalyst surface, thereby enhancing selectivity for desired products. By carefully designing the atomic arrangement, HEAs can be tailored to preferentially bind certain reactants while excluding others, thus controlling the overall reaction pathway. Ligand effects in HEAs can sometimes lead to alterations in the reaction mechanism itself, enabling new reaction pathways that are not accessible with traditional catalysts. This ability to influence the mechanism can result in more efficient reaction routes, reduced

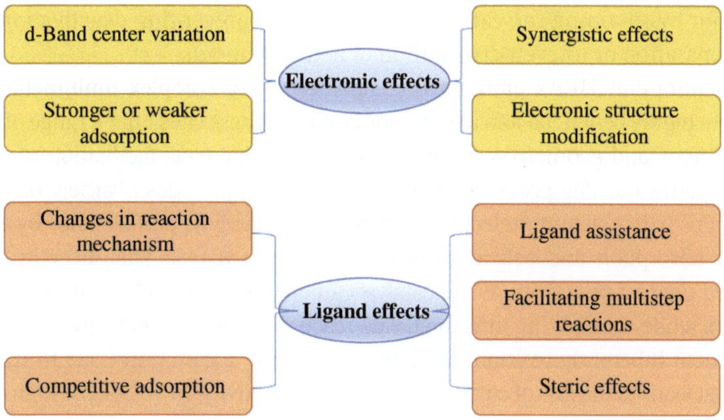

Figure 7.19 Unique electronic and ligand effects of HEAs.

energy barriers, and improved product selectivity, making HEAs a powerful platform for catalytic innovation (Figure 7.19).

The presence of multiple elements in HEAs can facilitate complex, multistep reactions by providing distinct active sites for each reaction step. This capability enhances the overall catalytic efficiency, as each element within the alloy can be tuned to optimize a specific part of the reaction mechanism, leading to smoother and more efficient conversions from reactants to products.

The electronic and ligand effects arising from the multicomponent nature of HEAs contribute significantly to their catalytic performance. By understanding and controlling these effects, it is possible to design HEAs with tailored properties for specific catalytic applications. The interplay between electronic modification, ligand interactions, and structural configuration provides HEAs with a level of versatility that surpasses traditional catalysts. However, further research is needed to fully elucidate the complex relationships between the composition, structure, and catalytic function of HEAs, as these relationships are key to unlocking their full potential in industrial catalysis.

7.4.3 Entropy's Role in Stabilizing Catalytic Active Phases

In HEAs, entropy plays a crucial role in stabilizing active phases and enabling the formation of complex multimetallic sites. The contribution of entropy is significant in catalysis, as it directly influences the performance, selectivity, and stability of HEA catalysts. The high degree of disorder associated with a homogeneous solid solution enhances the stability of the alloy, preventing the segregation of individual elements. This stable, disordered state is critical for maintaining the consistency of catalytic activity over time. The entropy of mixing (ΔS_{mix}) is a driving force behind the formation of HEAs. When the entropy of mixing is sufficiently high, it can overcome positive enthalpy barriers (ΔH_{mix}) that would otherwise lead to phase segregation. This entropy-driven mixing ensures that the different elements remain uniformly distributed throughout the alloy, contributing to the stability of the active catalytic phases. This uniform distribution is crucial for maintaining the catalytic efficiency and preventing deactivation due to phase separation or uneven distribution of active elements.

High entropy in HEAs promotes the stabilization of complex multimetallic sites, which are highly beneficial for catalytic applications. These sites offer a range of adsorption strengths and geometries, allowing for the simultaneous facilitation of multiple reaction pathways. The presence of multiple elements provides a variety of potential interactions with reactant molecules, enhancing the catalytic performance by enabling selective adsorption and activation of specific intermediates. This diversity leads to improved product selectivity, as the catalyst can be fine-tuned to favor desired reaction pathways while minimizing unwanted side reactions. The high-entropy state of HEAs makes them inherently resistant to phase segregation, a common issue in traditional alloys that leads to the loss of catalytic activity and selectivity. In HEAs, the high configurational entropy discourages the formation of separate phases, ensuring that the catalyst surface maintains a stable and uniform composition. This stability is vital for maintaining active site structure and preserving catalytic performance over extended periods, especially under harsh reaction conditions where traditional catalysts may degrade.

The high entropy of HEAs can lead to unique and enhanced surface properties, such as increased surface area and a higher density of under-coordinated surface atoms. These features contribute to an increase in the number of active sites available for catalysis, thereby enhancing catalytic activity. The high density of under-coordinated atoms provides highly reactive sites for reactant adsorption, while the increased surface area ensures better accessibility of the catalyst surface to reactant molecules, improving overall catalytic efficiency. The entropy-driven stabilization of active phases in HEAs also provides a platform for fine-tuning catalytic properties by adjusting the alloy's composition. By varying the proportions of different elements, researchers can modulate the electronic structure, surface properties, and overall catalytic performance of the alloy. This tunability allows for the design of catalysts with tailored properties for specific reactions, offering a significant advantage over traditional catalysts that have more rigid and less adaptable structures.

HEAs often exhibit excellent stability under the harsh conditions typically encountered in catalytic reactions, such as high temperatures and pressures. The high-entropy state contributes to this stability by discouraging structural transformations or the formation of unwanted phases that could lead to catalyst deactivation. This resilience is crucial for industrial applications, where catalysts are required to maintain high activity and selectivity over long operational periods (Figure 7.20).

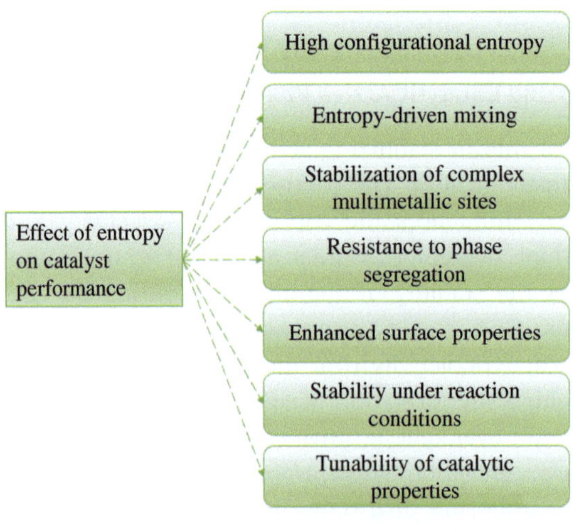

Figure 7.20 Effect of entropy on catalyst performance.

Entropy plays a key role in stabilizing active phases in HEA catalysts, enabling the formation of complex multimetallic sites and enhancing overall catalytic performance. The high-entropy state of HEAs promotes a homogeneous distribution of elements, resists phase segregation, and provides a tunable platform for designing catalysts with tailored properties for specific applications. By leveraging the entropy-driven stability and unique properties of HEAs, it is possible to develop more efficient, robust, and adaptable catalysts for a wide range of catalytic processes.

7.4.4 Mechanistic Insights Through Theoretical Calculations and Spectroscopy

Computational methods and spectroscopic techniques are essential tools for understanding the mechanisms of HEA catalysis. These approaches provide valuable insights into the atomic-scale structure, electronic properties, and active sites of HEAs, enabling the rational design and optimization of HEA catalysts for various catalytic applications.

DFT is a key computational method used to investigate the electronic structure of HEAs. DFT helps determine the electron distribution, d-band center, and density of states, which influence how reactants interact with the catalyst surface. By calculating adsorption energies for reactants, intermediates, and products, DFT provides insights into catalytic activity and selectivity. Additionally, DFT can model energy barriers for different reaction pathways, aiding in the identification of the most favorable catalytic routes and helping to predict the optimal composition of HEAs for specific reactions.

Spectroscopic techniques complement computational studies by offering experimental data on the local atomic and electronic structure of HEA catalysts. X-ray absorption spectroscopy, including X-ray Absorption Near Edge Structure (XANES) and Extended X-ray Absorption Fine Structure (EXAFS), provides information on the oxidation states and coordination environment of the elements in HEAs, revealing how they interact within the alloy. X-ray photoelectron spectroscopy (XPS) is used to analyze the surface electronic structure and track changes in oxidation states and electronic properties during catalysis. Mössbauer spectroscopy, for elements like iron, gives insights into oxidation states, coordination geometry, and magnetic properties, all crucial for understanding the catalytic process.

By combining DFT calculations with spectroscopic data, researchers gain a comprehensive understanding of HEA catalysis, from adsorption and intermediate formation to reaction pathways and energy barriers. This integrated approach enables the design of HEAs with optimized compositions and structures tailored for specific catalytic applications. The synergy between computational predictions and experimental validation is vital for advancing the development of HEA catalysts, ultimately improving their performance in a wide range of chemical transformations [52].

7.4.5 Designing HEA Catalysts for Specific Reactions

Designing HEA catalysts tailored for specific reactions involves a multifaceted approach that leverages the unique properties of HEAs. By adjusting their composition, creating specific nanostructures, and applying surface engineering techniques, researchers can optimize the activity and selectivity of HEA catalysts for targeted applications. The following are some key strategies for designing HEA catalysts for specific reactions. The multiprincipal element composition of HEAs can be fine-tuned to enhance catalytic performance. By manipulating the types and proportions of elements, it is possible to significantly influence the electronic structure, active site availability, and overall catalytic behavior.

Substituting one element with another can alter the electronic structure of the HEA, change the availability of active sites, and affect the interaction between the catalyst and reactant molecules. For example, replacing a base metal with a noble metal might enhance selectivity or increase resistance to deactivation. Adjusting the ratios of the constituent elements can influence the formation of specific phases, the distribution of elements on the surface, and the overall catalytic performance. Fine-tuning these ratios allows for precise control over the catalytic properties, enabling the optimization of activity, selectivity, and stability. Adding small amounts of promoters or dopants can significantly enhance the catalytic activity and selectivity of HEAs. Promoters can modify electronic properties, create new active sites, or even help stabilize intermediates, thereby boosting the catalyst performance.

The physical form of HEAs, particularly at the nanoscale, plays a critical role in determining their catalytic properties. Engineering the nanostructure of HEAs can greatly enhance their effectiveness as catalysts. Synthesizing HEA nanoparticles with controlled sizes can maximize surface area and expose more active sites, leading to increased catalytic activity. Smaller nanoparticles often have a higher density of reactive sites, which can significantly enhance reaction rates. The shape and morphology of nanoparticles (e.g., spheres, rods, cubes) can be controlled to optimize catalytic activity and selectivity. Different shapes expose different crystallographic facets, which can have varying catalytic properties, thus allowing for the tuning of reaction pathways. Creating hierarchical nanostructures, which feature multiscale porosity, can improve mass transport and facilitate reactant access to active sites. This is particularly beneficial for catalytic reactions involving large or bulky molecules, where efficient diffusion is critical. Manipulating the surface properties of HEAs is a powerful strategy for enhancing their interaction with reactants and optimizing catalytic performance.

Functionalizing the surface with specific groups or molecules can enhance the adsorption of reactants and improve selectivity. For example, adding functional groups that preferentially bind to certain reactants can increase the rate of desired reactions while suppressing unwanted side reactions. Deliberately introducing surface defects or vacancies can create additional active sites and influence reaction pathways. Defects often serve as high-energy sites that can enhance the adsorption and activation of reactant molecules, improving catalytic efficiency. Applying stress to the surface of HEAs can modify the electronic structure and reactivity of surface atoms. Controlled surface stress can lead to changes in adsorption strength, which can be exploited to enhance selectivity or reduce energy barriers for certain reaction steps.

Computational methods play a crucial role in the design and optimization of HEA catalysts, allowing researchers to predict catalytic performance before synthesis. DFT calculations are used to model the interaction of reactants with the catalyst surface, providing insights into activity and selectivity. DFT can predict which compositions are likely to exhibit optimal catalytic properties, reducing the need for extensive trial-and-error experimentation.

Machine learning algorithms can analyze large datasets of HEA compositions and catalytic performance, identifying trends and suggesting promising candidates. By leveraging data-driven approaches, machine learning can accelerate the discovery of effective HEA catalysts. High-throughput experimentation enables the rapid screening of a vast array of compositions and conditions to identify promising HEA catalysts for specific reactions. Robotic systems can be used to rapidly synthesize and test large numbers of HEA compositions, allowing researchers to quickly identify those with the best catalytic properties. Techniques such as X-ray diffraction and XPS can be used in situ to provide real-time data on the catalyst's structure and performance. This allows for immediate feedback and optimization during the experimentation process.

Ensuring the long-term stability of HEA catalysts under reaction conditions is crucial for their practical application in industry. Strategies to enhance stability include: Designing HEAs with high oxidation resistance is essential for improving their stability in reactions involving air or oxygen. The addition of certain elements, such as chromium or aluminum, can form protective oxide layers that prevent further oxidation. Developing HEAs that are resistant to coking is important for extending their lifespan in reactions prone to carbon buildup, such as hydrocarbon reforming. Elements that inhibit coke formation can be incorporated into the HEA to maintain catalyst activity over prolonged periods. The design of HEA catalysts for specific reactions is a complex and iterative process that involves a combination of compositional tuning, nanostructure control, surface engineering, and the use of advanced computational and experimental techniques. Each of these strategies can be applied individually or in combination to achieve the desired catalytic performance. By leveraging the unique properties of HEAs and employing a systematic approach, it is possible to develop catalysts with optimized activity, selectivity, and stability for a wide range of catalytic applications.

References

1 Sun L, Wen K, Li G, et al. High-entropy alloys in catalysis: progress, challenges, and prospects. *ACS Materials Au*, 2024, 4: 547–556. https://doi.org/10.1021/acsmaterialsau.4c00080

2 Wang H-X, Toh W L, Tang B Y, et al. Metal surfaces catalyse polarization-dependent hydride transfer from H_2. *Nature Catalysis*, 2023, 6: 351–362. https://doi.org/10.1038/s41929-023-00944-1

3 Cai J, Ruffieux P, Jaafar R, et al. Atomically precise bottom-up fabrication of graphene nanoribbons. *Nature*, 2010, 466: 470–473. https://doi.org/10.1038/nature09211

4 Sinha N, Yaltseva P, Wenger O S. The nephelauxetic effect becomes an important design factor for photoactive first-row transition metal complexes. *Angewandte Chemie International Edition*, 2023, 62: e202303864 (2023). https://doi.org/10.1002/anie.202303864

5 Gao S, Hao S, Huang Z, et al. Synthesis of high-entropy alloy nanoparticles on supports by the fast moving bed pyrolysis. *Nature Communications*, 2020, 11: 2016. https://doi.org/10.1038/s41467-020-15934-1

6 Li H, Han Y, Zhao H, et al. Fast site-to-site electron transfer of high-entropy alloy nanocatalyst driving redox electrocatalysis. *Nature Communications*, 2020, 11: 5437. https://doi.org/10.1038/s41467-020-19277-9

7 Glasscott M W, Pendergast AD, Goines S, et al. Electrosynthesis of high-entropy metallic glass nanoparticles for designer, multi-functional electrocatalysis. *Nature Communications*, 2019, 10: 2650. https://doi.org/10.1038/s41467-019-10303-z

8 Feng G, Ning F, Song J, et al. Sub-2 nm Ultrasmall high-entropy alloy nanoparticles for extremely superior electrocatalytic hydrogen evolution. *Journal of the American Chemical Society*, 2021, 143: 17117–17127. https://doi.org/10.1021/jacs.1c07643

9 Wu D, Kusada K, Nanba Y, et al. Noble-metal high-entropy-alloy nanoparticles: atomic-level insight into the electronic structure. *Journal of the American Chemical Society*, 2022, 144: 3365–3369. https://doi.org/10.1021/jacs.1c13616

10 Jia Z, Yang T, Sun L, et al. A novel multinary intermetallic as an active electrocatalyst for hydrogen evolution. *Advanced Materials*, 2020, 32: 2000385. https://doi.org/10.1002/adma.202000385

11 Qiao Y, Yao Y, Liu Y, et al. Thermal shock synthesis of nanocatalyst by 3D-printed miniaturized reactors. *Small*, 2020, 16: e2000509. https://doi.org/10.1002/smll.202000509

12 Song J-Y, Kim C, Kim M, et al. Generation of high-density nanoparticles in the carbothermal shock method. *Science Advances*, 2021, 7: eabk2984. https://doi.org/10.1126/sciadv.abk2984

13 Yao Y, Huang Z, Xie P, et al. Carbothermal shock synthesis of high-entropy-alloy nanoparticles. *Science*, 2018, 359: 1489–1494. https://doi.org/10.1126/science.aan5412

14 Dong Q, Yao Y, Cheng S, et al. Programmable heating and quenching for efficient thermochemical synthesis. *Nature*, 2022, 605: 470–476. https://doi.org/10.1038/s41586-022-04568-6

15 Yao Y, Huang Z, Xie P, et al. High temperature shockwave stabilized single atoms. *Nature nanotechnology*, 2019, 14: 851–857. https://doi.org/10.1038/s41565-019-0518-7

16 Jiao M, Yao Y, Pastel G, et al. Fly-through synthesis of nanoparticles on textile and paper substrates. *Nanoscale*, 2019, 11: 6174–6181. https://doi.org/10.1039/c8nr10137j

17 Yao Y, Huang Z, Hughes L A, et al. Extreme mixing in nanoscale transition metal alloys. *Matter*, 2021, 4: 2340–2353. https://doi.org/10.1016/j.matt.2021.04.014

18 Feng J, Chen D, Pikhitsa P V, et al. Unconventional alloys confined in nanoparticles: building blocks for new matter. *Matter*, 2020, 3: 1646–1663. https://doi.org/10.1016/j.matt.2020.07.027

19 Wang X, Huang Z, Yao Y, et al. Continuous 2000 K droplet-to-particle synthesis. *Materials Today*, 2020, 35: 106–114. https://doi.org/10.1016/j.mattod.2019.11.004

20 George E P, Raabe D, Ritchie R O. High-entropy alloys. *Nature Reviews Materials*, 2019, 4: 515–534. https://doi.org/10.1038/s41578-019-0121-4

21 Hsu W-L, Tsai C-W, Yeh A-C, et al. Clarifying the four core effects of high-entropy materials. *Nature Reviews Chemistry*, 2024, 8: 471–485. https://doi.org/10.1038/s41570-024-00602-5

22 Chen W, Hilhorst A, Bokas G, et al. A map of single-phase high-entropy alloys. *Nature Communications*, 2023, 14: 2856. https://doi.org/10.1038/s41467-023-38423-7

23 Zhou Y, Shen X, Qian T. A review on the rational design and fabrication of nanosized high-entropy materials. *Nano Research*, 2023, 16(5): 7874–7905. https://doi.org/10.1007/s12274-023-5419-2

24 Lu Z, Chen Z W, Singh C V. Neural network-assisted development of high-entropy alloy catalysts: decoupling ligand and coordination effects. *Matter*, 2020, 3: 1318–1333. https://doi.org/10.1016/j.matt.2020.07.029

25 Batchelor T A, Pedersen J K, Winther S H, et al. High-entropy alloys as a discovery platform for electrocatalysis. *Joule*, 2019, 3: 834–845. https://doi.org/10.1016/j.joule.2018.12.015

26 Sun Y, Dai S. High-entropy materials for catalysis: a new frontier. *Science Advances*, 2021, 7: eabg1600. https://doi.org/10.1126/sciadv.abg1600

27 Yeh J W, Chen S K, Lin S J, et al. Nanostructured High-Entropy Alloys with Multiple Principal Elements: Novel Alloy Design Concepts and Outcomes. *Advanced Engineering Materials*, 2004, 6: 299–303. https://doi.org/10.1002/adem.200300567

28 Li H, Lai J, Li Z, et al. Multi-Sites Electrocatalysis in High-Entropy Alloys. *Advanced Functional Materials*, 2021, 31: (47):2106715. https://doi.org/10.1002/adfm.202106715

29 Huang X, Zhao Z, Cao L, et al. High-performance transition metal–doped Pt3Ni octahedra for oxygen reduction reaction. *Science*, 2015, 348: 1230–1234. https://doi.org/10.1126/science.aaa8765

30 Ding K, Cullen D A, Zhang L, et al. A general synthesis approach for supported bimetallic nanoparticles via surface inorganometallic chemistry. *Science*, 2018, 362: 560–564. https://doi.org/10.1126/science.aau4414

31 Wu D, Kusada K, Yamamoto T, et al. Platinum-group-metal high-entropy-alloy nanoparticles. *Journal of the American Chemical Society*, 2020, 142(32): 13833–13838. https://doi.org/10.1021/jacs.0c04807

32 Mori K, Hashimoto N, Kamiuchi N, et al. Hydrogen spillover-driven synthesis of high-entropy alloy nanoparticles as a robust catalyst for CO(2) hydrogenation. *Nature Communications*, 2021, 12: 3884. https://doi.org/10.1038/s41467-021-24228-z

33 Bell T E, Torrente-Murciano L. H_2 production via ammonia decomposition using non-noble metal catalysts: a review. *Topics in Catalysis*, 2016, 59: 1438–1457. https://doi.org/10.1007/s11244-016-0653-4

34 Xie P, Yao Y, Huang Z, et al. Highly efficient decomposition of ammonia using high-entropy alloy catalysts. *Nature Communications*, 2019, 10: 4011. https://doi.org/10.1038/s41467-019-11848-9

35 Xin L, Yongqiang H, Husheng J. Pt-Rh-Pd alloy group gauze catalysts used for ammonia oxidation. *Rare Metal Materials and Engineering*, 2017, 46: 339–343. https://doi.org/10.1016/S1875-5372(17)30091-7

36 Liu H, Zhang L, Wang K, et al. Composition-dependent near-surface structure of high-entropy alloy catalysts for the semihydrogenation of alkynes. *The Journal of Physical Chemistry C*, 2023, 127: 7211–7219. https://doi.org/10.1021/acs.jpcc.3c00456

37 Liu H, Zhang Y, Zhang L, et al. Unveiling atomic-scaled local chemical order of high-entropy intermetallic catalyst for alkyl-substitution-dependent alkyne semihydrogenation. *Journal of the American Chemical Society*, 2024, 146: 20193–20204. https://doi.org/10.1021/jacs.4c05295

38 Li T, Yao Y, Huang Z, et al. Denary oxide nanoparticles as highly stable catalysts for methane combustion. *Nature Catalysis*, 2021, 4: 62–70. https://doi.org/10.1038/s41929-020-00554-1

39 Nakaya Y, Hayashida E, Asakura H, et al. High-entropy intermetallics serve ultrastable single-atom pt for propane dehydrogenation. *Journal of the American Chemical Society*, 2022, 144: 15944–15953. https://doi.org/10.1021/jacs.2c01200

40 Xing F, Ma J, Shimizu K I. High-entropy intermetallics on ceria as efficient catalysts for the oxidative dehydrogenation of propane using CO_2. *Nature Communications*, 2022, 13: 5065. https://doi.org/10.1038/s41467-022-32842-8

41 Yao Y, Liu Z, Xie P, et al. Computationally aided, entropy-driven synthesis of highly efficient and durable multi-elemental alloy catalysts. *Science Advances*, 2020, 6: eaaz0510 https://doi.org/10.1126/sciadv.aaz0510

42 Feng D, Dong Y, Zhang L, et al. Holey Lamellar High-Entropy Oxide as an Ultra-High-Activity Heterogeneous Catalyst for Solvent-free Aerobic Oxidation of Benzyl Alcohol. *Angewandte Chemie International Edition*, 2020, 59: 19503–19509. https://doi.org/10.1002/anie.202004892

43 Li Y, Bai X, Yuan D, et al. Challenging thermodynamics: combining immiscible elements in a single-phase nano-ceramic. *Nature Communications*, 2024, 15: 1167. https://doi.org/10.1038/s41467-024-45413-w

44 Sun Y, Yu L, Xu S, et al. Battery-driven N(2) electrolysis enabled by high-entropy catalysts: from theoretical prediction to prototype model. *Small*, 2022, 18: e2106358. https://doi.org/10.1002/smll.202106358

45 Dong Q, Hong M, Gao J, et al. Rapid synthesis of high-entropy oxide microparticles. *Small*, 2022, 18: e2104761. https://doi.org/10.1002/smll.202104761

46 Li Y, Bai X, Yuan D, et al. Cu-based high-entropy two-dimensional oxide as stable and active photothermal catalyst. *Nature Communications*, 2023, 14: 3171. https://doi.org/10.1038/s41467-023-38889-5

47 Wang T, Chen H, Yang Z, et al. High-entropy perovskite fluorides: a new platform for oxygen evolution catalysis. *Journal of the American Chemical Society*, 2020, 142: 4550–4554. https://doi.org/10.1021/jacs.9b12377

48 Ding Y, Wang Z, Liang Z, et al. Monolayer high-entropy layered hydroxide frame for efficient oxygen evolution reaction. *Advanced Materials*, 2023, 6: 2302860. https://doi.org/10.1002/adma.202302860

49 Hao J, Zhuang Z, Cao K, et al. Unraveling the electronegativity-dominated intermediate adsorption on high-entropy alloy electrocatalysts. *Nature Communications*, 2022, 13: 2662. https://doi.org/10.1038/s41467-022-30379-4

50 Zhang W, Liaw P K, Zhang Y. Science and technology in high-entropy alloys. *Science China Materials*, 2018, 61: 2–22. https://doi.org/10.1007/s40843-017-9195-8

51 Yao Y, Dong Q, Brozena A, et al. High-entropy nanoparticles: synthesis-structure-property relationships and data-driven discovery. *Science*, 2022, 376: eabn3103. https://doi.org/10.1126/science.abn3103

52 Oses C, Toher C, Curtarolo S. High-entropy ceramics. *Nature Reviews Materials*, 2020, 5: 295–309. https://doi.org/10.1038/s41578-019-0170-8

8

Clean Energy and Electrocatalysis

*Bing Zhu, Dongshuang Wu**

School of Materials and Engineering, Nanyang Technological University, Singapore

*Corresponding email: dongshuang.wu@ntu.edu.sg

8.1 Introduction to Electrocatalysis and Its Importance

With the rapid growth of population and industrialization, global demand for fossil fuels has continuously increased, leading to severe environmental and climate crises. These issues include rising greenhouse gas emissions, global warming, and pollution, all of which pose serious threats to ecosystems and human health. Addressing these challenges requires an urgent transition to sustainable energy systems that rely on renewable and environmentally friendly resources. Achieving sustainable energy and environmental protection on a global scale are two of the most significant challenges of this century. In this context, advanced electrochemical processes have emerged as a promising solution. These processes leverage the earth's abundant and renewable resources such as water, carbon dioxide, and nitrogen to produce high-purity chemical fuels. Electrocatalytic reactions play a crucial role in these processes, including hydrogen production via water electrolysis, electrochemical reduction of nitrate to synthesize ammonia, electrochemical reduction of CO_2 to generate carbon-based value-added products, and the electrochemical oxygen reduction reactions in fuel cells.

By enabling these transformative processes, electrocatalysis is not only reducing reliance on fossil fuels but also paving the way for a circular carbon economy, where waste carbon dioxide is converted into useful products. Continued advancements in this field will be critical for achieving the dual goals of energy sustainability and environmental preservation, making electrocatalysis a cornerstone in the evolution of new energy technologies.

8.2 Application of High-Entropy Alloys in Key Reactions of Electrocatalysis

High-Entropy Alloys as Electrocatalyst. Electrocatalytic reactions will occur at the electrode–electrolyte interface and require energy input to overcome inherent barriers in electrochemical processes. The potential difference between the applied

Functional High Entropy Alloys and Compounds, First Edition. Edited by Yong Zhang, Yonggang Yao and Haojie Liu.

potential and the thermodynamic potential, known as the overpotential, is needed to drive these reactions. Catalysts can reduce the overpotential and accelerate reaction rates, directly impacting the efficiency of electrocatalytic reactions. Since most electrocatalytic reactions involve interdependent multistep processes, designing catalysts with uniformly distributed active sites on their surfaces is essential to break these limitations and enhance catalytic performance.

The d-band model, proposed nearly 30 years ago by Bjørk Hammer and Jens Nørskov, is a foundational theory in electrochemical catalysis. It explains how the electronic structure of transition metal catalysts is directly linked to their catalytic activity. According to this theory, the position and width of the metal's d-band, relative to the Fermi level, affect the binding strength of adsorbed reactants on the catalyst's surface. A catalyst with an optimally positioned d-band can achieve a balance between adsorption and desorption of reactant molecules, enhancing reaction rates without overly stabilizing or destabilizing intermediates.

The unique features of high-entropy alloys (HEAs) catalysts can be understood through the d-band profile, which describes variations in binding energies across different catalyst compositions (Figure 8.1). For simple unary catalysts, all binding sites are identical, resulting in a single binding energy. Adding a second element or creating core/shell structures introduces additional peaks due to distinct interactions between the elements. In intermetallic phases, each element has a unique, consistent binding energy. However, in solid-solution phases, the range of neighboring atoms varies. In a quinary alloy, at least five distinct bond strengths exist, resulting in five peaks in the d-band profile. This broadens the profile, creating an almost continuous range of binding energies. Such a continuous profile enables a diverse array of binding energies, facilitating the formation of active sites with near-optimal binding strengths. Consequently, HEA can activate elements that are otherwise inert, offering a multitude of unique surface binding sites. This nearly continuous distribution of adsorption energies provides a basis for fine-tuning binding energies, allowing for precise control over reaction properties.

Water Splitting. Molecular hydrogen (H_2) is one of the most attractive fuels due to its high gravimetric energy density and its only combustion by-product being clean water. This makes hydrogen an excellent energy carrier and a potential candidate for future

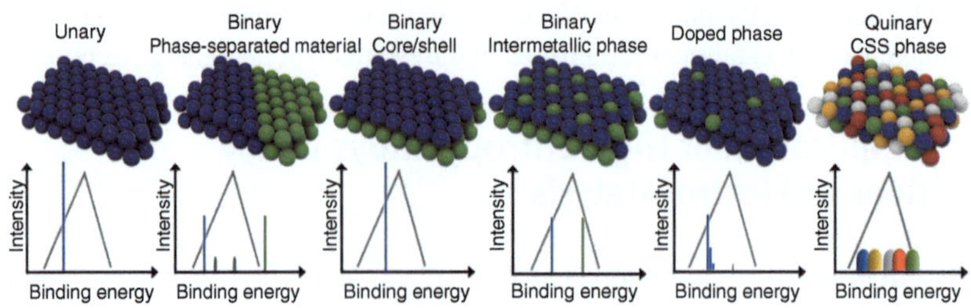

Figure 8.1 Illustration of d-band profile in different catalysts [1] / with permission of John Wiley & Sons

low-carbon energy systems. As one of the ideal green energy sources, effectively obtaining hydrogen is of significant practical importance. Currently, hydrogen is primarily produced from steam reforming of fossil fuels, a process that generates substantial carbon dioxide emissions. Therefore, electrocatalytic water splitting powered by renewable energy is key to successfully achieving a hydrogen economy [2, 3].

Water electrolysis is a clean method of hydrogen production that uses electrical energy to split water molecules into hydrogen and oxygen. First reported in 1789, the core reaction is:

$$H_2O\ (1) \rightarrow H_2\ (g) + 1/2\,O_2\ (g)$$

In the process of water electrolysis, the hydrogen evolution reaction (HER) is the key cathodic reaction for hydrogen generation. The initial step of HER, known as the Volmer reaction, involves the transfer of electrons to the catalyst surface, where they interact with protons or water molecules (depending on the pH of the electrolyte) to form adsorbed hydrogen atoms (H^*). This step is often the rate-limiting step for HER. The efficiency of HER catalysts depends on optimizing the hydrogen adsorption energy during this Volmer step. A ΔG_{H*} value near zero is considered ideal because weak adsorption limits interaction with protons, while overly strong adsorption hinders the release of H_2. After the formation of H^*, H_2 is generated through either the Heyrovsky or Tafel reactions. In acidic conditions, the reaction benefits from hydronium ions, which promote fast kinetics. In alkaline media, water dissociation and the management of hydroxyl adsorption play crucial roles in determining the overall activity of HER.

Due to their unique electronic structures, HEA can rationally tune water dissociation and hydrogen adsorption by incorporating earth-abundant, weakly active transition metals—even completely inert elements—achieving comparable or even superior reaction performance. Wu et al. revealed that HEAs exhibit a wider distribution of ε_d values and richer diversity in the local density of states (LDOS) on their surface atoms (Figure 8.2) [4]. This allows inert elements such as Os, Au, and

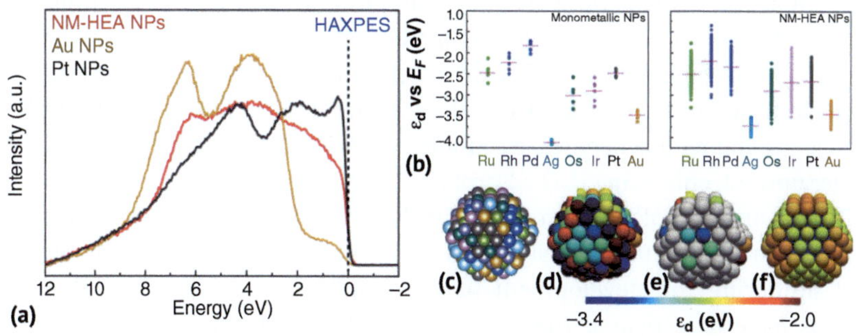

Figure 8.2 (a) Valence Band spectra obtained by X-ray photoelectron spectroscopy (XPS). (b) The ε_d value of surface atoms in monometallic and noble-metal (NM) HEA-nanoparticles (NPs). (c) Density functional theory (DFT) model of NM-HEA-NP. (d) The ε_d value of surface atoms in NM-HEA-NP. (e) The ε_d of the surface Pt atoms in the NM-HEA-NP. (f) The ε_d value in Pt NP [4]. *Source:* With permission of American Chemical Society.

Ag within the HEA to possess LDOS profiles suitable for the HER. These elements can directly serve as active sites or indirectly act as beneficial neighboring atoms to certain reactive centers, thereby enhancing catalytic activity.

Based on the inspiration of this, Wei et al. designed a PtMoPdRhNi HEA with strong *d*–*d* electron interactions [5]. In this alloy, Pt and Ni sites accelerate water dissociation, while Pt and Rh sites promote the combination of H^* into H_2, maintaining high activity during HER. Li et al. synthesized PtNiFeCoCu HEA nanoparticles that exhibit higher HER performance in alkaline media compared to commercial Pt catalysts. Water molecules initially adsorb onto Fe sites, activating water dissociation and stabilizing OH^* intermediates, while nearby Ni and Co sites facilitate the stabilization of $^*H$, further promoting the HER process [6].

Reducing particle size to achieve extremely small dimensions and increasing the surface-to-volume ratio is another strategy to enhance catalytic activity. Minamihara et al. developed a continuous-flow reactor for large-scale liquid-phase reduction synthesis, producing approximately 1 nm ultrasmall HEA nanoparticles that achieved an overpotential of only 6 mV at a current density of 10 mA/cm^2 in acidic media [7]. Feng et al. prepared 2 nm NiCoFePtRh HEA nanoparticles, achieving an ultrahigh mass activity of 28.3 A mg^{-1} noble metal at −0.05 V, with no decay after 10,000 cycles [8].

In the anode of water electrolysis, the oxygen evolution reaction (OER) is a complex four-electron/four-proton process with a large energy barrier that severely limits the rate of water splitting. The most commonly used catalysts are Ru-based and Ir-based materials, which are constrained by high cost and a trade-off between stability and activity. As emerging materials, HEAs have shown great potential in enhancing OER activity by maximizing the utilization of NMs during catalysis and possessing the ability to adjust local coordination environments and electronic structures.

Many studies have focused on the strong local electronic interactions within HEAs. Mei et al. revealed a unique electron transfer phenomenon in FeCoNiMo HEA, where Mo acts as an electron donor, limiting OH^* adsorption on Fe, Co, and Ni sites, thus improving OER performance [9]. Zhu et al. introduced low-electronegativity Mn atoms into FeCoNiCu to form high-entropy FeCoNiCuMn HEA nanoparticles [10]. The high-entropy effect drives electron migration, activating the inert, high-electronegativity Cu sites. He et al. combined 4*d* Mo, 5*d* W, and 3*d* FeCoNi elements, generating a synergistic electronic coupling effect that shifts the Ni d-band center upward, enhancing H_2O adsorption [11].

During anodic polarization in the OER, metal dissolution and surface reconstruction occur simultaneously, often leading to significant performance decay and hindering the development of water-splitting devices. Several research groups have investigated the effects of metal dissolution and surface reconstruction in HEAs on OER performance. Yao et al. discovered a reconstructed core–shell structure in an IrRuCoNiMo HEA after 24 hours of electrolysis [12]. Through monitoring during 500 hours of water splitting with XPS and X-ray absorption fine structure (XAFS) analyses, they found that Ir in the 2 nm Ir-rich oxide shell captures electrons from other transition metals (Figure 8.3a). This

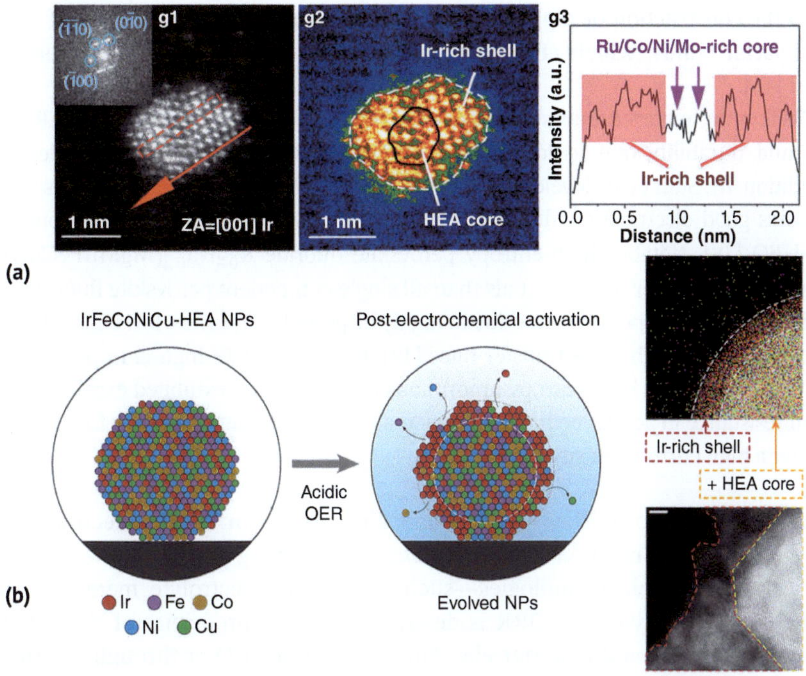

Figure 8.3 (a) Ir-rich shell in IrRuCoNiMo [12] / with John Wiley & Sons permission of Wiley. (b) Ir-rich shell in IrFeCoNiCu [13] / with permission of American Chemical Society.

electron transfer suppresses transition metal migration and prevents internal metal dissolution during electrolysis. Similarly, Maulana et al. found that after electrochemical activation of an IrFeCoNiCu HEA, the surface 3*d* transition metals significantly dissolved, forming a 2–6 nm thick surface layer with 70% Ir atomic content (originally 19%), while the core maintained a uniform elemental distribution (Figure 8.3b). This heterogeneous structure exhibited superior OER stability in subsequent chronopotentiometry tests, with negligible dissolution [13]. Kwon et al. reported similar phenomena in a ZnNiCoIrMn HEA with low Ir content [14]. By optimizing the electronic structure through the high-entropy effect, the dissolution of 3*d* metals is confined to the surface, forming an Ir-rich layer. The undissolved HEA core influences the electronic structure of the Ir-rich surface, causing a downward shift of the d-band center and weakening the adsorption energy of reaction intermediates.

Recently, the enhancement of OER by HEA-derived electrocatalysts has garnered increasing attention. Various high-entropy oxides with different structures have been successfully synthesized and applied in OER, including rock-salt, spinel, perovskite, fluorite, and layered double hydroxide (LDH) structures. Wu et al. found that in a high-entropy perovskite hydroxide system, the d-band centers of Fe, Co, and Ni shift upward due to synergistic effects, leading to enhanced OER activity [15]. Liu et al. demonstrated the high OER stability of FeCoNiMg-LDH, attributing it to the construction of high configurational entropy and the formation of Mg—O interfacial bonds, which effectively mitigate phase segregation and transformation [16].

Baek et al. reported that spinel CoFeNiCrMn HEO exhibits optimized OER activity and durability due to local strain effects, surpassing low-entropy oxides [17]. Additionally, Suhr et al. developed (Ti—Co—Mo—Ta—W)N high-entropy nitrides, which possess good electrical conductivity and superior water oxidation activity compared to HEO [18]. Notably, high-entropy perovskite fluoride $K_{0.8}Na_{0.2}(MgMnFeCoNi)F_3$ demonstrated lower overpotentials than all single-component perovskite fluorides and exhibited excellent performance due to highly dispersed active sites, low charge transfer resistance, and high mass transfer rates [19]. CoFeNiMoWTe high-entropy chalcogenide glasses with a high-entropy amorphous microstructure exhibited excellent corrosion resistance in acidic media [20]. They maintained long-term stability for 100 hours under a high current density of 1 A/cm^2.

Fuel Cell. Due to its slow kinetics and high overpotential, the electrocatalytic oxygen reduction reaction (ORR) remains a challenging process in metal-air batteries and fuel cell technologies—such as polymer electrolyte membrane fuel cells. The pathway of the ORR is determined by the properties of the catalyst. It can proceed via a direct four-electron reduction to H_2O or through a competitive indirect route involving H_2O_2 as an intermediate. Although platinum-group NMs are effective catalysts, their low abundance in the earth's crust and high cost severely limit their widespread commercial application. Theoretically, HEA with low Pt content—or even Pt-free HEAs loaded with transition metals—can achieve catalytic performance comparable to commercial Pt catalysts [21]. He et al. developed a CrMnFeCoNi system whose synergistic combination achieved performance at least comparable to Pt under identical conditions, reaching lower overpotentials [22]. The removal of any single element resulted in a significant decrease in activity across all quaternary alloys. Yu et al. developed a PtPdFeCoNi HEA that achieved a mass activity of 1.23 A mg^{-1} Pt, greatly surpassing that of pure Pt [23]. Even after 50,000 accelerated durability tests, it maintained high stability, whereas pure Pt and PtCo nanoparticles suffered severe dissolution, segregation, or aggregation.

Developing long-range ordered high-entropy structures provides opportunities to optimize electronic structures and stabilize reaction intermediates. A PtFeCoNiCuZn high-entropy intermetallic (HEI) was developed, where charge transfer from non-NMs to Pt due to ligand effects shifted the overall d-band center downward [24]. Simultaneously, the non-NMs were activated as highly active sites, achieving an impressive total mass activity of 1.64 A mg^{-1}. A fuel cell assembled with PtPdFeCoNi-HEI/C as the cathode delivered an ultrahigh peak power density of 1.73 W cm^{-2} under 1.0 bar back pressure and operated continuously for 80 hours. Wang et al. developed long-range ordered PtPdFeCoNi HEA and HEI materials [25]. The HEI exhibited a more stable electronic structure during the ORR process and improved the strong adsorption of OH^*, thereby reducing the reaction energy barrier. A zinc–air battery using this catalyst operated continuously for 200 hours at 4.0 mA cm^{-2}, demonstrating excellent cycling stability. Feng et al. designed a PtIrFeCoCu HEI specifically for ORR, where the high-activity (001) facets provided the lowest activation energy barrier and an optimal *d*-band center [26].

Carbon Dioxide Reduction Reaction (CO_2RR). The CO_2RR enables the conversion of CO_2 into high-value-added chemical fuels and everyday chemicals, thereby achieving carbon recycling. The reaction yields a wide array of products, including C_1 compounds such as CO, HCOOH (formic acid), CH_3OH (methanol), and CH_4 (methane); C_2 compounds like C_2H_4 (ethylene), CH_3CHO (acetaldehyde), C_2H_5OH (ethanol), and CH_3COOH (acetic acid); as well as various C_2^+ products, including *n*-C_3H_7OH (*n*-propanol). These processes involve complex multielectron and proton coupling steps with sluggish kinetics. Catalysts based on single metal elements are no longer sufficient to meet these demands, making the development of efficient and highly selective catalysts an urgent necessity. Accommodating multistep reactions that require different adsorption energy levels is essential [27]. A high-entropy oxide $BiSbInCdSnO_4$ has been developed to effectively reduce CO_2 to formate. Achieving a Faradaic efficiency exceeding 95% within a potential window of −0.5 to −1.6 V, the catalyst remains stable for 100 hours at a current density of 150 mA cm^{-2}. DFT calculations indicate that long-range electron donor–acceptor interactions between indium (In) and bismuth (Bi) enhance CO_2 adsorption, thereby accelerating formate production [28].

Pedersen and colleagues investigated two HEA systems: CoCuGaNiZn and AgAuCuPdPt [29]. They discovered that modifying the alloy composition alters the distribution of adsorption energies, which in turn affects the catalyst's selectivity and activity. Nellaiappan et al. confirmed the high activity of the AgAuCuPdPt system, achieving a Faradaic efficiency of approximately 100% for gaseous hydrocarbon products at −0.3 V versus the reversible hydrogen electrode [30]. Cavin et al. developed two-dimensional high-entropy dichalcogenide alloys, specifically $(MoWVNbTa)S_2$, which exhibited an outstanding current density of 0.51 A cm^{-2} at −0.8 V and a turnover frequency of 58.3 s^{-1} [31].

Nitrate Reduction Reaction (NO_3RR). The nitrogen cycle has long been a crucial elemental cycle in industrial processes and biological activities. Nitrate pollution, primarily from agricultural runoff and industrial waste, poses serious environmental and health risks. Electrochemical reduction of nitrates offers a method to mitigate this pollution by converting harmful nitrates into less harmful substances. Ammonia is a key component in fertilizers and various chemicals. Traditional ammonia synthesis via the Haber–Bosch process is energy-intensive and relies heavily on fossil fuels. NO_3RR provides a more sustainable alternative by producing ammonia under milder conditions and potentially utilizing renewable energy sources.

The reaction mechanism involves multiple proton and electron transfer steps, leading to the formation of ammonia. The overall reaction can be represented as follows:

$$NO_3^- + 9H^+ + 8e^- \rightarrow NH_3 + 3H_2O$$

Due to the complexity of its mechanisms, the high diversity of products, and the inevitable competitive reactions (e.g., HER), multiple possible reaction pathways exist. Developing efficient and selective catalysts is crucial for enhancing the performance of NO_3RR. Recent research has focused on various materials, including transition metals, alloys, and metal oxides, to improve activity and selectivity toward ammonia production. For instance, copper-based catalysts have shown promise due to their favorable binding energies with reaction intermediates, leading to high selectivity for ammonia.

Zhang et al. introduced an $L1_2$ + FCC dual-phase FeCuNiAlTi HEA to enhance the electrochemically active surface area and charge transfer efficiency [32]. DFT calculations confirmed the multisite catalytic activity involving Fe, Co, and Ni, with the Co site exhibiting the lowest energy barrier for nitrate reduction. Qi et al. discovered in ultrathin high-entropy Fe-based spinel oxide nanosheets that incorporating high-entropy components significantly reduces the band gap and increases the density of states near the Fermi level [33]. This adjustment enhances electronic conductivity and modulates the adsorption strength of Fe sites toward O and N species.

8.3 Advanced *Ex/In Situ* Characterization Techniques

Microscopy. Researchers employ microscopy to examine microstructure, geometric morphology, and compositional distribution, providing insights into the growth and reaction mechanisms within materials. Advanced techniques like electron microscopy and tomography enable three-dimensional reconstruction, allowing researchers to visualize internal structures with high spatial resolution. This facilitates a deeper understanding of the spatial relationships and distribution of phases or defects within the material, crucial for elucidating underlying mechanisms in materials science and enhancing design and performance in applications.

Electron microscopy, one of the most widely used microscopy techniques, includes transmission electron microscopy (TEM), scanning electron microscopy (SEM), and scanning transmission electron microscopy (STEM), each serving distinct purposes. TEM enables atomic-level imaging of internal structures, ideal for studying crystallography and fine internal features, while SEM provides detailed surface topography and compositional data, making it suitable for examining surface structures over larger areas. STEM combines aspects of TEM and SEM, offering atomic-resolution imaging alongside compositional analysis through spectroscopic techniques, making each method uniquely valuable for specific applications in materials science and related fields.

The oxidation behavior of FeCoNiCuPt HEA nanoparticles at 400 °C in air has been investigated using *in situ* gas-phase TEM [34]. The results indicated that the oxidation of HEA nanoparticles followed logarithmic rate constants, governed by the Kirkendall effect. During oxidation, Fe, Co, Ni, and Cu were segregated to the surface, while Pt remained in the core region. Notably, the oxidation rate of HEA nanoparticles was significantly slower than that of monometallic nanoparticles. The researchers further applied the same method to study the reduction behavior of oxidized FeCoNiCuPt HEA nanoparticles in H_2 at 400 °C. During reduction, outward diffusion of transition metals caused expansion of the oxide layer; Cu was reduced to Cu nanoparticles, while Fe, Co, and Ni remained in oxidized states.

The high-entropy effect, a critical characteristic of HEAs, reflects how the increase in component diversity raises configurational entropy, enhancing the system's stability. *In situ* TEM at elevated temperatures has directly demonstrated the thermal stability of HEA nanoparticles, contrasting with monometallic Pd nanoparticles, which undergo Ostwald ripening at temperatures above 700 °C (Figure 8.4) [35].

TEM is effective for verifying surface reconstruction phenomena following electrochemical tests, revealing structural evolution associated with abundant defect formation and the emergence of single-atom (SA) sites, as well as compositional segregation due to the dissolution of non-precious metals. With the advent of spherical aberration correction, the spatial resolution and sensitivity of STEM have reached the SA scale. High-resolution STEM can reveal defects such as point defects, dislocations, bulk defects, and twins after cyclic voltammetry (CV) activation, as well as SA sites formed through metal dissolution and redeposition. These observations highlight the critical role of reconstruction in increasing the number of active sites and enhancing surface area.

The development of advanced electron microscopy techniques enables dynamic tracking of the structural evolution of HEAs during electrochemical processes, providing visual insights into localized processes at the microscopic scale. *In situ* electrochemical TEM holders equipped with nanoelectrodes have been developed, allowing repeated access to individual nanoparticles at the same location as CV progresses. Before CV cycling, the pristine particles exhibit small attached catalyst

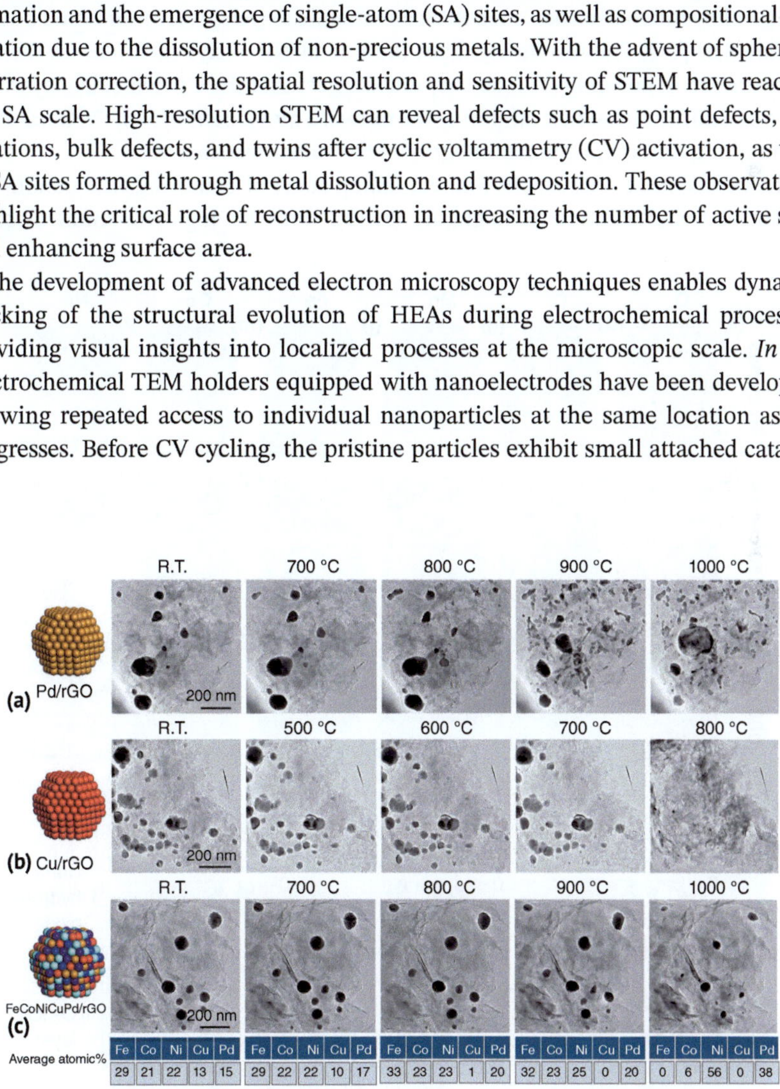

Figure 8.4 *In situ* TEM of FeCoNiCuPd HEA under different temperatures [35] / with permission of the American Chemical Society.

particles and partially amorphous carbon-covered surfaces, while post-cycling, the surfaces become smoother, exposing areas initially covered by carbon.

Atomic-resolution tomography has also been successfully applied in HEA electrocatalyst research. Atom probe tomography (APT), a mass spectrometry-based technique, detects ionized atoms ejected from the specimen, providing both spatial and chemical information. APT can identify elemental composition and distribution, including light elements, at subnanometer spatial resolution and down to parts-per-million levels, offering invaluable insights into HEA structural and compositional characteristics.

The compositional changes of reconstruction in different surface layers of a Cantor alloy microelectrode after 70 electrochemical cycles were monitored by APT (Figure 8.5) [36]. The results highlight the role of NiFe oxyhydroxides on the topmost surface in boosting OER activity and suggest that the Cr-rich oxide layer may serve as a diffusion barrier, inhibiting the outward movement of Mn, Fe, Co, and Ni, maintaining the compositional integrity of the subsurface layers.

The precision of APT is still limited, falling short of achieving true atomic resolution. Atomic electron tomography (AET), with its high-resolution electron

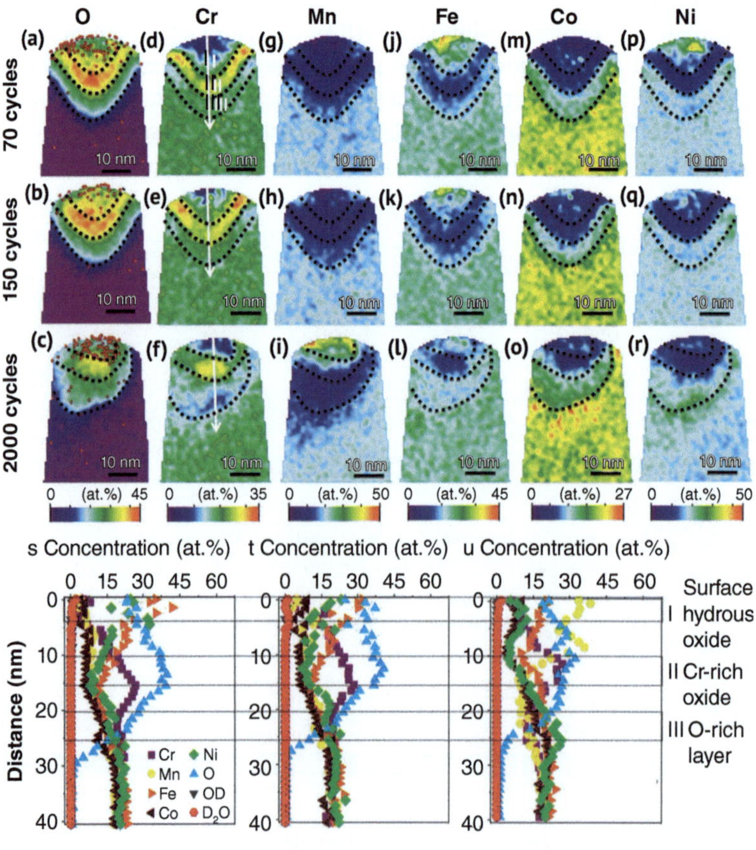

Figure 8.5 2D APT profiles under OER conditions [36] / American Chemical Society.

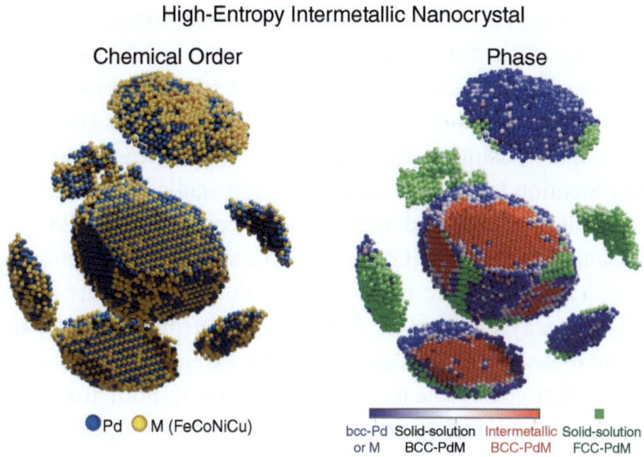

Figure 8.6 AET images of PdFeCoNiCu HEI [38] / American Chemical Society.

transmission and optimized iterative algorithms, addresses this challenge. AET can resolve chemical ordering and structural information within extremely small regions, even within individual nanoparticles [37]. By analyzing 3D atomic distributions, AET can extract local atomic coordination environments—such as coordination number (CN), bond length distributions, and strain fields. However, its chemical sensitivity is relatively lower, primarily providing structural information and distinguishing only elements with substantial atomic number differences. For instance, AET has revealed the 3D atomic configuration of PdFeCoNiCu HEI, showing that surface Pd atom electron density is modulated by the local chemical ordering within the nanoparticle core, further influencing alkyl substitution-dependent alkyne semi-hydrogenation (Figure 8.6) [38].

Due to the complexity of HEAs, a more thorough understanding of their electrocatalytic processes requires advanced *in situ* microscopy techniques that offer high spatial and temporal resolution for structural and compositional information. In addition to the developments in microscopy techniques mentioned earlier, other advanced characterization methods such as scanning electrochemical cell microscopy and *in situ* electrochemical atomic force microscopy are expected to be applied in HEA electrocatalyst research [39]. These techniques will facilitate effective catalyst screening and provide insights into local electrochemical activity, reaction kinetics, and surface morphology changes, advancing our understanding of catalytic mechanisms in HEAs.

Diffraction. X-ray diffraction (XRD) can determine a material's phase structure, grain size, and crystallinity based on the positions and intensities of Bragg peaks. Since single-phase solid-solution alloys are a critical consideration in defining the HEA, XRD coupled with Rietveld refinement serves as a primary decisive method for their characterization. *In situ* XRD has been developed as a pivotal technique to monitor the relationship between structure, activity, and stability in electrochemical environments. Within the initial hours of the HER, the original cubic phase of $NiSe_2$ rapidly transforms into a hexagonal NiSe phase, which exhibits enhanced catalytic activity and serves as the active phase under HER conditions [40]. The phase transformation behaviors of iridium and ruthenium oxides under HER conditions have also been investigated using *in situ* surface XRD studies. In alkaline environments, these oxides transition to hydroxides; however, in acidic conditions (0.5 M H_2SO_4), such transformations are not evident; nor do they convert to metallic phases. Instead, lattice expansion occurs under HER conditions, attributed to proton incorporation.

Traditional crystallographic methods may not fully capture the complexities of HEAs. High-energy synchrotron X-rays and neutron scattering are utilized for total scattering measurements. Total scattering extends conventional powder diffraction by simultaneously measuring and analyzing both Bragg scattering and diffuse scattering. The Fourier transform of total scattering data yields a pair distribution function (PDF), which is a weighted histogram of interatomic distances within the irradiated volume. The positions and shapes of peaks in the PDF offer information about the local environment surrounding each atom, rather than the average structural information contained in Bragg data. Lattice distortion (strain fields) is one of the four significant characteristics of HEAs. The atomic size mismatch among constituent elements can lead to local strain fields within HEAs. This can modulate the energy barriers for molecular adsorption, activation, and transformation in electrocatalytic reactions, thereby significantly influencing the materials' electrocatalytic performance. Therefore, investigating local lattice distortion is crucial for understanding the catalytic properties of HEAs. While long-range lattice distortion can be determined through the broadening of XRD peaks, local atomic-scale lattice distortion does not cause such peak broadening. Instead, the peak positions and widths in the PDF can determine the distribution of interatomic distances, directly reflecting the impact of lattice distortion on the actual local atomic arrangement in the lattice. Neutron radiation was used to study the total scattering measurements of CrMnFeCoNi HEA [41]. The HEA exhibited broader PDF peaks compared to binary and ternary alloys, possibly indicating the highest level of lattice strain.

Characterization techniques that provide atomic-level insights into local structures are particularly important in the study of HEAs for electrochemical applications. In HEA structures, the increase in the number of mixed elements and the reduction in nanoparticle size may result in poor crystallinity of synthesized HEAs. Additionally, self-reconstruction is a significant feature in electrochemical reactions such as the OER. In OER, disordered or low-crystallinity (oxy)hydroxides formed on the surfaces of nanoparticles are often considered to have high activity. This poses resolution challenges for traditional XRD, which relies on ordered structures that produce Bragg diffraction. PDF analysis varied adjustment of the atomic coordination environment in FeCoNiCrMo HEA by varying the Mo content [42]. Due to Mo's relatively large atomic size, HEAs with high Mo content have the smallest nearest-neighbor CNs, leading to a more loosely arranged surface, narrower band distribution, higher *d*-band centers, and stronger adsorption energies of the catalyst. This is speculated to be the reason for its excellent OER catalytic activity.

Selected area electron diffraction (SAED), a technique within TEM, enables the examination of specific regions within a sample, providing detailed information about the crystal structure and phase composition at a higher spatial resolution compared to XRD. Furthermore, electron pair distribution function (ePDF) analysis, derived from electron diffraction data, offers insights into both local and medium-range order within materials. By analyzing the distribution of interatomic distances, ePDF can reveal subtle structural features that are not discernible through traditional diffraction techniques. This method is especially useful for characterizing amorphous or poorly crystalline regions, which are often present in HEAs due to their complex compositions and processing histories. In the context of HEAs, combining SAED and ePDF techniques allows for a comprehensive understanding of their local structural variations.

Spectroscopy. Spectroscopic techniques are analytical methods used to investigate material composition and structure by examining the interactions of light or particles at various energies with matter. Based on the type of radiation source and detection principle employed, spectroscopic techniques can be categorized into several types, including electron spectroscopy, X-ray spectroscopy, and inductively coupled plasma spectroscopy. As the application of HEA in electrocatalysis advances, precise characterization of their complex compositions and active sites has become essential. Spectroscopy, particularly at atomic-scale spatial resolution, provides comprehensive chemical insights into HEA electrocatalysts, elucidating the relationship between their structure and activity.

Currently, the precision in synthesizing HEA electrocatalysts has reached the subnanometer level. With the growing complexity of compositions, the demand for atomic-level characterization has intensified. In electrocatalysis, the integration of electron spectroscopy with electron microscopy, using high-energy electron beams interacting with samples, offers ultrahigh-resolution imaging of surface and structural features, achieving spatial resolutions at the subnanometer or even atomic level. This multidimensional information on spatial and chemical characteristics of HEAs enables accurate characterization of their microstructures, which is pivotal for understanding the origins of their catalytic properties and optimizing synthesis methods. This combination not only reveals surface morphology but also enables in-depth analysis of atomic-scale compositional distribution and chemical states.

X-Ray Spectroscopy. Due to the complexity of structure–activity relationships in HEA electrocatalysts, structural morphology and compositional information alone are insufficient for designing new HEA catalysts. Given the critical impact of electronic structure on intrinsic catalytic activity, more detailed characterization techniques are required. X-ray spectroscopy has become an indispensable tool for studying HEA electrocatalysts, offering in-depth insights into phase structures, oxidation states, and electronic configurations.

In situ and operando X-ray spectroscopy techniques have been developed to study HEA electrocatalysts under actual operating conditions. In HEA electrocatalyst research, X-ray spectroscopy, combined with in situ/operando setups, has enabled the precise characterization of complex, multielemental compositions and the detection of subtle electronic structure and chemical environment change during catalytic processes. This capability is crucial for understanding transient intermediate species and reaction mechanisms. Avoid these potential artifacts by maintaining the material in its reaction environment, preserving its true structure and composition. Meantime, synchrotrons produce highly intense and tunable X-ray beams, allowing for deep analysis of materials. Recent advancements in synchrotron technology have significantly enhanced both resolution and detection speed for X-ray analysis.

Energy-dispersive X-ray spectroscopy (EDS) is one of the most widely used X-ray spectroscopic techniques combined with STEM. EDS uses an incident electron beam of known energy to excite atomic elements within the sample, generating characteristic X-ray signals. Specifically, when the electron beam strikes the sample, it excites inner-shell electrons, creating vacancies that are filled by outer-shell electrons, releasing X-rays with characteristic energies. EDS uses these X-ray signals to determine the types and concentrations of elements in the local structure of the sample.

Because most compositional analysis tools lack spatial resolution (primarily targeting bulk materials), EDS elemental mapping has become indispensable for analyzing compositional distributions in heterogeneous nanostructures. Heterogeneity within particles has become a critical factor in electrocatalyst design. Core–shell HEA nanoparticles as representative materials are widely applied in electrochemical reactions such as EOR, ORR, and HER, where shell thickness reduction demands increasingly high spatial resolution. Recently, the construction of subnanometer PtRuFeCoNi atomic layers on Pd NP surfaces demonstrated atomic-resolution EDS mapping and corresponding line scan profiles, verifying the spatial distribution of elements [43].

Local structural heterogeneity, such as domain segregation or clustering, within nanoparticles also necessitates atomic-resolution EDS mapping. Through atomic-resolution EDS mapping (Figure 8.7), it was found that, in PtPdFeCoM (M = Ru/Rh) HEAs, the concentration of Ru and Rh can significantly alter atomic-level chemical order [39]. The inclusion of Rh reduces the formation energy of a uniformly mixed HEA phase, while Ru increases system energy. Elevated Ru concentrations promote the clustering of Ru in local structures. This local configuration difference generates heterogeneous metal interfaces that can modulate interfacial electric fields, with a pivotal influence on LDOS distribution, underscoring its potential in electrochemical applications.

However, with the increasing interest in SA catalysis and the pursuit of maximizing atomic utilization efficiency, atomically dispersed sites have also been integrated into HEA. Currently, spatial identification of these isolated atomic sites primarily depends on Z-contrast in STEM, leveraging differences in atomic numbers (Z) for contrast. Nevertheless, achieving atomic-level mapping of these SA sites remains challenging due to limitations in signal-to-noise ratio, which restricts the acquisition of accurate information. Thus, a comprehensive analysis combining multiple characterization techniques is essential for reliable identification and validation.

X-Ray Absorption Spectroscopy (XAS). XAS is typically divided into two regions: X-ray absorption near-edge structure (XANES) and extended X-ray absorption

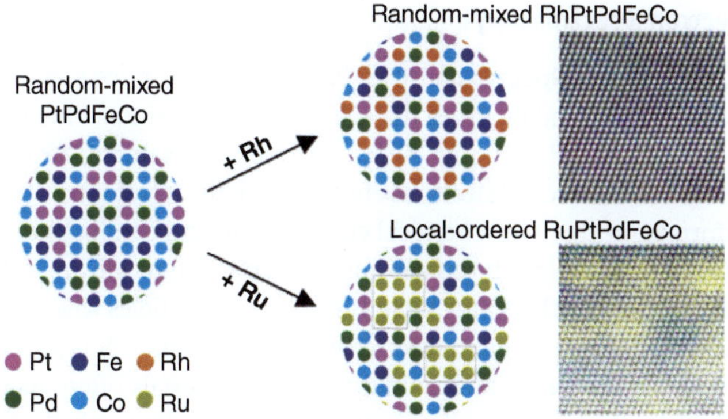

Figure 8.7 Atomic-resolution EDS mapping for PtPdFeCoM (M = Ru/Rh) HEAs [44] / with permission of American Chemical Society.

fine structure (EXAFS). XANES refers to the fine structure within approximately 50 eV around the absorption edge and primarily reflects the local electronic structure and chemical environment around the absorbing atom, providing information on valence state, CN, and symmetry. Due to its sensitivity to changes in electronic states, XANES is commonly used to investigate oxidation states and local symmetry. EXAFS, which occurs within the range of 50 to 1000 eV beyond the absorption edge, consists of oscillations that arise from interference between the photoelectron excited by the X-rays and those scattered by surrounding atoms. EXAFS provides information on the distance between the absorbing atom and neighboring atoms, CNs, and disorder, making it highly suitable for studying atomic arrangements in both crystalline and amorphous materials.

For example, *in situ* XAFS was used to study the most electroactive sites of Ni within a FeCoNiMnCr HEA during the OER [45]. The CN of Ni-metal significantly decreased from 5.6 to 3.1, accompanied by an increase in Ni—O coordination, suggesting that a substantial proportion of Ni atoms were oxidized from the HEA phase. Simultaneously, the absorption edges of other metals shifted to higher energies, indicating that higher-valence metal sites facilitated the OER process. Quasi-*in situ* XAFS deciphered the effects of reactive oxygen adsorption on metal sites in the high-entropy metal selenide $(FeCoNiCrCu)Se_4$ during OER. Results revealed that the pre-edge features shifted to higher photon energies, attributed to increased electron transfer from metal atoms to adsorbed oxygen species. The enhancement and shift of the first-shell peak represent the rearrangement of the local coordination structure due to adsorbed oxygen.

Traditionally, collecting a full XAS spectrum on standard beamlines requires several to tens of minutes, significantly limiting time resolution for *in situ* reaction monitoring. Recent developments in quick-XAS, utilizing specially designed continuously scanning monochromators and rapid data acquisition, have reduced time resolution to the scale of seconds, greatly shortening collection times.

8.3.1 Electron Spectroscopy

X-Ray Photoelectron Spectroscopy (XPS). XPS is a surface-sensitive analytical technique that detects and analyzes the elemental composition, chemical state, and electronic state of materials. Using X-rays to irradiate a sample, XPS measures the kinetic energy and number of electrons emitted from the top atomic layers, providing detailed information about the chemical environment and bonding state of elements on a material's surface. Since it probes only the surface (typically up to 10 nm deep), XPS is ideal for studying thin films, coatings, and nanoparticles where surface chemistry plays a critical role. In nano-electrocatalyst research, XPS is particularly valuable. The catalytic efficiency and stability of nanomaterials often depend on their surface composition, oxidation states, and atomic bonding, which directly affect reaction pathways and the overall performance of catalytic processes. With XPS, researchers can characterize active sites, monitor changes in oxidation states during reactions, and analyze the interactions between nanoparticles and support materials.

XPS effectively demonstrated that, in FeCoNiMnRu, due to the high electronegativity of Ru, electrons transfer from Fe, Co, Ni, and Mn atoms to Ru atoms. This charge redistribution modifies the electronic structure of each metal element, thereby enhancing the adsorption of various intermediates in the OER [46]. Beyond detecting metal valence states, XPS is also essential for analyzing the chemical state of oxygen. In the Au SA-MnFeCoNiCu LDH, XPS measurements confirmed the presence of oxygen vacancies introduced by SA Au, verifying the improvement in intrinsic catalytic activity [47].

Electron Energy Loss Spectroscopy (EELS). EELS is another advanced electron spectroscopy technique, frequently used alongside EDS, and is particularly valuable for obtaining chemical and bonding information at atomic resolution. EELS measures the energy lost by electrons as they pass through a thin sample. When electrons interact with the atoms in the sample, they can lose energy through various inelastic scattering processes, which correspond to excitations in the sample, such as inner-shell electron excitations, plasmon resonances, or phonon vibrations. By analyzing the spectrum of these energy losses, EELS provides detailed information about the sample's chemical composition, oxidation states, and even electronic structure.

In the context of HEA electrocatalysts, EELS complements EDS by adding another layer of chemical specificity. For example, EELS can detect subtle variations in bonding environments, such as changes in oxidation states or coordination chemistry, which are crucial for understanding the active sites in catalytic processes. Unlike EDS, which is more sensitive to heavier elements, EELS can detect light elements such as oxygen, making it particularly useful for studying complex oxides and other light-element-rich materials. EELS can thus reveal how elemental distributions and oxidation states change at the atomic level, contributing to a deeper understanding of catalytic mechanisms and material stability under reaction conditions.

For instance, Miao et al. used EELS to reveal substantial differences in the oxidation states and bonding environments of Co and O within HEO thin films

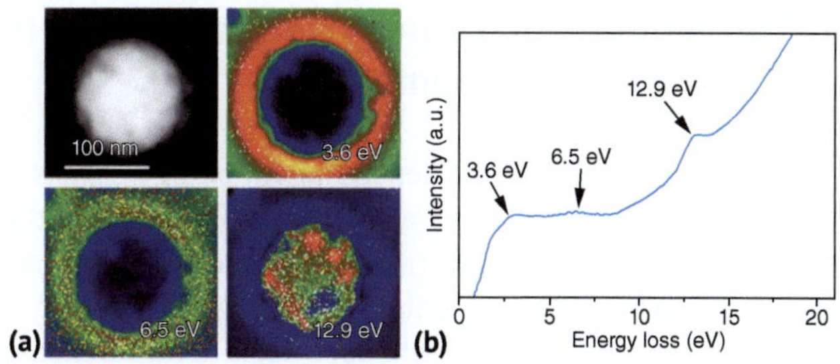

Figure 8.8 EELS mapping and profile of GaInAuCuPtPd high-entropy liquid metal [49] / John Wiley & Sons / CC BY 4.0.

synthesized under varied conditions (e.g., temperature and composition), while the chemical environments of other elements remained unchanged [48]. This finding offers valuable guidance for tuning kinetic parameters in the synthesis of HE materials. Similarly, EELS was employed to analyze the complex liquid-solid mixed layered microstructure of high-entropy liquid metal alloy nanoparticles [49]. EELS maps, along with spectra processed to remove the zero-loss peak (Figure 8.8), exhibited multiple peaks linked to plasmon resonances, bulk plasmon states, and oxidation states, providing a detailed characterization of the alloy's intricate structure.

Inductively Coupled Plasma (ICP). *Although not a form of electron spectroscopy, ICP is often discussed alongside advanced spectroscopic techniques due to its crucial role in elemental analysis and its complementarity to surface-sensitive electron spectroscopy methods.* In ICP, the sample is introduced into a high-temperature plasma, where atoms or ions are excited or ionized, generating characteristic spectral signals. Depending on the detection method, ICP spectroscopy is divided into two main techniques: ICP-optical emission spectroscopy (ICP-OES) and ICP-mass spectrometry (ICP-MS). ICP-OES determines the types and quantities of elements by detecting characteristic spectra emitted from excited atoms, suitable for multielement quantitative analysis with high sensitivity and accuracy. ICP-MS, on the other hand, measures the mass-to-charge ratio of ionized atoms, achieving high-sensitivity detection of trace elements with detection limits down to the ppt level, making it ideal for ultratrace element analysis.

Compared to EDS, ICP offers advantages in quantitative accuracy. EDS signals can exhibit relatively low intensity, especially for elements with atomic numbers below 30 ($Z < 30$), reducing detection efficacy and potentially leading to discrepancies between EDS analysis and actual composition. In contrast, ICP, through bulk measurements, delivers exceptional quantitative analysis with higher accuracy and reliability. However, while ICP has high sensitivity and accuracy for elemental analysis, it has limitations in sample preparation and spatial resolution. First, sample preparation for ICP is complex; it requires samples to be in ionic form, often necessitating the dissolution of solid samples into liquid, adding complexity to sample preparation. Additionally, ICP provides only bulk compositional information, lacking the ability to reveal the spatial distribution of elements within the microregions of the sample.

8.4 High-Throughput and Data-Driven Techniques in High-Entropy Catalyst Development

High-entropy catalysts have garnered extensive attention in catalysis research due to their multielemental alloy structures and unique physicochemical properties. However, the complexity of their compositional space and diverse structural features pose challenges for traditional experimental approaches in exploring and optimizing high-entropy catalysts. Traditional methods struggle to comprehensively sample these materials; even selecting combinations of five elements from a set of 10 results in nearly 300 possible configurations, not to mention variations in element ratios. To address this, high-throughput (HT) and data-driven approaches have been introduced to accelerate the discovery and design of high-entropy catalysts.

High-Throughput. HT techniques, by leveraging parallelization and automation, allow for rapid screening of large numbers of samples, significantly accelerating experimental workflows [50, 51]. Applications of HT methods in HEA catalyst development span synthesis, testing, and characterization.

HT Computation. HT computation in high-entropy catalyst research largely centers on DFT calculations. DFT, a quantum mechanical modeling method, can estimate intermediate states and product selectivity in catalytic processes, revealing the electronic states of various metal atoms under high-entropy conditions and thus facilitating rational composition design. However, the number of theoretically synthesizable catalysts is vast and too expansive to explore solely through experimental methods. To accelerate catalyst discovery, first-principles computational chemistry, particularly DFT, is increasingly applied.

Machine learning (ML) models trained on extensive databases generated by HT DFT calculations can predict atomic structures with near-DFT accuracy, significantly reducing computational costs. This is especially useful for high-dimensional systems such as HEAs. In materials modeling, deep learning potential (DLP) surface models, which approximate potential energy surfaces using deep neural networks and describe atomic local environments with symmetry functions, provide an efficient alternative with high accuracy across structural variations. Various ML approaches, such as multilayer perceptrons, Gaussian approximation potentials, spectral neighbor analysis potential, and many-body tensor representation, have proven effective. DLPs achieve near-DFT accuracy with significantly improved computational efficiency, allowing large-scale global optimizations up to four orders of magnitude faster than pure DFT methods, enabling more efficient structural searches across diverse conditions. Techniques such as stochastic surface walking, genetic algorithms, and grand canonical Monte Carlo allow DLPs to generate a wide range of optimized structures, aiding in understanding structural evolution and identifying low-energy surfaces.

The combination of DLPs with DFT is especially valuable in HEA design. For HEAs, with their complex multicomponent structures, pure DFT calculations are insufficient for efficient exploration of material properties. DLPs enable rapid screening of stable low-energy structures among many candidate configurations, followed by DFT analysis of optimal candidates for precise electronic structure evaluations. This hybrid approach is also applicable to phase diagram prediction and stability analysis of HEAs: thermodynamic and structural data generated by DLPs, combined with DFT validation, allow phase diagram construction and stability prediction across different compositional combinations. A typical HEA catalyst design workflow might begin with DFT to generate data on representative HEA configurations, utilize DLPs for HT screening to identify promising catalytic sites, and then refine electronic structure predictions for these candidates with DFT. This approach reduces computational load and enhances efficiency, significantly advancing HEA catalyst research. In summary, combining DLPs with DFT offers robust computational support for rapid screening and optimization of HEAs and other complex materials, expediting the design of efficient catalysts and stability assessments.

HT Synthesis. HT synthesis methods, such as sputter deposition and inkjet printing, enable rapid preparation of multielement alloy combinations, particularly suited for thin films or small sample preparation. Vapor deposition allows for the efficient production of HEO films, offering advantages in control precision and sample uniformity. In HT applications, these methods allow for the rapid generation of thin films with diverse compositional combinations. Additionally, combining printed arrays of precursor materials with rapid pulsed heating enables the preparation of multicomponent HEAs. For example, 70 Pt-based HEA element combinations have been synthesized, providing a comprehensive platform for investigating the catalytic performance of ORR across various compositions [52]. HT wet-chemical or pulsed high-temperature synthesis methods, using liquid handling robots and automated pipetting systems, allow for flexible adjustments of solution compositions in line with experimental designs. This is especially effective in the synthesis of colloidal nanocrystals, enabling rapid preparation of nanoparticles with varying compositions and sizes. This method has been applied to the preparation of various high-entropy materials by combining different precursors, solutions, and reaction conditions to quickly tune alloy compositions and structures. With the advancement of AI-driven robotics, autonomous exploration of chemical research tasks has become possible, significantly reducing experimental and decision-making time [53–55]. This capability offers accelerated pathways for the development of HEA catalysts, enabling rapid discovery and optimization processes that were previously time-intensive.

HT Testing. HT testing enables researchers to rapidly evaluate the electrochemical performance of multicomponent electrocatalysts. Microelectrodes allow for rapid electrochemical testing, making large-scale screening feasible and enabling quick assessments of catalyst performance in specific electrochemical reactions (e.g., oxygen reduction, hydrogen evolution). Through parallelized testing, vast electrocatalyst combinations can be evaluated in a short time, providing data support for identifying optimal catalyst compositions.

HT Characterization. HT characterization relies on automated spectroscopic and microscopic techniques, enabling batch characterization of catalyst structures,

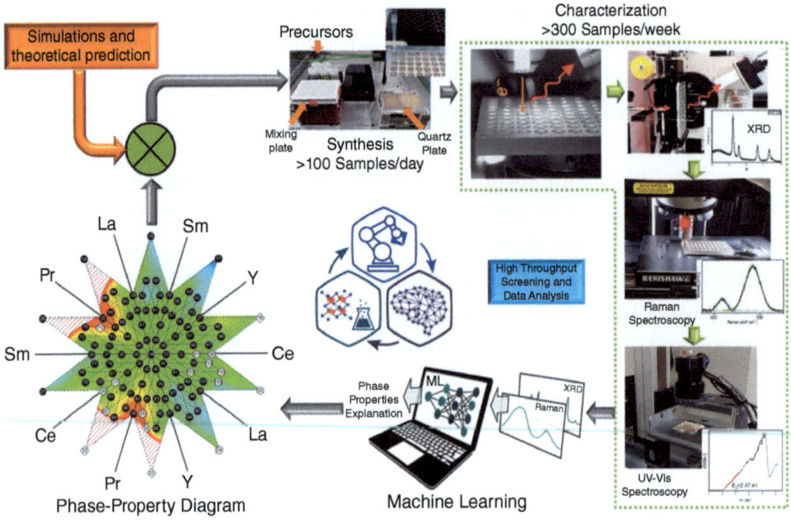

Figure 8.9 ML-assisted automated synthesis and characterization platforms [56] / with permission of John Wiley & Sons.

compositions, and surface properties (Figure 8.9). Automated XRD platforms can analyze material crystal structures, particularly useful for assessing single-phase solid solutions in HEAs. Coupled with ML-enhanced data analysis, automated XRD platforms can rapidly identify samples with appropriate structures from extensive characterization data, saving time on manual analysis. Automated scanning electron microscopy platforms with SEM-EDS can analyze morphology and elemental distributions, generating rapid compositional imaging for studying microstructure and surface composition. This platform, combined with HT XRD, provides comprehensive analyses of material phase compositions and microstructures. Automated spectroscopic techniques such as XPS and Raman spectroscopy further streamline surface chemical and electronic structure characterization, enabling rapid acquisition of surface composition and valence state information in HT screening processes.

Data-Driven Approaches. Data-driven approaches leverage the analysis of extensive experimental and computational data to uncover relationships between material properties and their composition and structure. In high-entropy catalyst research, large databases encompassing various element combinations, structural characteristics, and catalytic performances are constructed through HT experiments, sensors, and database integration. These data require thorough cleaning, storage, and management to ensure quality and usability [57–60].

Following data acquisition, feature engineering is conducted to preprocess, extract, and normalize features from the raw data, making them more representative and analyzable. Next, suitable ML or statistical models, such as regression, classification, or clustering models, are chosen to model and analyze the data. Finally, model performance is evaluated using metrics such as cross-validation, accuracy, and F1 scores to ensure accuracy and robustness. With these analyses, ML models can predict the catalytic performance of novel high-entropy alloys, guiding experimental design and reducing trial-and-error processes. Figure 8.10 illustrates a model that generates reactive sites on the (100) or (111) surfaces of a HEA by specifying the composition of a particular metal element. These sites are then input into different models for training to obtain the corresponding OH^* adsorption energies.

Bayesian optimization, an ML method, optimizes alloy composition progressively without fully traversing the design space, accelerating the identification of optimal catalyst formulations. Bayesian optimization often relies on DFT calculations or ML model predictions. For example, a DFT-based model was developed to predict catalytic activity for the ORR. Combining this model with Bayesian single-objective optimization, they identified optimal catalyst compositions within the Ag—Ir—Pd—Pt—Ru and Ir—Pd—Pt—Rh—Ru high-entropy alloy systems, pinpointing multiple efficient combinations in fewer than 50 experiments [62]. Similarly, Peng et al. highlighted the efficiency of Bayesian optimization in multicomponent catalyst composition spaces, using computational models and automated screening to significantly reduce screening time while reliably identifying top-performing compositions [63].

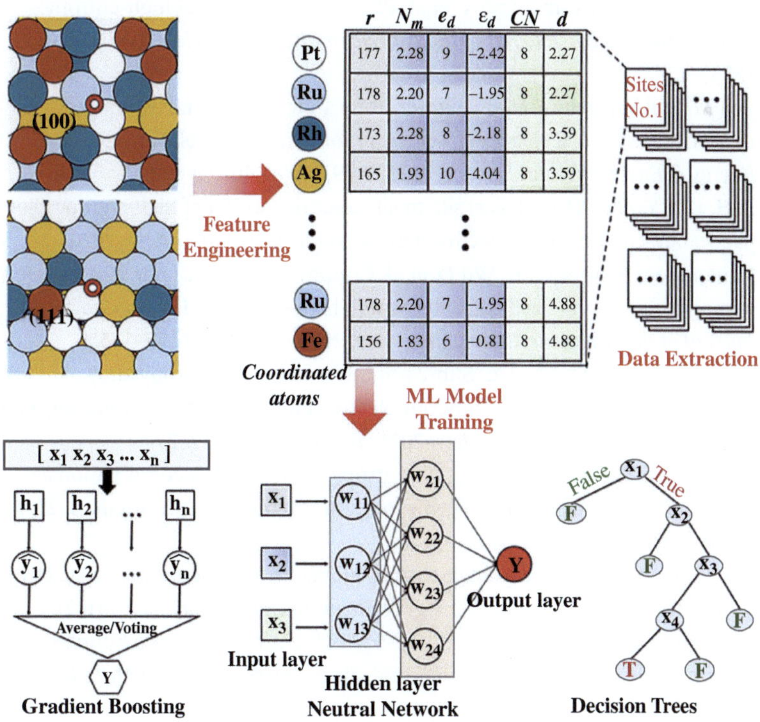

Figure 8.10 DFT-ML scheme for guidance of HEA catalyst synthesis [61] / Elsevier / CC BY 4.0.

While HT and data-driven methods hold tremendous potential in high-entropy catalyst research, challenges remain. First, the complexity of high-entropy alloys complicates data acquisition and analysis. Second, the predictive accuracy of ML models depends on high-quality, large-scale datasets, the construction of which requires extensive experimental and computational resources. As computational power and experimental techniques advance, HT and data-driven approaches are expected to play an increasingly significant role in high-entropy catalyst research. Through multidisciplinary collaboration that combines experimental, computational, and data science expertise, we can accelerate the discovery and application of high-performance catalysts.

References

1 Löffler T, Ludwig A, Rossmeisl J, et al. What makes high-entropy alloys exceptional electrocatalysts? *Angewandte Chemie International Edition*, 2021, 60(52): 26894–26903.

2 Hunter B M, Gray H B. and Müller A M. Earth-abundant heterogeneous water oxidation catalysts. *Chemical Reviews*, 2016, 116(22): 14120–14136.

3 Quan L, Jiang H, Mei G, et al. Bifunctional electrocatalysts for overall and hybrid water splitting. *Chemical Reviews*, 2024, 124(7): 3694–3812.

4 Wu D, Kusada K, Yamamoto T, et al. Platinum-group-metal high-entropy-alloy nanoparticles. *Journal of the American Chemical Society*, 2020, 142(32): 13833–13838.

5 Wei M, Sun Y, Zhang J, et al. High-entropy alloy nanocrystal assembled by nanosheets with d–d electron interaction for hydrogen evolution reaction. *Energy & Environmental Science*, 2023, 16(9): 4009–4019.

6 Li H, Han Y, Zhao H, et al. Fast site-to-site electron transfer of high-entropy alloy nanocatalyst driving redox electrocatalysis. *Nature Communications*, 2020, 11(1): 5437.

7 Minamihara H, Kusada K, Wu D, et al. Continuous-flow reactor synthesis for homogeneous 1 nm-sized extremely small high-entropy alloy nanoparticles. *Journal of the American Chemical Society*, 2022, 144(26): 11525–11529.

8 Feng G, Ning F, Song J, et al. Sub-2 nm ultrasmall high-entropy alloy nanoparticles for extremely superior electrocatalytic hydrogen evolution. *Journal of the American Chemical Society*, 2021, 143(41): 17117–17127.

9 Mei Y, Feng Y, Zhang C, et al. High-entropy alloy with Mo-coordination as efficient electrocatalyst for oxygen evolution reaction. *ACS Catalysis*, 2022, 12(17): 10808–10817.

10 Zhu H, Sun S, Hao J, et al. A high-entropy atomic environment converts inactive to active sites for electrocatalysis. *Energy & Environmental Science*, 2023, 16(2): 619–628.

11 He R, Yang L, Zhang Y, et al. A 3d-4d-5d High entropy alloy as a bifunctional oxygen catalyst for robust aqueous zinc–air batteries. *Advanced Materials*, (2023a), 35(46): 2303719.

12 Yao L, Zhang F, Yang S, et al. Sub-2 nm IrRuNiMoCo high-entropy alloy with iridium-rich medium-entropy oxide shell to boost acidic oxygen evolution. *Advanced Materials*, 2024, 36(25): 2314049.

13 Maulana A L, Chen P.-C, Shi Z, et al. Understanding the structural evolution of IrFeCoNiCu high-entropy alloy nanoparticles under the acidic oxygen evolution reaction. *Nano Letters*, 2023, 23(14): 6637–6644.

14 Kwon J, Sun S, Choi S, et al. Tailored electronic structure of ir in high entropy alloy for highly active and durable bifunctional electrocatalyst for water splitting under an acidic environment. *Advanced Materials*, 2023, 35(26): 2300091.

15 Wu H, Lu Q, Li Y, et al. Structural framework-guided universal design of high-entropy compounds for efficient energy catalysis. *Journal of the American Chemical Society*, 2023, 145(3): 1924–1935.

16 Liu D, Yan X, Guo P, et al. Inert Mg incorporation to break the activity/stability relationship in high-entropy layered hydroxides for the electrocatalytic oxygen evolution reaction. *ACS Catalysis*, 2023, 13(11): 7698–7706.

17 Baek J, Hossain M D, Mukherjee P, et al. Synergistic effects of mixing and strain in high entropy spinel oxides for oxygen evolution reaction. *Nature Communications*, 2023, 14(1): 5936.

18 Suhr E, Krysiak O A, Strotkötter V, et al. High-throughput exploration of structural and electrochemical properties of the high-entropy nitride system (Ti–Co–Mo–Ta–W) N. *Advanced Energy Materials*, 2023, 25(22): 2300550.

19 Wang T, Chen H, Yang Z, et al. High-entropy perovskite fluorides: a new platform for oxygen evolution catalysis. *Journal of the American Chemical Society*, 2020, 142(10): 4550–4554.

20 Jo S, Kim M.-C, Lee K. B, et al. Nonprecious high-entropy chalcogenide glasses-based electrocatalysts for efficient and stable acidic oxygen evolution reaction in proton exchange membrane water electrolysis. *Advanced Energy Materials*, 2023, 13(35): 2301420.

21 Wu G. and Zelenay P. Nanostructured nonprecious metal catalysts for oxygen reduction reaction. *Accounts of Chemical Research*, 2013, 46(8): 1878–1889.

22 He R, Yang L, Zhang Y, et al. A CrMnFeCoNi high entropy alloy boosting oxygen evolution/reduction reactions and zinc-air battery performance. *Energy Storage Materials*, 2023, 58, 287–298.

23 Yu Y, Xia F, Wang C, et al. High-entropy alloy nanoparticles as a promising electrocatalyst to enhance activity and durability for oxygen reduction. *Nano Research*, 2022, 15(9): 7868–7876.

24 Chen T, Qiu C, Zhang X, et al. An ultrasmall ordered high-entropy intermetallic with multiple active sites for the oxygen reduction reaction. *Journal of the American Chemical Society*, 2024, 146(1): 1174–1184.

25 Wang Y, Zhang X.-Y, He H, et al. Ordered mesoporous high-entropy intermetallics for efficient oxygen reduction electrocatalysis. *Advanced Energy Materials*, 2024, 14(8): 2303923.

26 Feng G; Ning F; Pan Y, et al. Engineering Structurally Ordered High-Entropy Intermetallic Nanoparticles with High-Activity Facets for Oxygen Reduction in Practical Fuel Cells. *Journal of the American Chemical Society*, 2023, 145 (20): 11140–11150.

27 Dattila F, Seemakurthi R R, Zhou Y, et al. Modeling operando electrochemical CO_2 reduction. *Chemical Reviews*, 2022, 122(12): 11085–11130.

28 Cai W, Cao X, Wang Y, et al. Spatial structure of electron interactions in high-entropy oxide nanoparticles for active electrocatalysis of carbon dioxide reduction. *Advanced Materials*, 2024, 36(45): 2409949.

29 Pedersen J K, Batchelor T A A, Bagger A, et al. High-Entropy Alloys as Catalysts for the CO2 and CO Reduction Reactions. *ACS Catalysis*, 2020, 10 (3): 2169–2176.

30 Nellaiappan S, Katiyar N K, Kumar R, et al. High-entropy alloys as catalysts for the CO_2 and CO reduction reactions: experimental realization. *ACS Catalysis*, 2020, 10(6): 3658–3663.

31 Cavin J, Ahmadiparidari A, Majidi L, et al. 2D high-entropy transition metal dichalcogenides for carbon dioxide electrocatalysis. *Advanced Materials*, 2021, 33(31): 2100347.

32 Zhang R, Zhang Y, Xiao B, et al. Phase Engineering of High-Entropy Alloy for Enhanced Electrocatalytic Nitrate Reduction to Ammonia. *Angewandte Chemie International Edition*, 2024, 63(35): e202407589.

33 Qi S, Lei Z, Huo Q, et al. Ultrathin High-Entropy Fe-Based Spinel Oxide Nanosheets with Metalloid Band Structures for Efficient Nitrate Reduction toward Ammonia. *Advanced Materials*, 2024, 36(27): 2403958.

34 Song B, Yang Y, Rabbani M, et al. In situ oxidation studies of high-entropy alloy nanoparticles. *ACS Nano*, 2020, 14(11): 15131–15143.

35 Hashimoto N, Mori K, Yoshida H, et al. Thermal stability of high-entropy alloy nanoparticles evaluated by in situ TEM observations. *Nano Letters*, 2024, 24(23): 7063–7068.

36 Luan C, Escalera-López D, Hagemann U, et al. Revealing dynamic surface and subsurface reconstruction of high-entropy alloy electrocatalysts during the oxygen evolution reaction at the atomic scale. *ACS Catalysis*, 2024, 14(17): 12704–12716.

37 Moniri S, Yang Y, Ding J, et al. Three-dimensional atomic structure and local chemical order of medium- and high-entropy nanoalloys. *Nature*, 2023, 624(7992): 564–569.

38 Liu H, Zhang Y, Zhang L, et al. Unveiling atomic-scaled local chemical order of high-entropy intermetallic catalyst for alkyl-substitution-dependent alkyne semihydrogenation. *Journal of the American Chemical Society*, 2024, 146(29): 20193–20204.

39 Shi Y, Collins L, Balke N, et al. In-situ electrochemical-AFM study of localized corrosion of AlxCoCrFeNi high-entropy alloys in chloride solution. *Applied Surface Science*, 2018, 439, 533–544.

40 Zhai L, Benedict Lo T W, Xu Z.-L, et al. In situ phase transformation on nickel-based selenides for enhanced hydrogen evolution reaction in alkaline medium. *ACS Energy Letters*, 2020, 5(8): 2483–2491.

41 Owen L R, Pickering E J, Playford H. Y, et al. An assessment of the lattice strain in the CrMnFeCoNi high-entropy alloy. *Acta Materialia*, 2017, 122, 11–18.

42 Zhou X, Zhu H, Fu S, et al. (2024) Atomic structure amorphization and electronic structure reconstruction of FeCoNiCrMox high-entropy alloy nanoparticles for highly efficient water oxidation. *Small*, 2023, 20(47): 2405596.

43 Wu C.-Y, Hsiao Y.-C, Chen Y, et al. A catalyst family of high-entropy alloy atomic layers with square atomic arrangements comprising iron- and platinum-group metals. *Science Advances*, 2024, 10(30): eadl3693.

44 Huang Z, Li T, Li B, et al. Tailoring local chemical ordering via elemental tuning in high-entropy alloys. *Journal of the American Chemical Society*, 2024, 146(3): 2167–2173.

45 Hu J, Guo T, Zhong X, et al. In situ reconstruction of high-entropy heterostructure catalysts for stable oxygen evolution electrocatalysis under industrial conditions. *Advanced Materials*, 2024, 36(14): 2310918.

46 Hao J, Zhuang Z, Cao K, et al. Unraveling the electronegativity-dominated intermediate adsorption on high-entropy alloy electrocatalysts. *Nature Communications*, 2022, 13(1): 2662.

47 Wang F, Zou P, Zhang Y, et al. Activating lattice oxygen in high-entropy LDH for robust and durable water oxidation. *Nature Communications*, 2023, 14(1): 6019.

48 Miao L, Sivak J T, Kotsonis G, et al. Chemical environment and structural variations in high entropy oxide thin film probed with electron microscopy. *ACS Nano*, 2024, 18(23): 14968–14977.

49 Allioux F-M, Nazari S, Ghasemian M B, et al. Atomic dispersion via high-entropy liquid metal alloys. *Small Structures*, 2024, 5(12): 2400294.

50 Batchelor T A A, Löffler T, Xiao B, et al. Complex-solid-solution electrocatalyst discovery by computational prediction and high-throughput experimentation. *Angewandte Chemie International Edition*, 2021, 60(13): 6932–6937.

51 Xu W, Diesen E, He T, et al. Discovering high entropy alloy electrocatalysts in vast composition spaces with multiobjective optimization. *Journal of the American Chemical Society*, 2024, 146(11): 7698–7707.

52 Pan Y, Shan X, Cai F, et al. Accelerating the discovery of oxygen reduction electrocatalysts: high-throughput screening of element combinations in Pt-based high-entropy alloys. *Angewandte Chemie International Edition*, 2024, 63(37): e202407116.

53 Coley C W, Thomas III D A, Lummiss J. A, et al. A robotic platform for flow synthesis of organic compounds informed by AI planning. *Science*, 2019, 365(6453): eaax1566.

54 Ha T, Lee D, Kwon Y, et al. AI-driven robotic chemist for autonomous synthesis of organic molecules. *Science Advances*, 2023, 9(44): eadj0461.

55 Zhao H, Chen W, Huang H, et al. A robotic platform for the synthesis of colloidal nanocrystals. *Nature Synthesis*, 2023, 2(6): 505–514.

56 Velasco L, Castillo J S, Kante M. V, et al. Phase–property diagrams for multicomponent oxide systems toward materials libraries. *Advanced Materials*, 2021, 33(43): 2102301.

57 Wang Q, Velasco L, Breitung B, et al. High-entropy energy materials in the age of big data: a critical guide to next-generation synthesis and applications. *Advanced Energy Materials*, 2021, 11(47): 2102355.

58 Banko L, Krysiak O A, Pedersen J. K, et al. Unravelling composition–activity–stability trends in high entropy alloy electrocatalysts by using a data-guided combinatorial synthesis strategy and computational modeling. *Advanced Energy Materials*, 2022, 12(8): 2103312.

59 Chen L, Chen Z, Yao X, et al. High-entropy alloy catalysts: high-throughput and machine learning-driven design. *Journal of Materials Informatics*, 2022, 2(4): 19.

60 Banko L, Tetteh E B, Kostka A, et al. Microscale combinatorial libraries for the discovery of high-entropy materials. *Advanced Materials*, 2023, 35(9): 2207635

61 Wan X, Zhang Z, Yu W, et al. Machine-learning-assisted discovery of highly efficient high-entropy alloy catalysts for the oxygen reduction reaction. *Patterns*, 2022, 3(9): 100553.

62 Pedersen J K, Clausen C M, Krysiak O A, et al. Bayesian optimization of high-entropy alloy compositions for electrocatalytic oxygen reduction. *Angewandte Chemie International Edition*, 2021, 60(45): 24144–24152.

63 Peng J, Damewood J K, Karaguesian J, et al. Navigating multimetallic catalyst space with Bayesian optimization. *Joule*, 2021, 5(12): 3069–3071.

9

High-Entropy Materials for Energy Storage Applications

*Yuwei Chen, Wei Luo**

Institute of New Energy for Vehicles, School of Materials Science and Engineering, Tongji University, Shanghai, China

*Corresponding Email: weiluo@tongji.edu.cn

9.1 Introduction to High-Entropy Materials for Battery Applications

High-entropy materials (HEMs), characterized by their composition of multiple principal elements in nearly equimolar ratios, have emerged as a transformative concept in material design, particularly in the field of energy storage. Unlike traditional materials that rely on a few key elements, HEMs harness the synergistic effects of multielement combinations to achieve superior properties. These include enhanced thermal and electrochemical stability, higher mechanical strength, and improved ionic conductivity, all of which are critical for advancing battery technologies (Figure 9.1) [1]. By leveraging these unique features, HEMs offer innovative solutions to challenges such as limited capacity, poor cycle life, and safety concerns in existing battery systems.

The application of HEMs in batteries spans a wide range of components, from electrodes to electrolytes. High-entropy oxides and high-entropy alloys (HEAs), for instance, have shown remarkable potential as electrode materials, delivering high specific capacities and outstanding cycling stability. These materials benefit from the high configurational entropy that stabilizes their crystal structures and mitigates phase separation during repeated charge and discharge cycles. Similarly, high-entropy solid-state electrolytes (HE SSEs) have gained attention for their exceptional ionic conductivity and wide electrochemical stability windows, making them suitable for all-solid-state batteries. These solid-state systems promise higher safety and energy density compared to conventional liquid electrolyte-based batteries.

Recent advances in the field also highlight the growing interest in high-entropy liquid electrolytes. By incorporating a diverse range of solutes, these electrolytes

Functional High Entropy Alloys and Compounds, First Edition. Edited by Yong Zhang, Yonggang Yao and Haojie Liu.

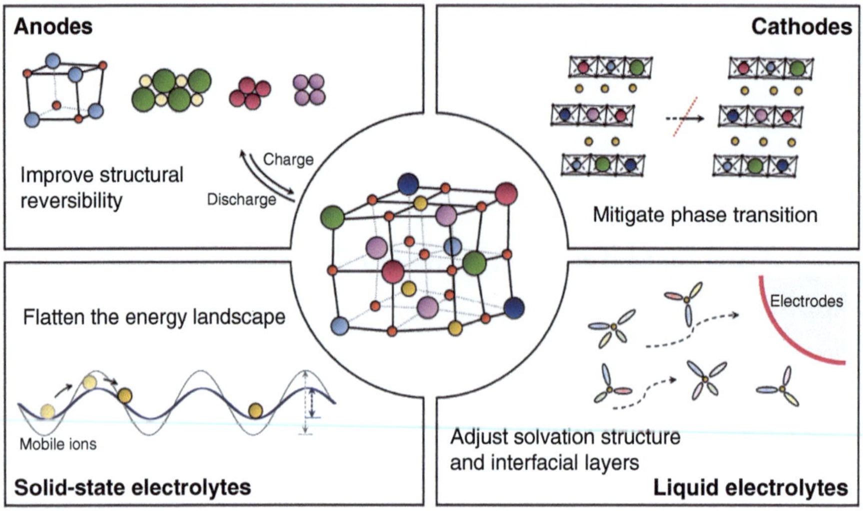

Figure 9.1 Schematic diagram of HEMs' application in batteries as anodes, cathodes, SSEs, and liquid electrolytes for energy storage.

exhibit improved thermal and chemical stability, reduced side reactions, and enhanced ionic transport. This innovation opens up opportunities for batteries operating under extreme conditions or requiring high-power outputs. Furthermore, the flexibility of high-entropy design allows for tailoring material properties to meet the specific demands of emerging battery chemistries, such as lithium-sulfur, lithium-air, and sodium-ion systems.

The rapid progress in HEM research is being driven by the integration of computational tools and experimental techniques. Machine learning (ML) and high-throughput screening are accelerating the discovery of optimal compositions, while advanced characterization methods provide deeper insights into the structure–property relationships of HEMs. These efforts are not only broadening the scope of HEMs but also paving the way for their commercialization in energy storage devices.

In conclusion, HEMs represent a paradigm shift in the design and development of battery components. Their unique properties and versatility offer significant advantages in terms of performance, safety, and sustainability, addressing the growing demands for efficient and reliable energy storage systems. As research continues to evolve, HEMs are expected to play a pivotal role in shaping the future of battery technology, contributing to a sustainable and electrified society.

9.2 High-Entropy Anode Materials

High-entropy anode materials have emerged as a promising approach to address the limitations of conventional anode materials in lithium-ion batteries (LIBs). Traditional transition metal oxides, widely used as anodes, achieve lithium storage through reversible conversion reactions between transition metal cations and lithium ions, offering a high theoretical specific capacity exceeding 600 mAh g^{-1}. However, these materials suffer from significant drawbacks, including drastic volume changes and poor reaction reversibility during charge–discharge cycling, which lead to rapid capacity fading and limited lifespan [2]. The high-entropy strategy provides a transformative solution by incorporating multiple principal elements into the anode structure, improving both reversibility and cycling stability.

The application of high-entropy principles to anode materials leverages the synergistic effects of multielement systems, including enhanced structural stability, tailored electronic properties, and resistance to mechanical degradation. High-entropy anodes, oxides [3], sulfides [4], phosphides [5], and carbides [6] have demonstrated superior electrochemical performance in recent studies. Among these, high-entropy oxides have garnered particular attention due to their diverse lattice structures, including rock salt, spinel, chalcocite, and fluorite. These structures enable the accommodation of large ionic and electronic transport pathways, reducing polarization and enhancing lithium storage efficiency. Furthermore, the high configurational entropy of these materials stabilizes their phase, preventing structural collapse during repeated cycling.

In addition to oxides, high-entropy sulfides and phosphides have shown promise as anode materials due to their unique electrochemical properties and potential for high capacity. Sulfides, for example, exhibit low lithium diffusion barriers and favorable redox kinetics, while phosphides offer high theoretical capacities and excellent conductivity. High-entropy carbides, characterized by their robustness and chemical stability, have also been explored as potential anodes, providing long-term cycling stability even under high current densities.

The design and optimization of high-entropy anodes are further supported by advancements in synthesis techniques and computational modeling. Methods such as high-temperature sintering, solvothermal processes, and ball milling enable precise control over the composition and microstructure of HEMs. Meanwhile, density functional theory (DFT) calculations and ML algorithms facilitate the prediction of optimal element combinations, guiding the rational design of high-performance anodes.

In conclusion, high-entropy anode materials represent a significant advancement in the quest for high-capacity, long-life batteries. By leveraging the unique benefits of multielement systems, these materials address the challenges associated with volume expansion, poor reaction reversibility, and structural instability in traditional anodes. As research progresses, high-entropy anodes are expected to play a crucial role in the development of next-generation energy storage systems, enabling the creation of more efficient and durable LIBs [7].

Traditional transition metal oxides are reduced to zero-valent metals and Li_2O during lithiation, making it difficult to fully return to the initial structure in the subsequent charging process. Fortunately, this challenge can be overcome by employing the high-entropy strategy. The complex components in the high-entropy anodes, typically comprising both active and inactive elements, serve distinct roles. The elements not involved in redox reactions usually act as a host matrix to maintain the framework during cycling. For instance, in the $(Co_{0.2}Cu_{0.2}Mg_{0.2}Ni_{0.2}Zn_{0.2})O$ anode, the inactive Mg helps to preserve the rock salt in the full voltage range, facilitating the reduced cations to diffuse back to the initial structure. Therefore, the configurational entropy stabilization of the lattice imparts exceptional cycling stability to the anode [8]. Chen et al. provided visual evidence using *in situ* X-ray diffraction (XRD) and *in situ* transmission X-ray microscopy (TXM), proving the pivotal role of the inactive Ti in conserving the structure stability in spinel-type $(Ni_{0.2}Co_{0.2}Mn_{0.2}Fe_{0.2}Ti_{0.2})_3O_4$ anode [9].

Intriguingly, the high-entropy anodes also show high reversibility though all the constituent elements are electroactive centers. Su et al. explored the mechanism of $(Co_{0.2}Ni_{0.2}Mn_{0.2}Zn_{0.2}Fe_{0.2})_3O_{3.2}$ anodes via *in situ* transmission electron microscopy (TEM). As shown in Figure 9.2a and b, during lithiation, the electrode undergoes conversion and alloying reactions to form Fe, Ni, Co, Mn, $Li_{0.2}Zn_{0.8}$, and Li_2O phases, which are uniformly distributed in the particles without obvious segregation. The original crystal structure can be recovered upon the charging process, and the stable structural reversibility can be maintained even after hundreds of cycles, which would undoubtedly enable specific capacity exceeding 600 mAh g^{-1} after 200 cycles, with a high coulomb efficiency of 98.6% [10].

However, an important issue of high-entropy anodes that should not be overlooked is the significant drop in specific capacity after the second cycle compared to the first. With the help of X-ray absorption spectroscopy (XAS) and TEM, Wang et al. discovered that Cu, Ni, and some Co are unable to participate in subsequent reactions after the initial cycle, incurring irreversible capacity loss. The mixed structure during the following cycles is illustrated in Figure 9.2c–f, which is composed of a metallic phase and an oxide phase [11]. As well, in the spinel $(CrMnFeNiCu)_3O_4$ anode, both Cu and some Ni undergo a unidirectional phase transition from ionic states to metallic states as revealed by *in situ* XAS analyses [12]. Every coin has two sides, and although the irreversible reaction impairs the capacity, these metallic Cu, Ni, and some Co penetrate throughout the microsized particles to form a 3D metallic network, as shown in Figure 9.2g, thus improving the electrical conductivity of the electrode. Combined with previous research [8, 13], the inactive MgO structural pillars within the 3D metallic network accommodate Li^+ ions in the discharged state and Co^{2+} and Zn^{2+} ions in the charged state, contributing to structural stability [11].

To date, numerous high-entropy anodes with similar or even identical compositions have been reported. Notwithstanding, the identical formula fails to guarantee comparable electrochemical performance due to the different microstructures resulting from the various synthesis methods [7b]. Currently, solid-state reactions, solution combustion syntheses, coprecipitation (CP) methods, and ultrafast Joule heating methods are commonly employed. Wet-chemical methods yield nanostructures such as nanowires, hollow nanospheres [14], and nanosheets. These nanostructures can shorten the ion migration path to improve diffusion kinetics and enhance the mechanical properties by reducing bulk strain. Consequently, they render the anodes modified rate performance and lifespan. Therefore, in addition to appropriate structural archetypes and well-designed composition ratios, morphology should be underscored to ensure comprehensive high performance.

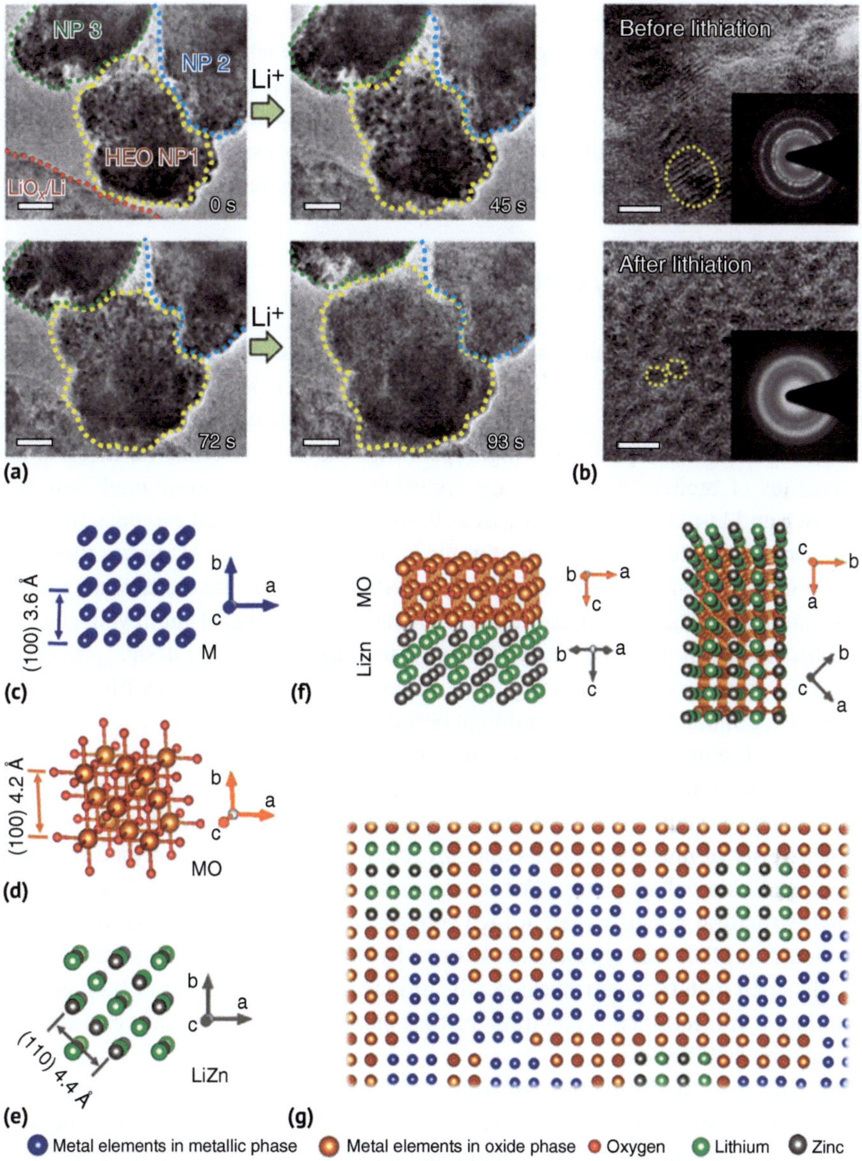

Figure 9.2 (a) TEM images of $(Co_{0.2}Ni_{0.2}Mn_{0.2}Zn_{0.2}Fe_{0.2})_3O_{3.2}$ at different lithiation processes. (b) High-resolution transmission electron microscopy (HRTEM) images and related electron diffraction patterns of $(Co_{0.2}Ni_{0.2}Mn_{0.2}Zn_{0.2}Fe_{0.2})_3O_{3.2}$ before and after lithiation. The scale bar is 5 nm [10]. Reproduced with permission from [10] / John Wiley & Sons. (c–e) Atomic structure sketches of the metal phase, metal oxide phase, and LiZn alloy phase. (f) Orientation relationship between oxide phase and LiZn alloy phase. (g) Schematic of all phases in the grain [11] / with permission of Springer Nature.

Although various types of materials have been extensively studied as high-entropy anodes, our understanding lags behind its significance where a systematic investigation to give the design criteria for tunable properties is lacking. Additionally, the reaction mechanisms of conversion reactions are intrinsically complex, and incorporating multiple elements further complicates the process. Thus, the development of more *in situ* and high-resolution instrumentation is essential to advance this field.

9.3 High-Entropy Cathode Materials

High-entropy cathode materials have emerged as a revolutionary approach in the development of next-generation batteries, addressing critical challenges associated with traditional cathode systems. From an energy density perspective, achieving both high specific capacity and high operating voltage is essential for meeting the demands of modern energy storage applications. Among conventional cathodes, Ni-rich and Li-rich layered oxides, as well as disordered rock-salt cathodes, are leading candidates due to their potential for high energy output. However, their performance is often hampered by complex phase transitions, severe capacity fading, and structural degradation, particularly when cycled at elevated cutoff voltages. The incorporation of a high-entropy strategy into cathode material design provides a promising avenue for mitigating these issues, delivering cathodes with enhanced stability, prolonged cycle life, and high performance.

The high-entropy approach introduces multiple principal elements into the cathode structure, leveraging the synergistic effects of configurational entropy, lattice distortion, and tailored electronic environments. This strategy significantly reduces the severity of phase transitions during cycling by stabilizing the crystal structure across a wide voltage range. For instance, high-entropy layered oxides maintain their structural integrity even at high voltages, minimizing the oxygen release and cation migration that typically cause capacity degradation in conventional Ni-rich and Li-rich materials. Furthermore, the unique multielement composition enables tuning of the redox activity, optimizing the balance between reversible capacity and long-term stability.

In the case of disordered rock-salt cathodes, the high-entropy design proves particularly beneficial. These materials, characterized by their inherently random atomic arrangements, already exhibit remarkable electrochemical properties due to their ability to facilitate multielectron redox reactions. Incorporating a high-entropy framework further enhances their cycling stability and mitigates issues such as the formation of inactive phases or irreversible structural changes. Studies have shown that high-entropy disordered rock-salt cathodes deliver excellent rate performance and retain their capacity over extended cycling, making them strong contenders for high-power and long-life applications.

The synthesis of high-entropy cathode materials involves advanced techniques to achieve uniform elemental distribution and precise control over microstructure. Methods such as solid-state reactions, sol–gel processes, and high-temperature sintering are commonly employed to ensure homogeneity and optimize particle morphology. Additionally, computational tools, including DFT and ML algorithms, play a pivotal role in predicting optimal compositions and guiding experimental efforts. These approaches enable the identification of element combinations that maximize configurational entropy while maintaining desirable electrochemical properties.

In conclusion, high-entropy cathode materials represent a transformative innovation in battery technology. By addressing the intrinsic limitations of traditional cathode systems, they pave the way for the development of high-energy-density batteries with superior cycling stability and reliability. As research continues to refine synthesis methods and deepen our understanding of structure–property relationships, high-entropy cathodes are poised to become a cornerstone of next-generation energy storage solutions.

In 2020, Ceder's group [15] designed a high-entropy disordered rock-salt-structured cathode containing six elements, denoted as $Li_{1.3}Mn^{2+}_{0.1}Co^{2+}_{0.1}Mn^{3+}_{0.1}Cr^{3+}_{0.1}Ti_{0.1}Nb_{0.2}O_{1.7}F_{0.3}$ (TM6), following their pioneering work on such materials in 2014 [16]. As the number of TM elements in the rock-salt structure increases from two elements (TM2) to four elements (TM4) and six elements (TM6), the randomly distributed heterogeneous elements suppress cation short-range order, evidenced by the square-like diffuse scattering patterns in TEM electron diffraction tests (Figure 9.3a–c). This enhanced cation disorder facilitates long-range Li percolation. Consequently, this cathode delivers a high specific capacity of 307 mAh g^{-1} when

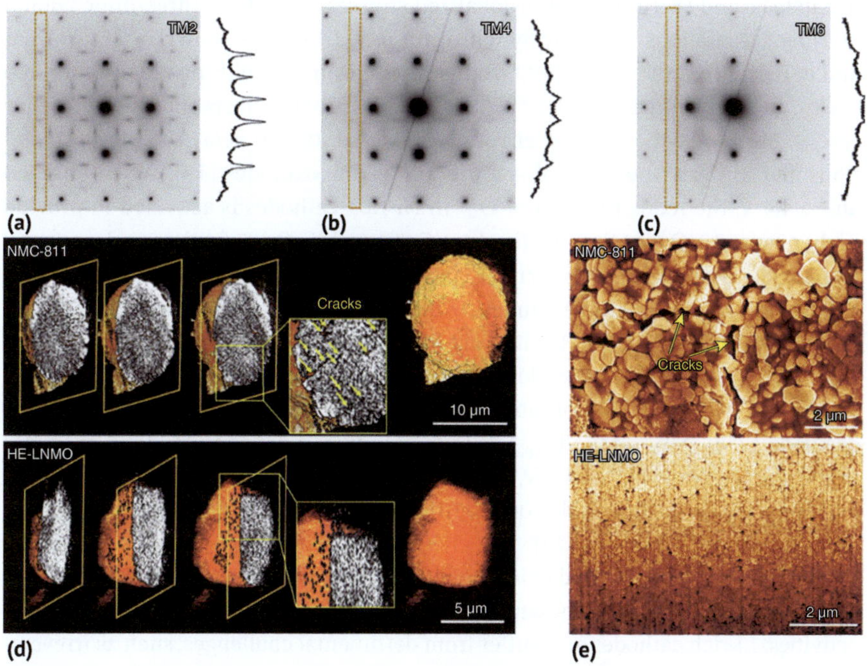

Figure 9.3 (a–c) TEM electron diffraction patterns of the TM2, TM4, and TM6 along the [100] zone axis [15]. Reproduced with permission from [15] / Springer Nature. (d) Synchrotron-based TXM tomography images of the long-cycled NMC811 and HE-LNMO secondary particles. (e) Representative scanning electron microscopy (SEM) cross-section images showing severely cracked NMC811 and dense HE-LNMO secondary particles [19]. Reproduced with permission from [19] / Springer Nature.

cycled between 1.5 and 4.7 V, outperforming other Mn-based disordered rock-salt cathodes with low entropy [15]. In addition, the regulation of the rock-salt lattice structure by these heterovalent elements contributes to the reduction of irreversible lattice oxygen loss, promoting the high-voltage stability of the high-entropy cathode [17].

Unlike disordered rock-salt cathodes that accommodate multiple electrochemically active elements, as an ideal platform for high-entropy engineering, the capacity of layered oxide cathodes predominantly depends on a single element. Introducing an excess of electrochemically inert elements can dilute the electroactive components, inevitably reducing the materials' theoretical capacity. To balance capacity and cycling performance, researchers have proposed an alternative method known as compositionally complex doping, or high-entropy doping, based on a multicomponent strategy where electroactive elements remain predominant [18].

Zhang et al. [19] applied the complex doping method to introduce a variety of heterovalent elements in place of the Co element in $LiNi_{0.8}Co_{0.1}Mn_{0.1}O_2$ (NCM811). Notably, the structural evolution of the prepared $LiNi_{0.8}Mn_{0.13}Ti_{0.02}Mg_{0.02}Nb_{0.01}Mo_{0.02}O_2$ (HE-LNMO) is greatly altered. During the lithiation and delithiation process, the volume change of the material is restricted to −0.3%, classifying it as a 'zero strain' material. Such a small lattice change reduces lattice defects and cracks caused by localized stresses. Therefore, three-dimensional (3D) reconstructions tested by Synchrotron-based TXM, as shown in Figure 9.3d and e, reveal that the HE-LNMO particles remain intact with no cracks after 100 cycles, in stark contrast to the NCM811 particles with substantial intergranular cracks. Moreover, the pinning effects of the dopants also delay the phase transitions and mitigate the thermal stability of Ni-rich cathodes [19]. When charged above 4.4 V, the redox reaction of O^{2-} in Ni-rich cathodes is activated to provide higher capacity. O^{2-} is first oxidized to $(O_2)^{n-}$ and if $(O_2)^{n-}$ further undergoes an irreversible overoxidation to form O_2 and escapes from the lattice, the lattice will collapse, causing irreparable damage to electrochemical performance. Liang et al. [20] replaced Co in NCM811 with various transition metal elements of low electronegativity to enhance the electron density of O in the TM-O_6 configuration, ensuring a higher $(O_2)^{n-}$ content and triggering a higher specific capacity. The designed $LiNi_{0.8}Mn_{0.12}Al_{0.02}Ti_{0.02}Cr_{0.02}Fe_{0.02}O_2$ gives a capacity of 203.6 mAh g^{-1} in the voltage range of 2.5–4.5 V. More importantly, these elements anchor $(O_2)^{n-}$ and suppress its further oxidation, contributing to the superior structure stability of both cations and anions [20].

Another prominent layered oxide cathode with an even higher capacity is the layered Li-rich cathode. Since the capacity is derived from both anion and cation redox activities, Li-rich cathodes also suffer from detrimental challenges, such as irreversible redox reaction of O^{2-}, related structural damage, activation of low-potential redox reactions, and generation of heterogeneous phases with Li_2MnO_3-like composition [21]. In the eight-element composed $Li_{1.0}(Li_{0.15}Mn_{0.50}Ni_{0.15}Co_{0.10}Fe_{0.025}Cu_{0.025}Al_{0.025}Mg_{0.025})O_2$, the introduced elements effectively stabilize the TM interlayer structure by increasing local structural diversity and distortion energy of Mn^{4+}. This improvement thus enhances the local structural adaptability and overall structural stability. Moreover, the phase transition from layered phase to spinel phase at low voltage results in a voltage decay of less than 0.02% per cycle [22].

Considering the ever-growing battery market, low-cost and resource-rich sodium batteries offer a solution to the constraints posed by limited and expensive lithium resources. The commonly used sodium cathode, layered oxides, and polyanionic cathode materials provide significant compositional flexibility due to the abundance of active elements they can incorporate. There is extensive research on high-entropy cathodes for sodium batteries, and their marked performance and underlying mechanisms are summarized by numerous publications [23].

During the development of $NaMO_2$ (where M represents transition elements), previous investigators have concluded that co-doping of several different elements can increase the average potential and suppress multiple phase transitions, thereby enhancing the capacity reversibility and cycling stability of the materials [24]. In 2015, Yue et al. introduced Ni, Co, Fe, Mn, and Ti into $NaMO_2$ to obtain a quintuple $NaNi_{1/4}Co_{1/4}Fe_{1/4}Mn_{1/8}Ti_{1/8}O_2$ cathode. This complex cathode demonstrates excellent structural stability, with capacity retention as high as 96.12% after 200 cycles at 1 C [25]. In the same year, the concept of high-entropy oxides was proposed, and Zhao et al. [26] applied this strategy to synthesize a high-entropy cathode incorporating nine transition elements, $NaNi_{0.12}Cu_{0.12}Mg_{0.12}Fe_{0.15}Co_{0.15}Mn_{0.1}Ti_{0.1}Sn_{0.1}Sb_{0.04}O_2$, in 2019. Successfully, the phase transition from the O3 phase to the P3 phase is delayed and the capacity maintains 83% after 500 cycles at a current density of 3 C. This groundbreaking study has formally initiated extensive research into high-entropy layered oxides for sodium batteries.

The energy density of the layered oxide cathodes lies in the available charge transfer number (ACTN) and the working potential. Hence, regulating the valence states of different elements and their electronic structure during the electrochemical process is crucial. On the base of $Na_xCu_{0.11}Ni_{0.11}Fe_{0.3}Mn_{0.48}O_2$, Ding et al. introduced low-valent Li^+ ions to modulate the electronic structure, which increased the relative ratio of Na^+ content to ACTN of TMs. This approach fully leverages the high-voltage transition metal redox reaction, and the capacity contribution of high-voltage transition metals is increased by 29% in the obtained $Na_{0.95}Li_{0.07}Cu_{0.11}Ni_{0.11}Fe_{0.3}Mn_{0.41}O_{1.97}F_{0.03}$ (LCNFMF) (Figure 9.4a and b), increasing the energy density efficiently [27]. The introduction of low-valent Li^+ ions in $Na_{2/3}Li_{1/6}Fe_{1/6}Co_{1/6}Ni_{1/6}Mn_{1/3}O_2$ also activates the redox of O, enabling the cathode with a high specific capacity of 171.2 mAh g^{-1} at 0.1 C in the voltage range of 2.0 to 4.5 V [28].

As the charging proceeds, the Na^+ ions in the layered sodium cathodes are continuously desodiated from the O3/P2 structure. This process elevates the valence of the active elements, enhancing the covalent bonding strength between transition metal–oxygen (TM—O) while weakening the oxygen-oxygen (O—O) bonding strength. These changes lead to the unfavorable phase transition from O3/P2 phase to P3/O2 phase, resulting in structural collapse and performance decay. Consequently, the high-entropy strategy tends to induce disorder among TMs to suppress the ordering of Na^+ with vacancies and limit the interlayer sliding, thus enhancing the cycling stability of the cathode [29]. Moreover, the Cu elements with high coordination energies act as pillars in O3-type $Na_{0.95}Li_{0.06}Ni_{0.25}Cu_{0.05}$

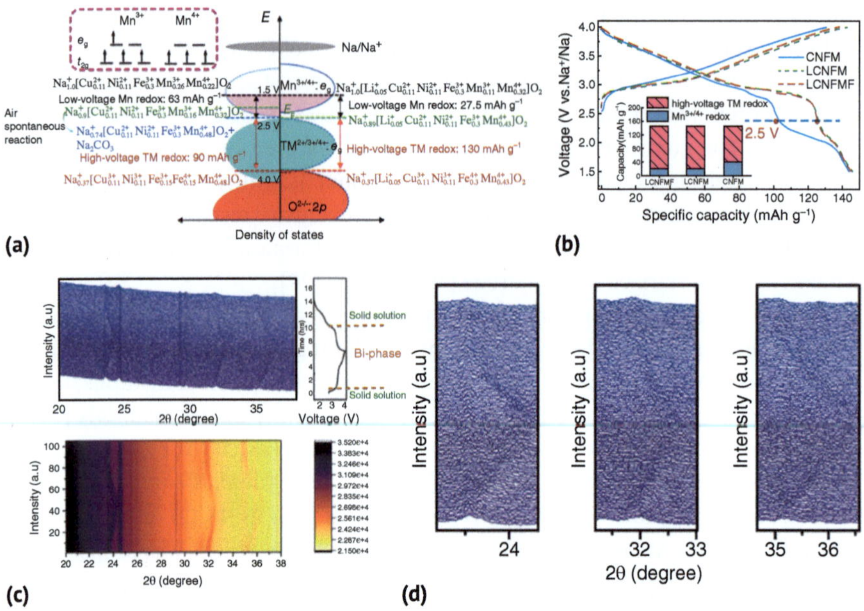

Figure 9.4 (a) Schematic of the electronic structure adjustment of O3-$Na_{0.9}Cu_{0.11}N$ $i_{0.11}F_{e0.3}Mn_{0.48}O_2$ (CNFM) and $Na_{0.89}Li_{0.05}Cu_{0.11}Ni_{0.11}Fe_{0.3}Mn_{0.43}O_2$ (LCNFM), with the dotted box indicating the electron occupancy in 3d orbitals of Mn^{3+} and Mn^{4+}. (b) Initial charge–discharge profiles of CNFM, LCNFM, and LCNFMF at 0.1 C across the voltage range of 1.5–4.0 V, with the inset depicting the quantitative analysis of reversible capacity contributions [27] / with permission of American Chemical Society. (c) *In situ* XRD analysis of $Na_4VFe_{0.6}Mn_{0.2}Cr_{0.1}(CoMgAl)_{0.1}(PO_4)_3$, and (d) zoomed-in patterns of observed peaks [36] / with permission of John Wiley & Sons.

$Fe_{0.15}Mn_{0.49}O_2$ Na to hinder the breakage of TM—O and O—TM—O bonds, ensuring structural stability under high voltage [30]. The difference in cation size brought about by the introduction of Li^+ ions in $Na_{2/3}Li_{1/6}Fe_{1/6}Co_{1/6}Ni_{1/6}Mn_{1/3}O_2$ leads to an ordered distribution between Li^+ ions and transition metal ions, forming a superlattice structure that helps relieve strain and maintain stability during the charging and discharging process [28].

Notably, the improved properties are not exclusively limited to structural stability, and additional beneficial properties beyond the prototype structure are also observed. The introduction of certain elements causes electron localization, and the resulting dense electron density can weaken the Na—O bond strength, promoting Na^+ ions migration [31]. Besides, the incorporation of multiple elements can narrow the energy band gap, which corresponds to enhanced electronic conductivity [32]. Explicitly, high ionic and electronic conductivities are directly related to reaction kinetics, significantly influencing the rate performance of the cathodes. What's more, the prominent air stability achieved through the high-entropy strategy provides valuable insights for the commercial production of layered Na-ion cathode materials [30, 31, 33].

Apart from the layered oxides, polyanionic-type cathodes hold valid promise for application. Similarly, this kind of cathode is also notable for its compositional flexibility, allowing elements such as V, Fe, Mn, Cr, and Ti to participate in redox reactions to supply capacity. Therefore, a high-capacity cathode with polycation redox reaction is enabled by the high-entropy strategy and the induced synergy effect of multiple elements inhibits irreversible phase transitions to ensure excellent cycling stability [34].

The redox couple of V^{4+}/V^{5+} occurs at a potential of about 4.0 V vs. Na^+/Na, which is conducive to satisfying the requirement of high energy density. However, the poor reversibility limits its cycling life, leading to severe capacity decay after only a few cycles. Fascinatingly, the introduced elements in the designed $Na_3VAl_{0.2}Cr_{0.2}Fe_{0.2}In_{0.2}Ga_{0.2}(PO_4)_3$ stabilize the ions at the Na1 site and help to maintain the structure above 4 V, greatly improving the reversibility of the V^{4+}/V^{5+} redox couple. Hence, the cathode exhibits an ultra-long lifespan of 5000 cycles at 20 C with a capacity retention of 86.8% and a minimal cell volume change of 1.1% [35]. When adjusting the elements to produce $Na_4VFe_{0.6}Mn_{0.2}Cr_{0.1}(CoMgAl)_{0.1}(PO_4)_3$, the corresponding $V^{3+}/V^{4+}/V^{5+}$ and Fe^{2+}/Fe^{3+} redox couples can contribute a specific capacity of 141.98 mAh g^{-1}. As shown in the *in situ* XRD patterns in Figure 9.4c, d, the excellent cycling stability is confirmed [36]. As mentioned above, the lattice and electronic structures of polyanionic-type cathodes can be regulated by the incorporation of various elements as well, reducing the Na^+ migration energy barrier to improve the charge transfer diffusion kinetics and enhancing the electronic conductivity. As a result, the cathodes demonstrate attractive rate performance and low-temperature performance. Operating at 0 °C, $Na_{3.5}V_{0.5}Mn_{0.5}Fe_{0.5}Ti_{0.5}(PO_4)_3$ cathode retains 93.45% of its initial capacity after 1000 cycles at 20 C [37].

Comprehensively, according to the reports, the mechanism related to the improved performance of the cathodes via the high-entropy strategy can be summarized as follows: (1) the interaction among multiple heterovalent elements can regulate the electronic structure and electrochemical behavior, potentially increasing the capacity of the cathode. (2) The local structure is distorted, with inactive elements acting as pillars to delay or mitigate the harmful phase transition during the charging and discharging process. (3) The modified lattice structure can expand channels for alkali metal ions diffusion, and the narrowed energy band gap can boost the electronic conductivity, resulting in improved reaction kinetics.

9.4 High-Entropy Solid-State Electrolytes

HE SSEs have emerged as a transformative approach to overcoming the limitations of traditional SSEs, which are integral to the development of safe and high-performance all-solid-state batteries. Conventional SSEs, while offering superior safety compared to liquid organic electrolytes by eliminating the risks of leakage and flammability, face challenges related to their relatively low ionic conductivity and poor interfacial compatibility with electrode materials. The introduction of high-entropy engineering into the design of SSEs provides a robust strategy to address these challenges by tuning the structural and chemical properties of the electrolyte materials, ultimately enhancing ion migration and interfacial stability.

One of the primary benefits of high-entropy engineering is its ability to optimize the local structure of SSEs, facilitating efficient ion migration. In SSEs, ionic conductivity is largely dictated by the migration energy barrier, which depends on the structural characteristics of the ionic channels. By incorporating multiple elements with varying ionic radii and electronegativity, high-entropy SSEs introduce significant local lattice distortions. These distortions modify bond lengths and angles, disrupt the periodicity of the crystal lattice, and alter the coulombic interactions between mobile ions and the surrounding framework. Such modifications can widen ionic channels or create new migration pathways, thereby reducing the migration energy barrier and improving ionic conductivity. For instance, in high-entropy oxide SSEs, the tailored local environment around lithium ions has been shown to enhance their mobility, enabling ionic conductivities comparable to or exceeding those of some liquid electrolytes.

In addition to improving ionic conductivity, high-entropy engineering addresses the critical issue of interfacial stability between SSEs and electrode materials. High-energy-density batteries require cathodes with high operating voltages and anodes with low potentials, which can drive undesired reactions at the electrode–electrolyte interface, leading to capacity degradation and impedance growth. By introducing multiple components, high-entropy SSEs can regulate their electronic structure, increasing their resistance to oxidation and reduction. For example, high-entropy halide SSEs exhibit improved electrochemical stability against high-voltage cathodes due to the formation of a passivating interfacial layer that mitigates further reactions.

The diversity of high-entropy SSEs, including oxides, sulfides, and halides, allows for tailored solutions depending on specific battery requirements. High-entropy sulfide SSEs, for instance, are known for their high ionic conductivity and soft lattice, which ensure excellent compatibility with electrodes. Conversely, high-entropy oxide SSEs offer exceptional chemical and thermal stability, making them suitable for long-term operation in extreme conditions. High-entropy halide SSEs provide a balance of ionic conductivity and electrochemical stability, with recent studies demonstrating their potential in lithium and sodium-ion solid-state batteries.

To advance the development of high-entropy SSEs, state-of-the-art synthesis techniques and computational modeling play critical roles. Advanced methods such as high-energy ball milling, spark plasma sintering, and CP enable the precise control of element distribution and microstructure. Computational tools, including DFT and molecular dynamics simulations, offer insights into ion migration mechanisms and guide the selection of optimal compositions. By combining experimental and theoretical approaches, researchers can systematically enhance the performance of high-entropy SSEs, tailoring their properties to meet the demands of various battery applications.

To date, many high-entropy electrolytes based on oxide, sulfide, and halide structural archetypes have been reported. Owing to the peculiarities of different types of electrolytes, the mechanisms of enhanced performance also differ from each other. Therefore, different types of HE SSEs will be analyzed separately in the following sections.

9.4.1 High-Entropy Oxide Solid-State Electrolytes

Among various oxide SSEs, garnet-type $Li_7La_3Zr_2O_{12}$ (LLZO) is of great potential for application due to its comparatively high ionic conductivity and wide electrochemical window. LLZO owns two phases, a thermodynamically stable tetragonal phase and a sub-stable cubic phase. During conventional sintering, the tetragonal phase gradually transforms to the cubic phase at a temperature above 800 °C. Although the cubic phase is fully formed at a temperature as high as 1230 °C [38], it may revert to the tetragonal phase during the cooling process. The ionic conductivity of tetragonal phase LLZO is two orders of magnitude lower than that of cubic phase LLZO (10^{-5}–10^{-6} S cm^{-1} versus 10^{-3}–10^{-4} S cm^{-1}) [39], far from application. To stabilize the cubic phase, Li vacancies are usually introduced through doping elements with higher valence [40]. Surprisingly, Jung et al. [41] designed two high-entropy garnets, $Li_7La_3Zr_{0.5}Hf_{0.5}Sc_{0.5}Nb_{0.5}O_{12}$ and $Li_7La_3Zr_{0.4}Hf_{0.4}Sn_{0.4}Sc_{0.4}Ta_{0.4}O_{12}$, where multiple heterovalent elements are introduced into the structure in equal molar ratios. Despite maintaining the Li content of 7 per formula unit, the structures are stabilized in the cubic phase. *In situ* XRD analyses (Figure 9.5a–f) reveal that high-entropy garnets with more than four elements at the Zr site lower the nucleation temperature of the cubic phase to 400 °C and prevent transition to the tetragonal phase during cooling, demonstrating the entropy effect on phase stabilization [41]. Similarly, Feng et al. and Fu et al. confirmed that increasing the number of elements in the system shifts the garnet structure from tetragonal to cubic phase and initiates formation at a lower temperature (Figure 9.5g–i) [42]. Though current studies indicate that multiple components in garnet-type oxides effectively stabilize the cubic phase, providing a prerequisite for oxides SSEs with high ionic conductivity. However, the underlying mechanisms of this entropy effect remain inadequately explained. The complexity of components alters the system's mixing entropy, encompassing configurational entropy, vibrational entropy, magnetic dipole entropy, and electronic random entropy. Yet, configurational entropy is sometimes overestimated. In designed garnets, configurational entropy appears similar for both cubic and tetragonal phases. Thus, further investigation is required to determine whether other entropy, such as electronic entropy or electronic configurational entropy, predominantly contributes to phase stabilization [41]. Compared with single-doped $Li_{6.75}La_3Zr_{1.75}Ta_{0.25}O_{12}$ (LLZT), high-entropy $Li_7La_3Zr_{0.5}Nb_{0.5}Ta_{0.5}Hf_{0.5}O_{12}$ presents uniform grain size of about 10 μm, regardless of sintering times. In contrast, LLZT grains grow rapidly with extended sintering, uncontrollably exceeding 100 μm after 20 h. It should be noted that fine and uniform ceramic grains are crucial for enhancing the mechanical properties of SSEs [42b]. Attractively, the ultrafast high-temperature sintering technique allows the synthesis of high-entropy garnet oxides with up to 10 elements at the Zr site in tens of seconds, significantly reducing trial-and-error time [42a].

In addition to the intact and continuous ionic channels in the cubic phase, mobile ions require high mobility for rapid migration within the electrolyte. Increasing the energy of mobile ions is an effective strategy to boost their mobility. Zeng et al. [43] employed ab initio modeling to introduce a standard deviation to the metal–oxygen bond length, thereby controlling structural distortions in SSEs. The approach alters the site energy for alkali ions. DFT calculations indicate that lattice distortions perturb the site energies, shifting from independent values to a certain distribution (Figure 9.6a,b). Consequently, the overlapped neighboring sites promote alkali ions hopping with low activation energy. When these sites with similar energies form a continuous network, mobile ions can realize macroscopic diffusion, thus

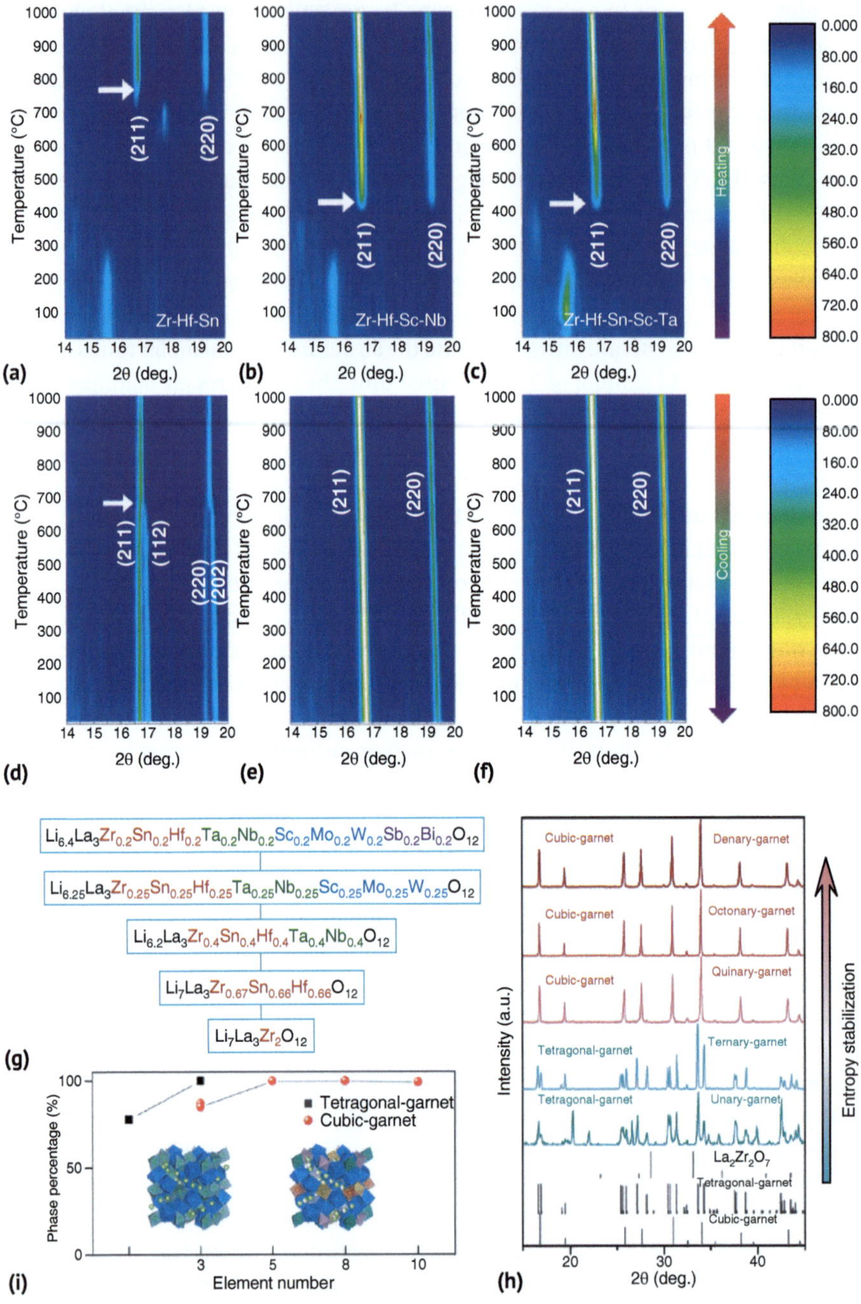

Figure 9.5 Contour plots of *in situ* XRD patterns during (a–c) heating and (d–f) cooling process in the range of 14–20° for $Li_7La_3Zr_{2/3}Hf_{2/3}Sn_{2/3}O_{12}$ (Zr—Hf—Sn), $Li_7La_3Zr_{0.5}Hf_{0.5}Sc_{0.5}Nb_{0.5}O_{12}$(Zr—Hf—Sc—Nb), and $Li_7La_3Zr_{0.4}Hf_{0.4}Sn_{0.4}Sc_{0.4}Ta_{0.4}O_{12}$(Zr—Hf—Sn—Sc—Ta). The white arrow denotes the temperature at which the cubic phase forms or transforms to the tetragonal phase [41] / Springer Nature / CC BY 4.0. (g) Designs of high-entropy garnets with different numbers of elements with (h) their respective XRD patterns. (i) The XRD phase distribution of garnets spanning from low entropy to high entropy [42a] / with permission of Elsevier.

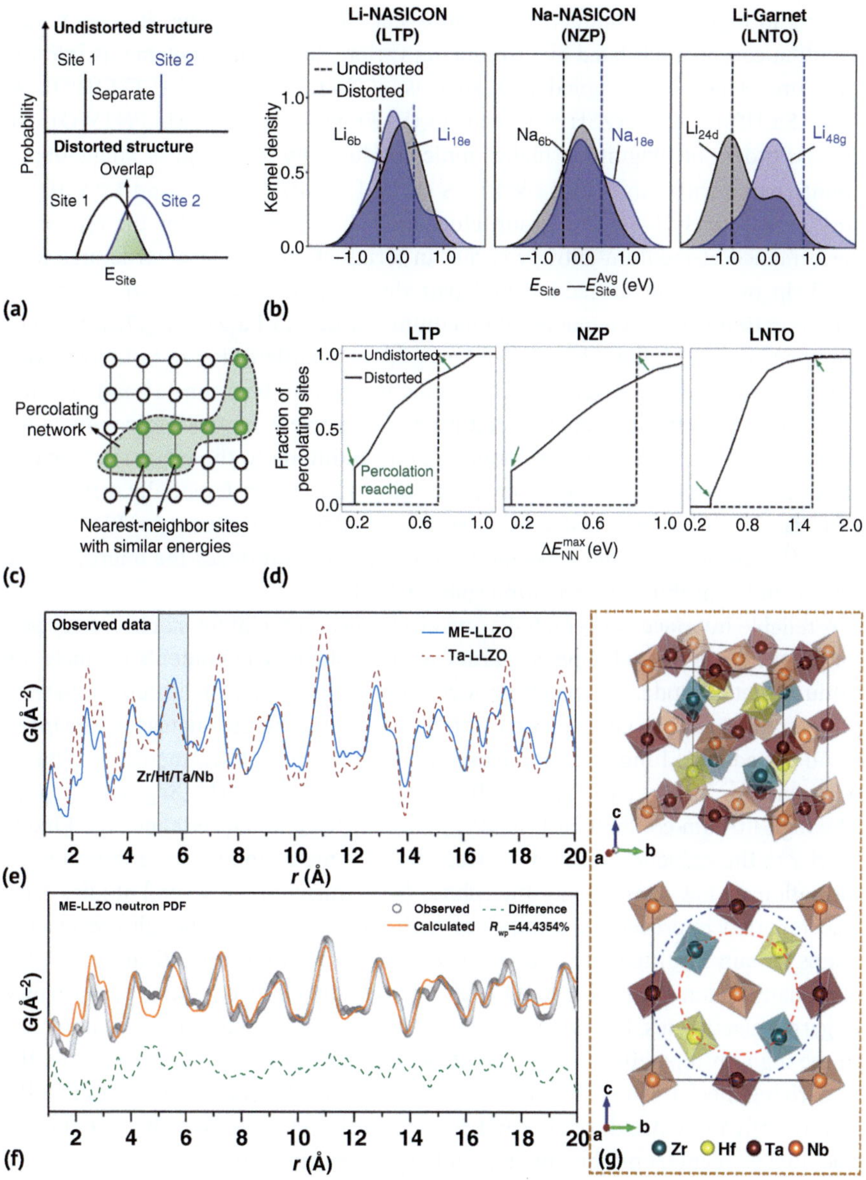

Figure 9.6 (a) Diagram of overlapping site-energy distributions created by local distortions. (b) Calculated site energies of the distorted $LiTi_2(PO_4)_3$, $NaZr_2(PO_4)_3$, and $Li_3Nd_3Te_2O_{12}$ materials. (c) Percolation network of sites where the nearest-neighbor sites have similar site energies. (d) Calculated proportions of percolating Li or Na sites as a function of maximum site-energy difference between neighboring sites (ΔE_{NN}^{max}) in both distorted (solid lines) and undistorted (dashed lines) structures[43] / with permission of American Association for the Advancement of Science. (e) Neutron PDF data for $Li_6La_3Zr_{0.5}Hf_{0.5}Ta_{0.5}Nb_{0.5}O_{12}$ (ME-LLZO) and $Li_{6.5}La_3Ta_{0.5}Zr_{1.5}O_{12}$ (Ta-LLZO) at the range of 1–20 Å for the atomic paired radial distribution. (f) PDF refinement of ME-LLZO with the lowest energy supercell model. (g) The corresponding occupancy of Zr^{4+}, Hf^{4+}, Ta^{5+}, and Nb^{5+} at the 16a site [44] / with permission of Elsevier.

leading to high ionic conductivity (Figure 9.6c,d). Accessibly, the site energy distribution can be modulated by introducing various heterovalent elements into the structure. Hence, the designed SSEs, such as $Li(Ti,Zr,Sn,Hf)_2(PO_4)_3$(LTZSHPO), $Na(Ti,Zr,Sn,Hf)_2(PO_4)_3$(NTZSHPO), and $Li_3(La,Pr,Nd)_3(Te,W)_2O_{12}$(LLPNTWO), all possess orders of magnitude higher ionic conductivity than that of single-doped counterparts, with values of 2.2×10^{-5} S cm^{-1}, 1.2×10^{-6} S cm^{-1}, and 1.7×10^{-6} S cm^{-1}, respectively [43]. Our group elucidated the mechanism for the enhanced performance of medium-entropy garnet in terms of specific local structures with the help of neutron diffraction and pair distribution function (PDF) methods. Incorporating various elements into medium-entropy $Li_6La_3Zr_{0.5}Hf_{0.5}Ta_{0.5}Nb_{0.5}O_{12}$ lengthens the Li—O bond lengths and expands the bottleneck size in Li-ion transport pathways. This weakens the coulombic interaction between Li^+ and O^{2-} and decreases the activation energy, resulting in relatively high ionic conductivity, despite a Li content of only 6 per formula unit. Additionally, these heterovalent elements induce local clustering, where pentavalent ions are preferentially surrounded by tetravalent ions, as revealed by supercell calculations and neutron PDF refinement (Figure 9.6e–g). Such specific local clustering minimizes the energy of the system and contributes to structural stability [44].

A reliable interface between SSEs and electrodes is crucial for stable battery performance. Once the cubic phase garnet with insufficient Li contents contacts the lithium metal anode, the resulting low chemical potential of lithium induces a phase transition from cubic phase to tetragonal phase, thereby increasing the interfacial impedance. Jung et al. compared the performance of symmetric cells based on cubic $Li_{6.6}La_3Zr_{0.4}Hf_{0.4}Sn_{0.4}Sc_{0.2}Ta_{0.6}O_{12}$ and $Li_7La_3Zr_{0.4}Hf_{0.4}Sn_{0.4}Sc_{0.4}Ta_{0.4}O_{12}$. The high lithium contents in $Li_7La_3Zr_{0.4}Hf_{0.4}Sn_{0.4}Sc_{0.4}Ta_{0.4}O_{12}$ enhance the kinetic barrier to the reduction reaction between lithium metal and garnet, inhibiting the formation of low-conductivity interphase [41]. Intriguingly, several studies have also reported improved air stability of high-entropy garnets, though the mechanisms remain insufficiently understood and warrant further investigation. In general, many factors contribute to the high ionic conductivity and stable performance of garnet electrolytes. It is premature to attribute performance enhancements solely to the high-entropy effect. Careful control of variables is essential to elucidate the underlying mechanisms. Moreover, the characteristics of grain boundaries in SSEs significantly influence durability and dendrite resistance, necessitating attention to potential elemental enrichment at grain boundaries and surfaces.

9.4.2 High-Entropy Sulfide Solid-State Electrolytes

Sulfide SSEs, typified by $Li_{10}GeP_2S_{12}$ and Li-argyrodites $Li_6PS_5X(X = Cl,Br,I)$, feature appropriate structural frameworks for fast ion migration. Their good mechanical ductility also helps to address interfacial contact issues, placing high hopes for future applications. Unlike oxides SSEs, sulfide SSEs can incorporate multiple anions, which significantly impacts the distribution and dynamics of Li^+ ions. Previous studies have demonstrated that regulating the S^{-2}/X^- site inversion via doping can potentially modify the Li sublattice [45]. As the high-entropy strategy gained popularity, researchers sought to find a more efficient path to boost ionic conductivity.

Strauss et al. [46] introduce various anions and cations to substitute parts of P, S, and Cl, either individually or in combination, in Li_6PS_5X (X = Cl, Br, I). According to the ^{31}P magic-angle spinning nuclear magnetic resonance (NMR) spectroscopy and neutron diffraction refinement, the local structural disorder in $Li_6P[S_{2.5}Se_{2.5}][Cl_{0.33}Br_{0.33}I_{0.33}]$ and $Li_{6.5}[Ge_{0.5}P_{0.5}][S_{2.5}Se_{2.5}][Cl_{0.33}Br_{0.33}I_{0.33}]$ is much higher than that in $Li_6PS_5[Cl_{0.33}Br_{0.33}I_{0.33}]$, leading to the decreased activation energy. However, the excessive softening of the lattice in $Li_{6.5}[Ge_{0.5}P_{0.5}][S_{2.5}Se_{2.5}][Cl_{0.33}Br_{0.33}I_{0.33}]$ instead slightly decreases the ionic conductivity compared to $Li_6P[S_{2.5}Se_{2.5}][Cl_{0.33}Br_{0.33}I_{0.33}]$ (1.1 vs. 1.3 mS cm^{-1}) [46]. It demonstrates that high-entropy engineering is a valuable strategy to improve ionic conductivity by regulating the local lattice structure. In the same year, they synthesized high-entropy $Li_{6.5}[P_{0.25}Si_{0.25}Ge_{0.25}Sb_{0.25}]S_5I$ with multicationic substitution at the P site to study the resulting changes in the Li sublattice. The randomly distributed cations give rise to a high S^{2-}/I^- site inversion of 11%, which favors Li^+ ion migration. The induced redistribution of Li^+ at the 48h (62%) and 24g sites (38%) lowers the activation energy for Li^+ diffusion. Such a large percentage of Li^+ occupying the 24g site leads to strong coulomb repulsion between two neighboring 48h sites, expanding the 48h–48h doublet jump distance from 1.25 Å in Li_6PS_5I to 1.39 Å in $Li_{6.5}[P_{0.25}Si_{0.25}Ge_{0.25}Sb_{0.25}]S_5I$. This change, in turn, shortens the intercage jump distance to 3.24 Å, effectively facilitating long-range lithium diffusion. Consequently, these structural superiorities endow $Li_{6.5}[P_{0.25}Si_{0.25}Ge_{0.25}Sb_{0.25}]S_5I$ with a low activation energy of 0.2 eV and high ionic conductivity of up to 13 mS cm^{-1} [47]. When introducing Si, Ge, Sn, and Cl to Li_6PS_5Br (LPSBr), these heterovalent elements increase the inversion between 4a and 4d sites in $Li_6P_{0.85}Si_{0.05}Ge_{0.05}Sn_{0.05}S_5Br$ (HE-LPSBr) and $Li_6P_{0.85}Si_{0.05}Ge_{0.05}Sn_{0.05}S_{4.5}BrCl_{0.5}$ (HE-LPSBrCl) and the disorder increased by a factor of 6.83-fold for HE-LPSBrCl. Ab initio molecular dynamics calculations and solid-state NMR further confirm the regulation of Li sublattice in HE-LPSBrCl. Lithium-ion cages begin to interconnect from their original isolated state, promoting intercage jumps and Li^+ diffusion in HE-LPSBrCl and effectively reducing the activation energy of ion migration (Figure 9.7a–c). Hence, a low activation energy of 0.18 eV is achieved and the ionic conductivity is increased by an order of magnitude from 0.85 mS cm^{-1} for LPSBr to 7.96 mS cm^{-1} for HE-LPSBrCl [48].

In contrast to the introduction of multiple cations, both Li et al. and Wang et al. adjusted the ratio of anions in Li_6PS_5X (X = Cl, Br) to probe the effect of anionic site inversion on ionic conductivity, obtaining high-entropy SSEs with conductivities exceeding 20 mS cm^{-1} [49]. Using a combination of XRD, neutron diffraction, and NMR measurements, Wang et al. [49b] unravel that the ratio of Cl and Br in $Li_{5.3}PS_{4.3}Cl_{1.7-x}Br_x$ ($0 \leq x \leq 1.7$) system affects the S and Cl/Br site inversion at the 4a and 4d sites (Figure 9.7d). As illustrated in Figure 9.7e, the similar proportion of Cl and Br induces a balanced distribution of anions at the 4d site, maximizing site disorder and reducing activation energy by lattice softening. As expected, the optimized anion frameworks increase the fraction of Li^+ at the 24g site and enhance the PS_4^{3-} rotational motion, paving the way for fast Li^+ transport. Figure 9.7f shows that Li^+ occupancy at the 24g site correlates with ionic conductivity. When $x = 1$, $Li_{5.3}PS_{4.3}Cl_{0.7}Br$ achieves the lowest activation energy of 0.155 eV and the highest ionic conductivity of 26 mS cm^{-1} [49b]. In the same vein, Li et al. [49a] also demonstrated the anion sublattice disorder positively affects Li^+ dynamics through

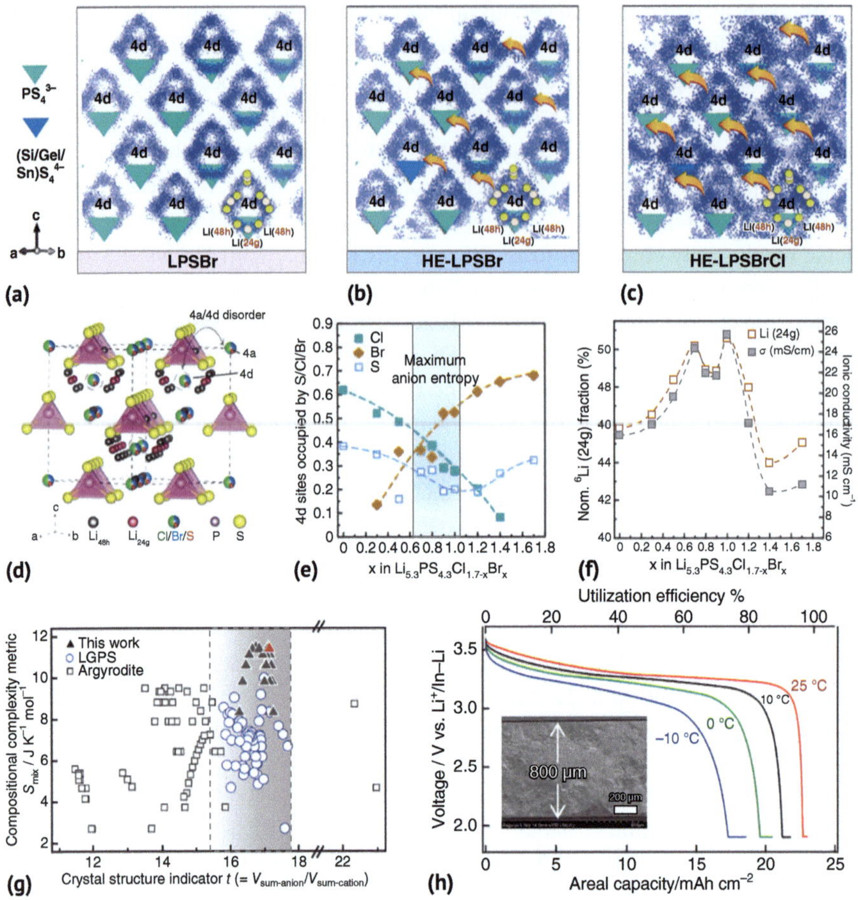

Figure 9.7 (a–c) AIMD simulations of the Li^+ probability densities at 900 K in LPSBr, HE-LPSBr, and HE-LPSBrCl, respectively [48] / with permission of John Wiley & Sons. (d) Crystal structure of $Li_{5.3}PS_{4.3}ClBr_{0.7}$. (e) The occupancy of Cl, Br, and S at Wyckoff 4d sites in $Li_{5.3}PS_{4.3}Cl_{1.7-x}Br_x$ based on the ^{31}P NMR results. (f) The fraction of Li^+ at Wyckoff 24g sites, inferred from the ^6Li NMR resonances [49b] / with permission of John Wiley & Sons. (g) The relationship between the crystal structure indicator (t) and compositional complexity metric (S_{mix}) for LGPS-type and argyrodite-type electrolytes. (h) Discharge curves for the all-solid-state cell under 0.025 C at temperatures ranging from 25 °C to −10 °C. The inset is the cross-sectional SEM image of the cathode (~800 μm) with a capacity loading of 23.5 mAh cm^{-2} [50]. Reproduced with permission from American Association for the Advancement of Science.

systematical exploration, with $Li_{5.5}PS_{4.5}Cl_{0.8}Br_{0.7}$ exhibiting a high ionic conductivity of 22 mS cm^{-1}. The remarkable property enables the solid-state cells assembled with single-crystal $LiNi_{0.9}Co_{0.06}Mn_{0.04}O_2$ cathode to cycle stably over 700 cycles [49a]. As discussed earlier, both cationic and anionic substitution can induce anionic site inversion. Anion mixing, in particular, creates a disordered distribution that further enhances an ultrahigh ionic conductivity by directly perturbing the site occupancies of neighboring Li^+.

Given the complexity of elements in the periodic table, establishing design rules for high-performance SSEs is urgently required for efficient materials research and development. Li et al. [50] integrated their experimental studies with reported LGPS-type and Li-argyrodites SSEs, elucidating the relationship between the crystal structure indicator t (volume ratio of anions to cations) and the compositional complexity metric S_{mix} (configuration entropy) as shown in Figure 9.7g. Based on this criterion, they designed an LGPS-type high-entropy electrolyte $Li_{9.54}[Si_{1-\delta}M_{\delta}]_{1.74}P_{1.44}S_{11.1}Br_{0.3}O_{0.6}$ ($M = Ge, Sn; 0 \le \delta \le 1$). Neutron diffraction and calculation results reveal that the highly disordered anions in the structure flatten the energy landscape and reduce the activation energy, endowing $Li_{9.54}[Si_{0.6}Ge_{0.4}]_{1.74}P_{1.44}S_{11.1}Br_{0.3}O_{0.6}$ with an unprecedented ionic conductivity as high as 32 mS cm^{-1}. This exceptional conductivity enables the SSEs to be coupled with high-loading cathodes. The all-solid-state cell, utilizing $LiNbO_3$-coated $LiCoO_2$ with a thickness of 800 μm and a theoretical surface capacity of 23.5 mAh cm^{-2}, delivers 22.7 mAh cm^{-2} when cycling at the cutoff voltage of 4.25 V at 25 °C (0.025 C rate). Even more impressive, when the temperature is lowered to −10 °C, the ionic conductivity of $Li_{9.54}[Si_{0.6}Ge_{0.4}]_{1.74}P_{1.44}S_{11.1}Br_{0.3}O_{0.6}$ remains at 9 mS cm^{-1}, retaining 75% of the discharge capacity observed at 25 °C, thus demonstrating excellent low-temperature performance (Figure 9.7h) [50].

Taken together, the high-entropy strategy is generally able to improve the anion disorder and optimize the hopping distance of Li^+ in sulfide SSEs, thereby reducing the activation energy and promoting Li^+ migration. However, the narrow electrochemical window poses challenges due to adverse redox reactions with both anodes and cathodes, hindering the development of sulfide SSEs. Therefore, ensuring compatibility between high-entropy sulfide SSEs and electrode materials merits attention to potentially achieve all-solid-state batteries with ultrahigh energy density, even at low operating temperatures.

9.4.3 High-Entropy Halide Solid-State Electrolytes

In analogy with sulfide SSEs, halide SSEs have garnered considerable attention because of their high ionic conductivity (10^{-3} S cm^{-1}) and good machinability. Among various structure types, UCl_3-type halide SSEs are notable for accommodating monovalent cations and providing migration pathways stemming from the abundant channel sites. Li et al. [51] demonstrated that partial substitution of La and Ce in $LaCl_3$ and $CeCl_3$ with other cations effectively enhances ionic conductivity while maintaining the skeleton structure. Introducing Zr^{4+}, Hf^{4+}, Al^{3+}, Ta^{5+}, and Nb^{5+} in different combinations induces disorder, successfully increasing conductivity by orders of magnitude. Interestingly, the incorporation of various elements also alters the microscopic morphology. The particle size of 1.5LiCl-($LaCl_3$·$CeCl_3$·$ZrCl_4$·$HfCl_4$·$TaCl_5$)$_{0.2}$ powder is about several hundred nanometers, with more than 50% amorphous content, which is speculated to enable internal or interfacial conduction and trigger fast ion migration. The large-sized UCl_3-type skeleton can also accommodate other mobile ions such as Na^+, K^+, Cu^{2+}, and Ag^+. Utilizing multication strategy, Cu-LaCeZrHfTa-Cl and Ag-LaCeZrHfTa-Cl achieve high ionic conductivity of 7.13×10^{-3} and 7.75×10^{-3} S cm^{-1}, respectively, verifying the broad applicability of this approach [51].

In the case of close-packed crystals, the introduction of different elements can lead to diverse structural outcomes. In contrast to traditional studies of halides, which often focus on the relationship between anion radius and phase stability, Wang et al. [52] also consider the charge of metal ions due to the impact on the interactions between the ions, which also influence structure stability. Hence, they defined ionic potential as $\Phi = n/R$, where n is the dimension-less ion charge number and R is the ion radius. As shown in Figure 9.8a, b, the plotted Li potential versus ionic potential ratio reveals a phase map that predicts the possible structure—hexagonal or orthogonal—of the designed halide SSEs. Following this criterion, the synthesized $Li_{2.8}In_{0.2}Sc_{0.2}Yb_{0.2}Lu_{0.2}Zr_{0.2}Cl_6$ owns a high ionic conductivity of 2.2 mS cm^{-1} due to the caused stacking faulted and disordered distribution of Li^+, which enhance three-dimensional diffusion in high-entropy halide SSE. Different from the lamellar stacking morphology of Li_3InCl_6 depicted in Figure 9.8c, $Li_{2.8}In_{0.2}Sc_{0.2}Yb_{0.2}Lu_{0.2}Zr_{0.2}Cl_6$ presents a spherical shape despite having a similar

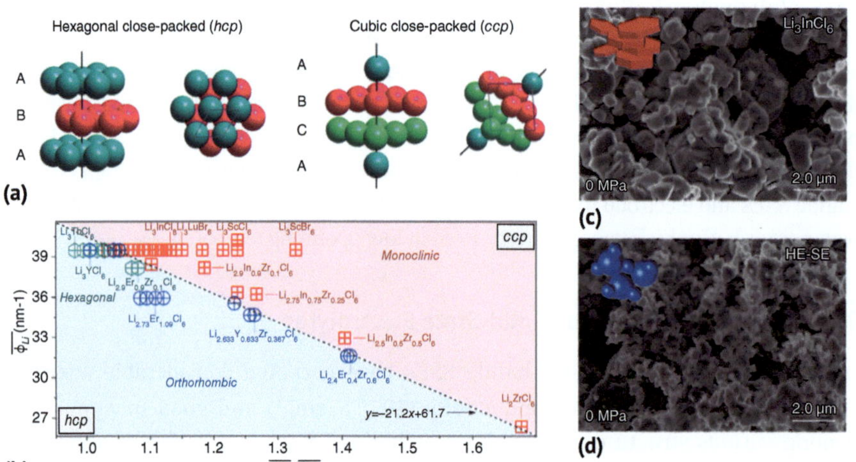

Figure 9.8 (a) Schematic of halide hexagonal close-packed (*hcp*) and cubic close-packed (*ccp*) arrangements. (b) Phase diagrams of the representative halide materials as a function of ionic potential ratio and Li potential [52] / Springer Nature / CC BY 4.0. (c, d) SEM images of Li_3InCl_6 and $Li_{2.8}In_{0.2}Sc_{0.2}Yb_{0.2}Lu_{0.2}Zr_{0.2}Cl_6$ SSEs [52]. Reproduced with permission from [52] / Springer Nature / CC BY 4.0 (e, f) The probability density isosurface of Cl^- in Li_3InCl_6 and HE-LIC supercells based on BVEL calculations. (g) CV curves of Li_3InCl_6 and HE-LIC-based solid-state cells within 2.7 and 5.0 V vs. Li^+/Li [53] / Springer Nature / CC BY 4.0.

layered crystal structure (Figure 9.8d). The spherical particle shape increases compaction density, which improves the overall conductivity of the SSE [52].

Another remarkable advantage is that the incorporation of diverse ions not only boosts the ionic conductivity of halide SSEs but also enhances their interfacial compatibility with high-voltage cathodes. Our group demonstrated that the chemical disorder-induced lattice distortion in $Li_{2.75}Y_{0.16}Er_{0.16}Yb_{0.16}In_{0.25}Zr_{0.25}Cl_6$ (HE-LIC) locally perturbs the site energies of neighboring Cl^-, as illustrated by the bond valence energy landscape (BVEL) calculations in Figure 9.8e, f. The inhomogeneous local energy distribution impedes the kinetics of Cl^- oxidation, raising the oxidation potential from 4.25 V in Li_3InCl_6 to 4.4 V in HE-LIC (Figure 9.8g). Thanks to the enhanced high-voltage stability, the HE-LIC-based full cell paired with $LiCoO_2$ cathode achieves a high capacity retention of 89.9% after 500 cycles within the voltage range of 2.5–4.2 V at 0.5 C, outperforming the Li_3InCl_6-based full cell, which retains 71.8%. Even when the cutoff voltage is further increased to 4.6 V, this full cell still exhibits excellent cycling stability, with a capacity retention of 91.6% after 50 cycles [53].

Though research on high-entropy halide SSEs is currently limited, various advantageous properties such as disorder-induced high ionic conductivity, the presence of amorphous components, altered particle morphology, and improved high-voltage stability have been detected. To fully leverage these benefits, comprehensive design strategies are required to guide the development of high-performance materials.

9.5 High-Entropy Liquid Electrolytes

Since 2020, the high-entropy strategy has been extensively applied to various types of liquid electrolytes, including both organic and aqueous systems and accommodating working ions such as Li^+, Na^+, Mg^{2+}, and Zn^{2+}. These approaches have been reviewed comprehensively in recent literature [1d, 1e]. Both increasing the solvent types or salt types can achieve high-entropy liquid electrolytes. The incorporation of multiple components can favorably modulate the solvation structure, tending to weaken the interaction strength between alkali metal ions and solvent molecules or salt anions, thus facilitating the migration kinetics of alkali metal ions. This results in great improvements in the electrochemical performance of batteries. Additionally, the complex components are anticipated to regulate the interfacial components between electrolytes and electrodes, rendering enhanced conductivity and good stability [54].

When lithium bis(fluorosulfonyl)imide (LiFSI), lithium bis(trifluorome thanesulfonyl)imide (LiTFSI), lithium difluoro(oxalato)borate (LiDFOB), and lithium nitrate ($LiNO_3$) are simultaneously introduced into a single solvent of dimethoxyethane, the resulting diverse solvation structures reduce the solvation strength of Li^+ with solvent and anions, facilitating Li^+ mobility and enhancing charge transfer. In the full cell assembled with lithium anode and NCM811 cathode, the formed solid–electrolyte interphase is much thinner and rich in inorganic materials. These advantages contribute to excellent high-voltage stability and rate performance, with a capacity retention higher than 80% after 600 cycles at 2.8–4.3 V [55]. Kim et al. [56]

succeeded in inhibiting the ionic aggregation in electrolytes through high solvation entropy stemming from the increased molecular diversity. The derived small clusters with anion-rich solvation structures enhance the ionic conductivity of weakly solvating electrolytes [56]. In addition to the improved Li^+ dynamics and the formation of a more stable interfacial layer due to the modified solvation structure, complex salt components broaden the operating temperature range of high-entropy electrolytes. The full cell with the electrolyte of 0.15 m $LiPF_6$/0.15 m LiFSI/0.15 m LiTFSI/0.15 m LiDFOB/0.15 m $LiNO_3$ in propylene carbonate/diethyl carbonate (1:1) with 5% fluoroethylene carbonate maintains 99% of its initial capacity after 60 cycles at −20 °C [57].

The success of multisolvent or multisalt approaches in enhancing the properties of liquid electrolytes underscores the flexibility and inclusivity of the high-entropy strategy. However, liquid systems significantly differ from solid materials with crystal structures, then the concept of high entropy is supposed to be adjusted appropriately. Therefore, it is imperative to refine the definition and design criteria of high-entropy liquid electrolytes through rigorous experiments and comprehensive analyses.

9.6 Perspectives on High-Entropy Battery Materials

HEMs represent a transformative strategy in the development of next-generation battery technologies, combining multiple elements in equimolar or near-equimolar proportions to create materials with unique structural and electrochemical properties. The flexibility inherent in the high-entropy approach offers unprecedented potential to tune a variety of material characteristics, leading to promising enhancements in battery performance. However, significant challenges remain in fully realizing the potential of HEMs for practical applications. In the following text, we address several key areas that require focused research and development to bring high-entropy battery materials closer to commercial viability (Figure 9.9).

Designing High-Performance HEMs: Empirical Rules Versus Systematic Approaches. Most current research on HEMs for batteries follows empirical guidelines derived from doping studies, often focusing on modifying existing structural archetypes or incorporating elements from neighboring groups in the periodic table. This trial-and-error approach, though successful in some cases, lacks generalizable design principles. In many instances, the choice of which elements to incorporate into the lattice—and how these elements interact with each other in both anionic and cationic sublattices—has not been systematically explored.

To overcome these limitations, systematic approaches need to be developed, focusing on a more rigorous understanding of the atomic interactions and structural stability of multicomponent systems. The potential of ML and computational screening could play a pivotal role in predicting optimal element combinations, identifying promising HEMs more efficiently, and even designing entirely new material structures. ML algorithms can be trained on large datasets of known materials to predict their electrochemical behavior, potentially accelerating the discovery of high-performance candidates.

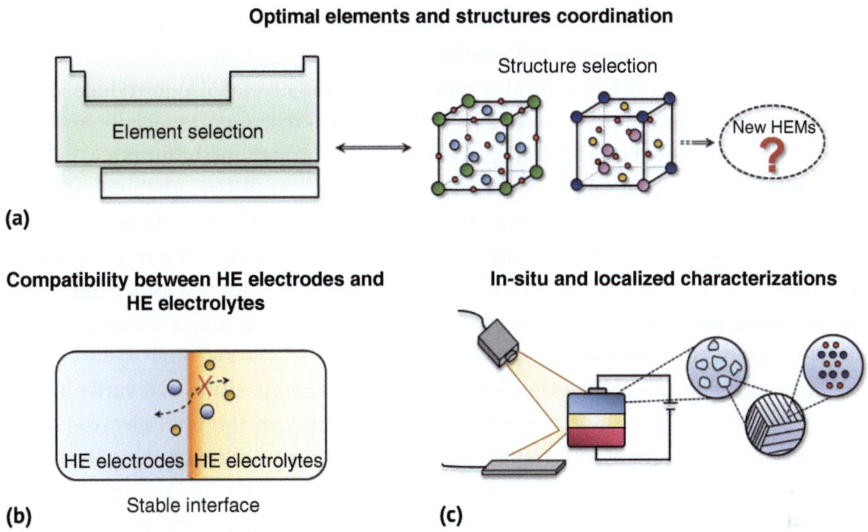

Figure 9.9 Schematic diagram of the key areas to advance the development of HEMs. (a) Select appropriate elements that align with the desired structure and develop new systems. (b) Ensure compatibility between HE electrodes and electrolytes to prevent detrimental side reactions. (c) Utilize *in situ* high-resolution and localized characterization techniques to explore the underlying mechanism of HEMs.

One promising area is HEAs for metallic anodes. Metals like Li, Na, and Mg exhibit high specific capacities, but their practical application is often hindered by large volume changes and poor reversibility during cycling. HEAs, by distributing strain more evenly across multiple elements, could mitigate these issues and enhance the mechanical properties of the anode material. Exploring multimetallic and multiphase alloys, which allow for better stress accommodation, is a critical avenue for further research.

In addition, the development of new structural archetypes—beyond those currently investigated in high-entropy oxides, phosphates, and sulfides—can provide new insights and materials platforms. This could lead to the identification of novel material families that extend the current understanding of high-entropy systems, including the exploration of non-oxide HEMs (e.g., carbides, nitrides, or borides).

Electrolyte–Electrode Compatibility: Challenges in Interfacial Stability and Lattice Distortions. One of the most significant challenges in high-entropy battery materials is ensuring compatibility between the electrolyte and the electrode. High-energy-density batteries require that both the anode and cathode materials have high specific capacities and stability across a wide range of potentials. However, the complex composition of HEMs, while beneficial for enhancing structural stability, can lead to unpredictable changes in electronic and ionic properties that affect the stability of the interface between the electrolyte and electrode.

Electrolyte–electrode interactions are particularly critical in systems where high-potential cathodes (e.g., those based on transition metal oxides or high-entropy oxides) are paired with low-potential anodes (e.g., lithium or sodium metal). In these systems, the electronic structure of the HEMs can be altered by the incorporation of multiple elements, potentially influencing the chemical and electrochemical stability of the electrolyte. The structural complexity of HEMs could also introduce lattice distortions that might hinder the kinetics of interfacial reactions, especially in solid–liquid interfaces where passivation layers or dendritic growth might form.

To mitigate these challenges, careful selection of elements for both the anode and cathode materials is necessary to balance the electrochemical and structural properties of each component. For instance, in high-entropy cathodes, one must consider the impact of multiple cations on the electrochemical stability and ionic conductivity. In high-entropy anodes, elements must be chosen to minimize volume expansion and promote better cycle reversibility.

Furthermore, gradient materials—materials with a composition that varies across the electrode—could help manage the transition between the bulk electrode and the electrolyte interface, improving the overall compatibility of the electrode–electrolyte system. For example, composite electrodes where HEMs are incorporated at the surface to enhance stability, while the core of the electrode consists of more traditional material, could be a promising strategy to improve interfacial compatibility.

Characterization of HEMs: Challenges and New Tools. The complexity of HEMs—due to their diverse elemental composition and potential for non-equilibrium phases—poses significant challenges for characterization. Most of the elements in high-entropy systems are often located in close proximity to the periodic table, making their differentiation via traditional characterization techniques such as XRD or SEM difficult. Moreover, the small particle sizes and potential amorphous morphologies of HEMs, especially those synthesized via solution-phase or ball-milling methods, complicate the analysis further.

A multimodal characterization approach using advanced techniques such as high-resolution electron microscopy (e.g., TEM), X-ray photoelectron spectroscopy (XPS), and *in situ* techniques (e.g., synchrotron-based XRD or neutron diffraction) will be necessary to gain a detailed understanding of the local atomic arrangements and phase behavior in these materials. *In situ* electrochemical characterization can also help track structural and electronic changes during battery cycling, providing critical insights into the stability and performance of high-entropy electrodes.

The integration of ML with spectroscopic data could enable more accurate prediction of material properties and facilitate the development of real-time monitoring systems for battery performance. This would help researchers correlate structural features with electrochemical behavior and identify the most promising materials for specific applications.

HEMs represent a promising frontier in battery research, offering a novel approach to addressing the challenges of energy density, cycling stability, and interfacial compatibility. While the high-entropy strategy has already demonstrated notable improvements over traditional materials, achieving practical applications will require overcoming several key obstacles, particularly in the areas of systematic design of multicomponent systems, electrolyte–electrode compatibility, and advanced characterization. With the continued development of computational tools, synthesis techniques, and characterization methods, HEMs are poised to make significant contributions to the next generation of high-performance, long-lifetime batteries.

As these materials mature, we anticipate that high-entropy batteries will offer a versatile platform for developing batteries that exceed the limitations of conventional materials—offering not only higher energy densities but also improved cycling performance, broader temperature stability, and better safety profiles for a wide range of applications from electric vehicles to large-scale energy storage systems.

References

1 (a) Zheng W, Liang G, Liu Q. The Promise of High-Entropy Materials for High-Performance Rechargeable Li-Ion and Na-Ion Batteries. *Joule*, 2023, 7: 2732–2748. https://doi.org/10.1016/j.joule.2023.10.016; (b) Ouyang B, Zeng Y. The Rise of High-Entropy Battery Materials. *Nature Communications*, 2024, 15: 973. https://doi.org/10.1038/s41467-024-45309-9; (c) Schweidler S, Botros M, Strauss F, et al. High-Entropy Materials for Energy and Electronic Applications. *Nature Reviews Materials*, 2024, 9: 266–281. https://doi.org/10.1038/s41578-024-00654-5; (d) Ren K F, Liu H, Guo J X, et al. Working Principles of High-Entropy Electrolytes in Rechargeable Batteries. *ACS Energy Letters*, 2024, 9(6): 2960–2980. https://doi.org/10.1021/acsenergylett.4c00621; (e) Zhao X, Fu Z, Zhang X. More is Better: High-Entropy Electrolyte Design in Rechargeable Batteries. *Energy & Environmental Science*, 2024, 17: 2406–2430. https://doi.org/10.1039/D3EE03821A.

2 (a) Zhang J, Yu, A. Nanostructured Transition Metal Oxides as Advanced Anodes for Lithium-Ion Batteries. *Science Bulletin*, 2015, 60: 823–838. https://doi.org/ 10.1007/s11434-015-0771-6; (b) Zhao Y, Li X, Yan B. Recent Developments and Understanding of Novel Mixed Transition-Metal Oxides as Anodes in Lithium Ion Batteries. *Advanced Energy Materials*, 2016, 6: 1502175. https://doi.org/10.1002/aenm.201502175.

3 (a) Nguyen T X, Patra J, Tsai C C, et al. Secondary-Phase-Induced Charge–Discharge Performance Enhancement of Co-Free High Entropy Spinel Oxide Electrodes for Li-Ion Batteries. *Advanced Functional Materials*, 2023, 33: 2300509. https://doi.org/10.1002/adfm.202300509; (b) Liu X, Xing Y, Xu K, et al. Kinetically Accelerated Lithium Storage in High-Entropy (LiMgCoNiCuZn)O Enabled by Oxygen Vacancies. *Small*, 2022, 18: 2200524. https://doi.org/10.1002/smll.202200524.

4 Cheng W, Liu J, Hu J, et al. Pressure-Stabilized High-Entropy $(FeCoNiCuRu)S_2$ Sulfide Anode Toward Simultaneously Fast and Durable Lithium/Sodium Ion Storage. *Small*, 2023, 19: 2301915. https://doi.org/10.1002/smll.202301915.

5 (a) Li W, Wang J H, Yang L, et al. High-Entropy Engineering of Cubic SiP with Metallic Conductivity for Fast and Durable Li-Ion Batteries. *Advanced Materials*, 2024, 36: 2314054. https://doi.org/10.1002/adma.202314054; (b) Wei Y, Yao R, Liu X. Understanding the Configurational Entropy Evolution in Metal-Phosphorus Solid Solution for Highly Reversible Li-Ion Batteries. *Advanced Science*, 2023, 10: 2300271. https://doi.org/10.1002/advs.202300271.

6 Du Z, Wu C, Chen Y, et al. High-Entropy Atomic Layers of Transition-Metal Carbides (MXenes). *Advanced Materials*. 2021, 33: 2101473. https://doi.org/10.1002/adma.202101473.

7 (a) Dong L, Tian Y, Luo C, et al. Porous High-Entropy Oxide Anode Materials for Li-Ion Batteries: Preparation, Characterization, and Applications. *Materials*, 2024, 17: 1542. https://doi.org/10.3390/ma17071542; (b) Li C, Chen X, Li J, et al. High Entropy Anodes in Batteries: from Fundamentals to Applications. *Energy Storage Materials*, 2024, 71: 103468. https://doi.org/10.1016/j.ensm.2024.103468; (c) Liu X, Li X, Li Y, et al. High-Entropy Oxide: a Future Anode Contender for Lithium-Ion Battery. *EcoMat*, 2022, 4: e12261. https://doi.org/10.1002/eom2.12261.

8 Sarkar A, Velasco L, Wang D. High Entropy Oxides for Reversible Energy Storage. *Nature Communications*, 2018, 9: 3400. https://doi.org/10.1038/s41467-018-05774-5.

9 Chen T-Y, Wang S-Y, Kuo C-H, et al. In Operando Synchrotron X-Ray Studies of a Novel Spinel $(Ni_{0.2}Co_{0.2}Mn_{0.2}Fe_{0.2}Ti_{0.2})_3O_4$ High-Entropy Oxide for Energy Storage Applications. *Journal of Materials Chemistry A*, 2020, 8: 21756–21770. https://doi.org/10.1039/D0TA06455F.

10 Su L, Ren J, Lu T. Deciphering Structural Origins of Highly Reversible Lithium Storage in High Entropy Oxides with In Situ Transmission Electron Microscopy. *Advanced Materials*, 2023, 35: 2205751. https://doi.org/10.1002/adma.202205751.

11 Wang K, Hua W, Huang X. Synergy of Cations in High Entropy Oxide Lithium Ion Battery Anode. *Nature Communications*, 2023, 14: 1487. https://doi.org/10.1038/s41467-023-37034-6.

12 Luo X-F, Patra J, Chuang W-T, et al. Charge–Discharge Mechanism of High-Entropy Co-Free Spinel Oxide Toward Li^+ Storage Examined Using Operando Quick-Scanning X-Ray Absorption Spectroscopy. *Advanced Science*, 2022, 9: 2201219. https://doi.org/10.1002/advs.202201219.

13 Wang S-Y, Chen T-Y, Kuo C-H, et al. Operando Synchrotron Transmission X-Ray Microscopy Study on (Mg, Co, Ni, Cu, Zn)O High-Entropy Oxide Anodes for Lithium-Ion Batteries. *Materials Chemistry and Physics*, 2021, 274: 125105. https://doi.org/10.1016/j.matchemphys.2021.125105.

14 Liu X, Yu Y, Li K. Intergrating Hollow Multishelled Structure and High Entropy Engineering Toward Enhanced Mechano-Electrochemical Properties in Lithium Battery. *Advanced Materials*, 2024, 36: 2312583. https://doi.org/10.1002/adma.202312583.

15 Lun Z, Ouyang B, Kwon D H. Cation-Disordered Rocksalt-Type High-Entropy Cathodes for Li-Ion Batteries. *Nature Materials*, 2021, 20: 214–221. https://doi.org/10.1038/s41563-020-00816-0.

16 Lee J, Urban A, Li X. Unlocking the Potential of Cation-Disordered Oxides for Rechargeable Lithium Batteries. *Science*, 2014, 343: 519–522. https://doi.org/10.1126/science.1246432.

17 Zhou S, Sun Y, Gao T. Enhanced Li^+ Diffusion and Lattice oxygen Stability by the High Entropy Effect in Disordered-Rocksalt Cathodes. *Angewandte Chemie International Edition*, 2023, 62: e202311930. https://doi.org/10.1002/anie.202311930.

18 (a) Zhang R, Wang C, Zou P, et al. Long-Life Lithium-Ion Batteries Realized by Low-Ni, Co-Free Cathode Chemistry. *Nature Energy*, 2023, 8: 695–702. https://doi.org/10.1038/s41560-023-01267-y; (b) Choi S, Feng W, Xia Y. High

Entropy and Co-Free High Nickel Based Layered $LiNi_{0.9}Mn_{0.1}O_2$ Cathode for Li-Ion Batteries. *ACS Applied Energy Materials*, 2024, 7: 3339–3346. https://doi.org/10.1021/acsaem.4c00095; (c) Li J, Yang H, Deng Q, et al. Stabilizing Ni-rich Single-Crystalline $LiNi_{0.83}Co_{0.07}Mn_{0.10}O_2$ Cathodes using Ce/Gd Co-doped High-entropy Composite Surfaces. *Angewandte Chemie International Edition*, 2024, 63: e202318042. https://doi.org/10.1002/anie.202318042.

19 Zhang R, Wang C, Zou P, et al. Compositionally Complex Doping for Zero-Strain Zero-Cobalt Layered Cathodes. *Nature*, 2022, 610: 67–73. https://doi.org/10.1038/s41586-022-05115-z.

20 Liang P, Qi K, Chen S, et al. Low-Electronegativity Cationic High-Entropy Doping to Trigger Stable Anion Redox Activity for High-Ni Co-Free Layered Cathodes in Li-Ion Batteries. *Angewandte Chemie International Edition*, 2024, 63: e202318186. https://doi.org/10.1002/anie.202318186.

21 (a) Hu S, Pillai AS, Liang G, et al. Li-Rich Layered Oxides and Their Practical Challenges: Recent Progress and Perspectives. *Electrochemical Energy Reviews*, 2019, 2: 277–311. https://doi.org/10.1007/s41918-019-00032-8; (b) He W, Guo W, Wu H, et al. Challenges and Recent Advances in High Capacity Li-Rich Cathode Materials for High Energy Density Lithium-Ion Batteries. *Advanced Materials*, 2021, 33: 2005937. https://doi.org/10.1002/adma.202005937.

22 Song J, Ning F, Zuo Y, et al. Entropy Stabilization Strategy for Enhancing the Local Structural Adaptability of Li-Rich Cathode Materials. *Advanced Materials*, 2023, 35: 2208726. https://doi.org/10.1002/adma.202208726.

23 (a) Gao H, Li J, Zhang F, et al. Revealing the Potential and Challenges of High-Entropy Layered Cathodes for Sodium-Based Energy Storage. *Advanced Energy Materials*, 2024, 14: 2304529. https://doi.org/10.1002/aenm.202304529; (b) Gao X, Zhang X, Liu X, et al. Recent Advances for High-Entropy based Layered Cathodes for Sodium Ion Batteries. *Small Methods*, 2023, 7: 2300152. https://doi.org/10.1002/smtd.202300152.

24 (a) Li X, Wu D, Zhou Y N, et al. O3-Type $Na(Mn_{0.25}Fe_{0.25}Co_{0.25}Ni_{0.25})O$: a Quaternary Layered Cathode Compound for Rechargeable Na Ion Batteries. *Electrochemistry Communications*, 2014, 49: 51–54. https://doi.org/10.1016/j.elecom.2014.10.003; (b) Sun X, Jin Y, Zhang CY, et al. $Na[Ni_{0.4}Fe_{0.2}Mn_{0.4-x}Ti_x]O_2$: a Cathode of High Capacity and Superior Cyclability for Na-Ion Batteries. *Journal of Materials Chemistry A*, 2014, 2: 17268–17271. https://doi.org/10.1039/C4TA03828B.

25 Yue J-L, Yin W-W, Cao M-H, et al. A Quinary Layer Transition Metal Oxide of $NaNi_{1/4}Co_{1/4}Fe_{1/4}Mn_{1/8}Ti_{1/8}O_2$ as a High-Rate-Capability and Long-Cycle-Life Cathode Material for Rechargeable Sodium Ion Batteries. *Chemical Communications*, 2015, 51: 15712–15715. https://doi.org/10.1039/C5CC06585B.

26 Zhao C, Ding F, Lu Y, et al. High-Entropy Layered Oxide Cathodes for Sodium-Ion Batteries. *Angewandte Chemie International Edition*, 2020, 59: 264–269. https://doi.org/10.1002/anie.201912171.

27 Ding F, Wang H, Zhang Q, et al. Tailoring Electronic Structure to Achieve Maximum Utilization of Transition Metal Redox for High-Entropy Na Layered Oxide Cathodes. *Journal of the American Chemical Society*, 2023, 145: 13592–13602. https://doi.org/10.1021/jacs.3c00879.

28 Yao L, Zou P, Wang C, et al. High-Entropy and Superstructure-Stabilized Layered Oxide Cathodes for Sodium-Ion Batteries. *Advanced Energy Materials*, 2022, 12: 2201989. https://doi.org/10.1002/aenm.202201989.

29 (a) Li R, Qin X, Li X, et al. High-Entropy and Multiphase Cathode Materials for Sodium-Ion Batteries. *Advanced Energy Material*, 2024, 14: 2400127. https://doi.org/10.1002/aenm.202400127; (b) Joshi A, Chakrabarty S, Akella S H, et al. High-Entropy Co-Free O3-Type Layered Oxyfluoride: A Promising Air-Stable Cathode for Sodium-Ion Batteries. *Advanced Materials*, 2023, 35: 2304440. https://doi.org/10.1002/adma.202304440; (c) Fu F, Liu X, Fu X, et al. Entropy and Crystal-Facet Modulation of P2-Type Layered Cathodes for Long-Lasting Sodium-Based Batteries. *Nature Communications*, 2022, 13: 2826. https://doi.org/10.1038/s41467-022-30113-0; (d) Wang B, Ma J, Wang K, et al. High-Entropy Phase Stabilization Engineering Enables High-Performance Layered Cathode for Sodium Ion Batteries. *Advanced Energy Materials*, 2024, 14: 2401090. https://doi.org/10.1002/aenm.202401090.

30 Cai T, Cai M, Mu J, et al. High-Entropy Layered Oxide Cathode Enabling High-Rate for Solid-State Sodium-Ion Batteries. *Nano-Micro Letters*, 2023, 16: 10. https://doi.org/10.1007/s40820-023-01232-0.

31 Wang H, Gao X, Zhang S, et al. High-Entropy Na-Deficient Layered Oxides for Sodium-Ion Batteries. *ACS Nano*, 2023, 17: 12530–12543. https://doi.org/10.1021/acsnano.3c02290.

32 Walczak K, Plewa A, Ghica C, et al. $NaMn_{0.2}Fe_{0.2}Co_{0.2}Ni_{0.2}Ti_{0.2}O_2$ High-Entropy Layered Oxide – Experimental and Theoretical Evidence of High Electrochemical Performance in Sodium Batteries. *Energy Storage Materials*, 2022, 47: 500–514. https://doi.org/10.1016/j.ensm.2022.02.038.

33 Feng J, Liu Y, Fang D, et al. Reusing the Steel Slag to Design a Gradient-Doped High-Entropy Oxide for High-Performance Sodium Ion Batteries. *Nano Energy*, 2023, 118: 109030. https://doi.org/10.1016/j.nanoen.2023.109030.

34 (a) Li H, Xu M, Long H. Stabilization of Multicationic Redox Chemistry in Polyanionic Cathode by Increasing Entropy. *Advanced Science*, 2022, 9: 2202082. https://doi.org/10.1002/advs.202202082; (b) Wu B, Hou G, Kovalska E, et al. High-Entropy NASICON Phosphates ($Na_3M_2(PO_4)_3$ and ($NaMPO_4O_x$, M = Ti, V, Mn, Cr, and Zr) for Sodium Electrochemistry. *Inorganic Chemistry*, 2022, 61: 4092–4101. https://doi.org/10.1021/acs.inorgchem.1c03861; (c) Zhu L, Wang M, Xiang S, et al. A Medium-Entropy Phosphate Cathode with Multielectron Redox Reaction for Advanced Sodium-Ion Batteries. *Advanced Energy Materials*, 2023, 13: 2302046. https://doi.org/10.1002/aenm.202302046.

35 Li M, Sun C, Ni Q, et al. High Entropy Enabling the Reversible Redox Reaction of V^{4+}/V^{5+} Couple in NASICON-Type Sodium Ion Cathode. *Advanced Energy Materials*, 2023, 13: 2203971. https://doi.org/10.1002/aenm.202203971.

36 Ahsan MT, Qiu D, Ali Z, et al. Unraveling the Fast Na Diffusion Kinetics of NASICON at High Voltage via High Entropy for Sodium-Ion Battery. *Advanced Energy Materials*, 2024, 14: 2302733. https://doi.org/10.1002/aenm.202302733.

37 Li M, Sun C, Yuan X, et al. A Configuration Entropy Enabled High-Performance Polyanionic Cathode for Sodium-Ion Batteries. *Advanced Functional Materials*, 2024, 34: 2314019. https://doi.org/10.1002/adfm.202314019.

38 (a) Bernstein N, Johannes M D, Hoang K. Origin of the Structural Phase Transition in $Li_7La_3Zr_2O_{12}$. *Physical Review Letters*, 2012, 109: 205702. https://doi.org/10.1103/PhysRevLett.109.205702; (b) Cussen E J. Structure and Ionic Conductivity in Lithium Garnets. *Journal of Materials Chemistry*, 2010, 20: 5167–5173. https://doi.org/10.1039/B925553B; (c) Geiger CA, Alekseev E, Lazic B, et al. Crystal Chemistry and Stability of "$Li_7La_3Zr_2O_{12}$" Garnet: A Fast Lithium-Ion Conductor. *Inorganic Chemistry*, 2011, 50: 1089–1097. https://doi.org/10.1021/ic101914e.

39 (a) Han J, Zhu J, Li Y, et al. Experimental Visualization of Lithium Conduction Pathways in Garnet-Type $Li_7La_3Zr_2O_{12}$. *Chemical Communications*, 2012, 48: 9840–9842. https://doi.org/10.1039/C2CC35089K; (b) Awaka J, Kijima N, Hayakawa H, et al. Synthesis and Structure Analysis of Tetragonal $Li_7La_3Zr_2O_{12}$ with the Garnet-Related Type Structure. *Journal of Solid State Chemistry*, 2009, 182: 2046–2052. https://doi.org/10.1016/j.jssc.2009.05.020.

40 (a) Buannic L, Orayech B, Lopez Del Amo JM, et al. Dual Substitution Strategy to Enhance Li^+ Ionic Conductivity in $Li_7La_3Zr_2O_{12}$ Solid Electrolyte. *Chemistry of materials*, 2017, 29: 1769–1778. https://doi.org/10.1021/acs.chemmater.6b05369; (b) Wang Y, Wu Y, Wang Z, et al. Doping Strategy and Mechanism for Oxide and Sulfide Solid Electrolytes with High Ionic Conductivity. *Journal of Materials Chemistry A*, 2022, 10: 4517–4532. https://doi.org/10.1039/D1TA10966A.

41 Jung SK, Gwon H, Kim H, et al. Unlocking the Hidden Chemical Space in Cubic-Phase Garnet Solid Electrolyte for Efficient Quasi-All-Solid-State Lithium Batteries. *Nature Communications*, 2022, 13: 7638. https://doi.org/10.1038/s41467-022-35287-1.

42 (a) Feng Y, Yang L, Yan Z, et al. Discovery of High Entropy Garnet Solid-State Electrolytes Via Ultrafast Synthesis. *Energy Storage Materials*, 2023, 63: 103053. https://doi.org/10.1016/j.ensm.2023.103053; (b) Fu Z, Ferguson J. Fine and Strengthened $Li_7La_3Zr_{0.5}Nb_{0.5}Ta_{0.5}Hf_{0.5}O_{12}$ High-Entropy Li-Garnet: A Comparison Study with Ta Doped Li-Garnet. *Ceramics International*, 2023, 49: 33981–33990. https://doi.org/10.1016/j.ceramint.2023.08.094.

43 Zeng Y, Ouyang B, Liu J, et al. High-Entropy Mechanism to Boost Ionic Conductivity. *Science*, 2022, 378: 1320–1324. https://doi.org/10.1126/science.abq1346.

44 Chen Y, Wang T, Chen H, et al. Local Structural Features of Medium-Entropy Garnet with Ultra-Long Cycle Life. *Matter*, 2023, 6: 1530–1541. https://doi.org/10.1016/j.matt.2023.03.002.

45 (a) Kraft MA, Ohno S, Zinkevich T, et al. Inducing High Ionic Conductivity in the Lithium Superionic Argyrodites $Li_{6+x}P_{1-x}Ge_xS_5I$ for All-Solid-State Batteries. *Journal of the American Chemical Society*, 2018, 140: 16330–16339. https://doi.org/10.1021/jacs.8b10282; (b) Minafra N, Culver SP, Krauskopf T, et al. Effect of Si Substitution on the Structural and Transport Properties of Superionic Li-Argyrodites. *Journal of Materials Chemistry A*, 2018, 6: 645–651. https://doi.org/10.1039/C7TA08581H; (c) Zhou L, Assoud A, Zhang Q, et al. New Family of Argyrodite Thioantimonate Lithium Superionic Conductors. *Journal of the American Chemical Society*, 2019, 141: 19002–19013. https://doi.org/10.1021/jacs.9b08357.

46 Strauss F, Lin J, Duffiet M, et al. High-Entropy Polyanionic Lithium Superionic Conductors. *ACS Materials Letters*, 2022, 4: 418–423. https://doi.org/10.1021/acsmaterialslett.1c00817.

47 Lin J, Cherkashinin G, Schäfer M, et al. A High-Entropy Multicationic Substituted Lithium Argyrodite Superionic Solid Electrolyte. *ACS Materials Letters*, 2022, 4: 2187–2194. https://doi.org/10.1021/acsmaterialslett.2c00667.

48 Li W, Chen Z, Chen Y, et al. High-Entropy Argyrodite-Type Sulfide Electrolyte with High Conductivity and Electro-Chemo-Mechanical Stability for Fast-Charging All-Solid-State Batteries. *Advanced Functional Materials*, 2024, 34: 2312832. https://doi.org/10.1002/adfm.202312832.

49 (a) Li S, Lin J, Schaller M, et al. High-Entropy Lithium Argyrodite Solid Electrolytes Enabling Stable All-Solid-State Batteries. *Angewandte Chemie International Edition*, 2023, 62: e202314155. https://doi.org/10.1002/anie.202314155; (b) Wang P, Patel S, Liu H, et al. Configurational and Dynamical Heterogeneity in Superionic $Li_{5.3}PS_{4.3}Cl_{1.7-x}Br_x$. *Advanced Functional Materials*, 2023, 33: 2307954. https://doi.org/10.1002/adfm.202307954.

50 Li Y, Song S, Kim H, et al. A Lithium Superionic Conductor for Millimeter-Thick Battery Electrode. *Science*, 2023, 381: 50–53. https://doi.org/10.1126/science.add7138.

51 Li X, Xu Y, Zhao C, et al. The Universal Super Cation-Conductivity in Multiple-cation Mixed Chloride Solid-State Electrolytes. *Angewandte Chemie International Edition*, 2023, 62: e202306433. https://doi.org/10.1002/anie.202306433.

52 Wang Q, Zhou Y, Wang X, et al. Designing Lithium Halide Solid Electrolytes. *Nature Communications*, 2024, 15: 1050. https://doi.org/10.1038/s41467-024-45258-3.

53 Song Z, Wang T, Yang H, et al. Promoting High-Voltage Stability Through Local Lattice Distortion of Halide Solid Electrolytes. *Nature Communications*, 2024, 15: 1481. https://doi.org/10.1038/s41467-024-45864-1.

54 (a) Wang S, Wang K, Zhang Y, et al. High-Entropy Electrolyte Enables High Reversibility and Long Lifespan for Magnesium Metal Anodes. *Angewandte Chemie International Edition*, 2023, 62: e202304411. https://doi.org/10.1002/anie.202304411; (b) Yang C, Xia J, Cui C, et al. All-Temperature Zinc Batteries with High-Entropy Aqueous Electrolyte. *Nature Sustainability*, 2023, 6: 325–335. https://doi.org/10.1038/s41893-022-01028-x.

55 Wang Q, Zhao C, Wang J, et al. High Entropy Liquid Electrolytes for Lithium Batteries. *Nature Communications*, 2023, 14: 440. https://doi.org/10.1038/s41467-023-36075-1.

56 Kim SC, Wang J, Xu R, et al. High-Entropy Electrolytes for Practical Lithium Metal Batteries. *Nature Energy*, 2023, 8: 814–826. https://doi.org/10.1038/s41560-023-01280-1.

57 Wang Q, Zhao C, Yao Z, et al. Entropy-Driven Liquid Electrolytes for Lithium Batteries. *Advanced Materials*, 2023, 35: 2210677. https://doi.org/10.1002/adma.202210677.

10

Future Trends and Concluding Regards

Yuxin Wen[2], *Yong Zhang*[1,2,*]

[1]*Key Laboratory of Silicon-based Materials, The Ministry of Education, and School of Materials Science and Engineering, Fuyao University of Science and Technology, Fuzhou, China*
[2]*School of Materials Science and Engineering, State Key Laboratory for Advanced Metals and Materials, University of Science and Technology Beijing, Beijing, P.R. China*

*Corresponding author: drzhangy@ustb.edu.cn/yongzhang@fyust.edu.cn

10.1 Key Challenges Presented in Functional High-Entropy Alloy and Compounds

The design, synthesis, and application of high-entropy materials (HEMs) represent a cutting-edge area of materials science due to their unique properties. However, while these materials hold great promise for a range of advanced applications, they also introduce several critical challenges. These challenges extend across various facets of materials development, from synthesis methods to characterization techniques and from theoretical modeling to real-world applications. The inherent complexities that arise from the entropic effects, phase stability, compositional flexibility, and the potential catalytic properties of HEMs make them both highly promising and exceptionally difficult to optimize. Let's further elaborate on these challenges:

10.1.1 Synthesis Challenges

The synthesis of HEMs is inherently more complicated than that of conventional alloys or compounds due to the large number of constituent elements and the unique entropy-driven stabilization mechanisms that govern their structure (Figure 10.1). The synthesis methods need to be carefully tailored to accommodate the distinct properties and behaviors of these multicomponent systems. Some of the key challenges include the following:

Phase Segregation and Immiscibility. HEMs often involve a combination of several elements with varying atomic sizes, bond strengths, and chemical reactivities.

Functional High Entropy Alloys and Compounds, First Edition. Edited by Yong Zhang, Yonggang Yao and Haojie Liu.

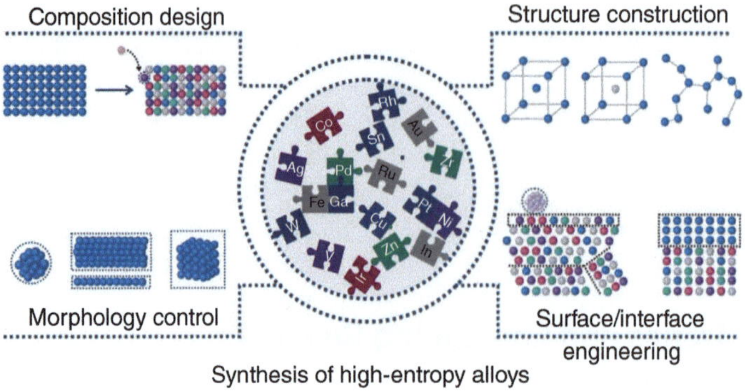

Figure 10.1 Schematic illustration of the controllable synthesis of HEAs [10] / with permission of Royal Society of Chemistry.

This increases the likelihood of phase segregation, where certain elements preferentially form separate phases rather than mixing uniformly. This is particularly problematic for HEMs intended for catalytic applications, as phase purity and uniformity directly influence the material's reactivity and efficiency. Many HEM systems exhibit high miscibility gaps where the individual elements have poor solubility in each other, which can result in the formation of immiscible phases. Overcoming these immiscibility issues while maintaining a stable high-entropy phase requires sophisticated nonequilibrium synthesis techniques like rapid cooling, flash sintering, or high-energy ball milling [1–8].

Controlling Nanostructure and Composition. Achieving fine control over the size and shape of HEMs is a complex task due to the competing thermodynamic forces present during their synthesis. The nanostructures of HEMs need to be precisely engineered to optimize their performance in various applications, such as catalysis, energy storage, and structural components. However, controlling the particle size and shape is challenging because the thermodynamic stability and kinetic factors often work in opposition. The reaction temperature, time, and precursor selection must be continuously optimized to balance these effects. For instance, increasing the temperature may promote the formation of larger particles due to enhanced diffusion, while lower temperatures may hinder atomic rearrangement, leading to nonuniform particle distribution. As a result, the synthesis process must be fine-tuned to ensure that the desired nanostructures are achieved without sacrificing material properties.

Another significant challenge lies in the compositional variability of HEMs, which is not only one of their key advantages but also a potential limitation. The ability to incorporate multiple elements in varying atomic ratios makes HEMs highly versatile, enabling the tailoring of properties for specific applications. However, this compositional flexibility makes it difficult to predict and reproduce material properties consistently. Small changes in elemental proportions can result in substantial differences in the material's behavior, making fine-tuning a slow and iterative process. For example, in electrocatalysis, adjusting the ratios of elements such as nickel, iron, and cobalt in a high-entropy alloy (HEA) can significantly influence catalytic activity, but achieving the optimal composition often requires extensive trial and error. This variability adds complexity to the design and synthesis process, particularly when scaling up the production of HEMs for commercial use.

Scaling up the synthesis of HEMs from laboratory-scale processes to industrial-scale production presents additional challenges. Methods that work well in small batches, such as solvothermal techniques or flash sintering, often face difficulties when scaled up. The precision and control required to maintain the quality of the material at larger scales are difficult to achieve. In high-energy processes such as ball milling, scaling up typically results in particle agglomeration, which compromises the material's properties and consistency. This poses a major hurdle in industrial applications where large quantities of high-quality material are needed. The challenge is to adapt these synthesis techniques to larger production scales without losing the precise control over size, shape, and structure that is crucial for HEM performance [9].

Cost-effectiveness is another critical consideration when producing HEMs on a large scale. Many of the advanced synthesis techniques required for HEM production, such as those involving high-energy inputs or specialized equipment, may not be economically viable for widespread industrial use. The cost of such processes must be carefully balanced against the performance of the final product. For example, while high-energy ball milling can produce finely dispersed particles, the high energy consumption and potential for particle agglomeration may make it less suitable for large-scale production. Similarly, solvothermal methods can be effective for synthesizing HEMs, but the need for specialized equipment and controlled conditions can drive up costs. As a result, finding cost-effective and scalable synthesis methods that do not compromise the material's properties remains a significant challenge for the industrial implementation of HEMs.

10.1.2 Characterization Challenges

Once synthesized, understanding the structure–property relationships in HEMs is crucial to optimize their performance. However, due to their complex and disordered nature, characterizing HEMs presents several challenges (Figure 10.2).

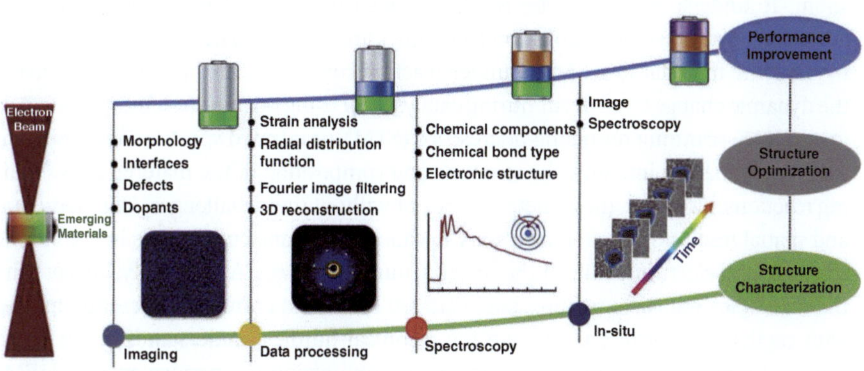

Figure 10.2 Schematic illustration of the analytical TEM techniques for emerging advanced materials [21] / with permission of Elsevier.

Atomic-Level Characterization. The complex atomic distributions inherent in HEMs introduce significant challenges in their characterization, primarily due to the disordered nature of their atomic structures. In HEMs, multiple elements are randomly distributed, which often leads to a lack of periodicity in the atomic arrangement. Traditional characterization techniques like X-ray diffraction (XRD), which rely on periodicity to generate diffraction patterns, are less effective in these materials. The disordered atomic configuration results in broad, poorly defined peaks in XRD spectra, making it difficult to extract detailed structural information. While XRD remains useful for phase identification, it fails to provide the high-resolution structural insights needed to understand the atomic-level arrangement of elements in HEMs. Consequently, advanced characterization methods are required to probe the unique features of these materials. Techniques such as atom probe tomography (APT) [11, 12], high-resolution transmission electron microscopy (HRTEM) [13, 14], and X-ray absorption spectroscopy (XAS) [15, 16] can offer higher resolution and more accurate structural details. APT, for example, provides 3D atomic-scale compositional mapping, while HRTEM allows for direct imaging of atomic arrangements. However, these techniques also have limitations when applied to highly disordered systems. They often require sophisticated equipment and, even then, may struggle to resolve the atomic distributions in HEMs with the level of precision required for comprehensive structural analysis. Furthermore, the interpretation of data from these methods can be complex, as the randomness of element distribution in HEMs adds an additional layer of difficulty in achieving a consistent and reproducible understanding of the material's atomic structure.

Another significant challenge arises in the identification of active sites within multimetallic HEMs, especially in catalytic applications such as electrocatalysis. In HEMs, where multiple elements are involved, the catalytic properties can be highly dependent on the local atomic environment of each element. The role of each individual component in catalysis is influenced not only by its identity but also by its bonding interactions with surrounding atoms, making it difficult to pinpoint which specific atoms or sites contribute most effectively to the reaction. This complexity is compounded by the heterogeneous nature of HEMs, where different regions of the material may exhibit distinct catalytic behaviors due to variations in elemental distribution, phase separation, or local strain. To understand the precise mechanisms governing catalysis in these materials, advanced operando characterization techniques are required. These techniques, which monitor the material in real-time under reaction conditions, are crucial for capturing the dynamic changes that occur during catalysis. By using operando techniques such as in-situ XAS, scanning tunneling microscopy (STM), or infrared spectroscopy, researchers can gain insight into how the structure and composition of the material evolve during reactions. However, these methods are not without their challenges. High temporal and spatial resolution is often necessary to capture the transient changes in the atomic structure and electronic state of the material during catalysis. Additionally, interpreting these complex data sets requires sophisticated models to correlate structural changes with catalytic performance. Despite the technical hurdles, understanding the active sites and their role in catalysis is essential for optimizing the performance of HEMs in industrial applications, and continued advancements in characterization techniques will be critical for overcoming these challenges.

Characterization Under Operating Conditions. One of the significant challenges in characterizing HEMs lies in performing operando studies, which involve monitoring their behavior under realistic operating conditions. This is particularly crucial for catalytic applications, where HEMs can undergo dynamic structural changes when exposed to reactive gases, high temperatures, or electrochemical environments. These transformations can include phase transitions, surface segregation, or the activation of new reaction pathways, all of which influence the material's catalytic performance. Traditional techniques such as XRD, Raman spectroscopy, and X-ray photoelectron spectroscopy (XPS) are commonly employed to monitor such changes in situ, but they have limitations. While XRD can detect phase changes and crystal structure modifications, it is often not capable of providing high spatial or temporal resolution to capture rapid or localized phenomena. Raman spectroscopy, which can provide valuable information on vibrational modes and bonding environments, may struggle with distinguishing subtle changes in complex HEM systems. XPS, on the other hand, offers insights into the chemical state of the surface atoms but typically only probes the very outermost layers of the material and often requires ultra-high-vacuum conditions, which may not fully replicate real reaction environments. Furthermore, these techniques often require sophisticated experimental setups, and their application can be limited by factors such as sample size, beam penetration, or the time scale of reactions.

The time-dependent behavior of HEMs under catalytic or electrochemical reaction conditions adds another layer of complexity. HEMs are often subjected to continuous changes in composition, structure, and surface properties during reaction cycles, and these changes can significantly affect long-term performance and stability. For example, phase segregation, which occurs when certain components of the alloy migrate to the surface or segregate into distinct phases, can alter the material's reactivity over time. Similarly, surface deactivation, due to the formation of noncatalytic phases or the loss of active sites, can drastically reduce catalytic efficiency. Tracking these processes in real time remains a major challenge in the study of HEMs. Advanced techniques such as in situ synchrotron-based X-ray spectroscopy, combined with time-resolved measurements, are increasingly being explored to monitor such time-dependent processes. However, capturing the full range of structural and chemical changes during long-term catalytic cycles remains an area of active research. Addressing these challenges requires innovative experimental approaches that combine multiple characterization techniques, each with its own strengths, to build a comprehensive understanding of the material's evolution. Ultimately, advancements in time-resolved operando studies will be critical to improving the design and optimization of HEMs for industrial applications, where durability and consistent performance are key [17–20].

10.1.3 Theoretical Challenges

Understanding the complex behavior of HEMs requires advanced theoretical frameworks that can accurately predict their properties, stability, and performance. However, several key theoretical challenges hinder the development of reliable predictive models for HEMs.

Phase Stability and Entropy-Driven Effects. Traditional thermodynamic models, which work well for simpler materials like binary alloys or oxides, often fail to predict the stability of HEMs. This is because HEMs are stabilized by configurational entropy, which these models don't consider. In conventional models, phase stability is determined by energy (enthalpy), but, for HEMs, entropy plays a major role in stabilizing their phases, especially since they contain multiple elements. This makes their phase diagrams difficult to define and leads to inaccurate predictions when applying traditional models to HEMs. To improve understanding, new thermodynamic models are needed that account for the unique stabilizing effects of entropy in these materials.

Additionally, while configurational entropy can stabilize HEMs at high temperatures, their behavior at lower temperatures can be more complex. Many HEMs are kinetically stable rather than thermodynamically stable at room temperature, meaning they can form metastable phases due to factors like slow atomic diffusion. These nonequilibrium phases may persist even when they aren't the most stable form in terms of energy. Understanding this requires models that consider both thermodynamic and kinetic factors, predicting not just stable phases but also how metastable phases might evolve under specific conditions [22] (Figure 10.3).

Catalytic Reaction Pathways. In catalytic applications, the adsorption and reaction mechanisms are intricately linked to the atomic arrangement of the material.

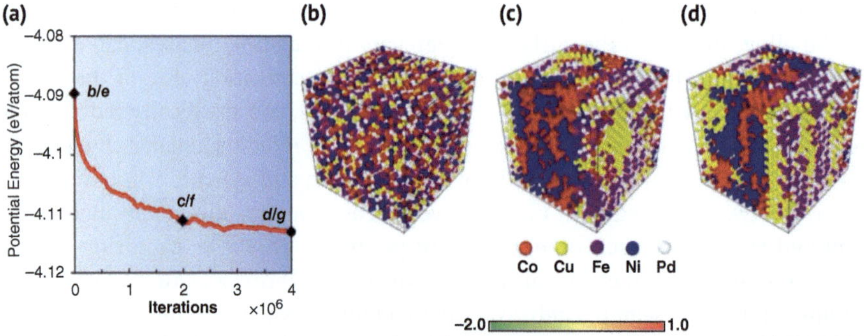

(e)

Neighboring Atom	Co	Cu	Fe	Ni	Pd
Pd	0.00	−0.01	0.00	0.00	0.01
Ni	0.01	−0.01	−0.02	−0.02	0.00
Fe	0.01	0.01	−0.04	0.02	0.00
Cu	0.00	0.01	0.01	−0.01	−0.01
Co	−0.02	0.00	0.01	0.01	0.00

(f)

Neighboring Atom	Co	Cu	Fe	Ni	Pd
Pd	0.70	0.05	−0.95	0.78	−0.58
Ni	−0.31	0.51	0.59	−1.57	0.78
Fe	0.45	0.36	−0.45	0.59	−0.95
Cu	0.51	−1.43	0.36	0.51	0.05
Co	−1.35	0.51	0.45	−0.31	0.70

(g)

Neighboring Atom	Co	Cu	Fe	Ni	Pd
Pd	0.66	0.20	−1.04	0.85	−0.67
Ni	−0.30	0.52	0.63	−1.70	0.85
Fe	0.39	0.58	−0.56	0.63	−1.04
Cu	0.64	−1.94	0.58	0.52	0.20
Co	−1.39	0.64	0.39	−0.30	0.66

(Columns: Central Atom)

Figure 10.3 Potential energy, atomic configurations, and Warren–Cowley parameters of the CoCuFeNiPd HEA during iterations at 300 K. (a) Variation of potential energy with iterations. Atomic configurations colored according to element types and tables of Warren–Cowley parameters at 0 (b and e), 2 × 106 (c and f), and 4 × 106 (d and g) iterations, as indicated in (a) [22] / Springer Nature / CC BY 4.0.

For HEMs, this becomes even more complex due to their highly disordered structures, where multiple elements with different atomic sizes, bond strengths, and chemical reactivities are present in the same phase. This complexity results in a highly variable environment on both the surface and subsurface of the material. The atomic arrangement can lead to a wide range of active sites with differing electronic properties, making the interaction with reaction intermediates unpredictable and difficult to model [23–25].

Traditional catalytic models, such as the Volcano plot and scaling relations, are based on the assumption that a catalyst's surface behaves in a relatively simple and predictable manner. These models typically work for materials with more ordered structures, where the interactions between intermediates and the surface are relatively straightforward. In the case of HEMs, however, the unique atomic configurations can cause reaction intermediates to bind in unexpected ways. The atomic environment at each active site may vary significantly due to the random distribution of elements, which alters the electronic and geometric characteristics of the site. This variability makes it challenging to predict the catalytic performance using existing models.

Moreover, traditional scaling relations that describe the adsorption energy of intermediates, like $^*OH$ or $^*OOH$ for oxygen evolution reactions, may not apply to HEMs because the relationship between these intermediates can differ drastically in HEMs compared to simpler materials. For instance, the adsorption energies of intermediates like $^*OH$ and $^*O$ in HEMs may not follow the usual scaling laws, due to the diverse atomic environments at the surface, which can lead to different binding modes for these intermediates. As a result, HEMs exhibit unique catalytic behaviors that do not align with the predictions of traditional catalytic models, requiring new models that take into account the complex atomic structure and dynamic surface behavior of these materials.

In addition, the multiple phases or local atomic environments within HEMs can lead to synergistic effects that are difficult to quantify using conventional models. These materials may also undergo dynamic changes in their structure during the reaction, further complicating the prediction of their catalytic properties. Consequently, to fully understand and optimize the catalytic performance of HEMs, it is essential to develop new, more sophisticated approaches that integrate both the structural complexity of HEMs and their dynamic response to reaction conditions. This will likely involve advances in computational modeling, in situ characterization techniques, and a deeper understanding of the interactions between materials and intermediates at the atomic level.

Machine Learning Integration. Machine learning (ML) models offer significant potential for optimizing HEMs, especially in predicting their composition, structure, and behavior. However, challenges arise due to the lack of robust descriptors that can capture the complex, disordered atomic arrangements of HEMs. Unlike traditional materials, which rely on well-established structural and compositional features, the unique variability in HEMs complicates the identification of key descriptors, particularly for properties like catalytic activity or stability. Additionally, training ML models on the multidimensional data required for HEMs is difficult, as

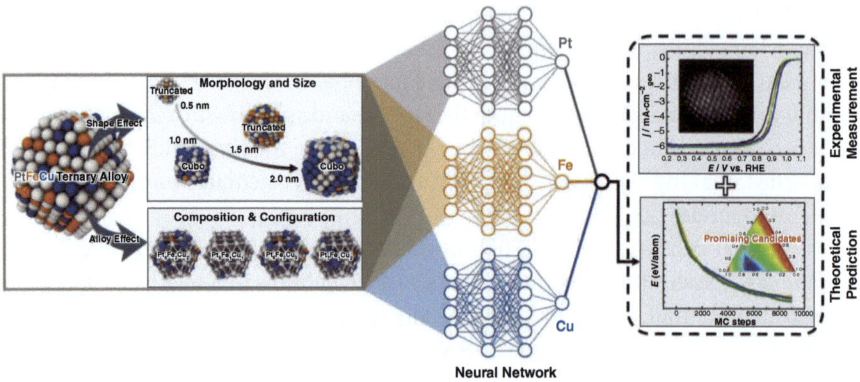

Figure 10.4 Schematic diagram of ternary alloy configuration search and theoretical predictions and experimental validation for ORR [26] / with permission of Elsevier.

these materials exhibit high compositional variability, with properties influenced by factors such as atomic size, electronegativity, and processing conditions. The data generated—encompassing phase stability, microstructure, and dynamic behavior—are highly multidimensional, requiring advanced data handling techniques. The behavior of HEMs is also sensitive to environmental factors like temperature, pressure, and reactant composition, making accurate predictions more challenging. Despite these hurdles, the field is progressing, with developments in descriptor identification, data collection, and computational models gradually overcoming these limitations. Techniques such as transfer learning and multi-fidelity models are helping improve predictions with fewer data points. As the understanding of HEMs advances, ML will play an increasingly important role in guiding the design of high-performance materials (Figure 10.4).

10.1.4 Application Challenges

HEMs show significant promise in a range of applications, from electrocatalysis and energy storage to structural materials for extreme environments. However, their application is fraught with challenges, particularly in terms of stability, performance, and scalability.

Catalytic Applications. In catalytic applications, HEMs have shown promising catalytic activity in a range of reactions, but optimizing their performance remains challenging due to issues with selectivity and stability. One of the key difficulties lies in balancing the material's catalytic activity with its selectivity. While HEMs can exhibit high reactivity due to their complex, multifaceted structure, they often suffer from unwanted side reactions or poor selectivity, especially when multiple elements are involved. The interaction between various atomic species can create a range of reaction pathways, making it difficult to optimize the material for a single, targeted reaction. Fine-tuning the number and composition of constituent elements to achieve a balance between these competing factors is a critical and ongoing challenge [27, 28].

Furthermore, the long-term stability of HEMs in catalytic environments is a significant concern. Over extended periods of operation, materials may experience surface segregation, phase transitions, or surface degradation, all of which can impair their catalytic performance. These stability issues often arise from the high compositional variability and dynamic behavior of the materials under reaction conditions, which can lead to structural changes that negatively affect their effectiveness. Understanding the mechanisms behind these degradations and developing strategies to mitigate them, such as modifying the synthesis process or surface treatment, is crucial for enabling the commercialization of HEMs for practical applications.

Industrial Scale-Up. The industrial scale-up of HEMs presents several challenges, particularly related to scalability and cost. While laboratory-scale synthesis methods have demonstrated significant potential, translating these techniques to larger scales is not a simple process. Many synthesis methods for HEMs rely on highly controlled environments and specialized equipment, which can be energy-intensive and may require significant infrastructure investment. This makes scaling up the production process both technically challenging and costly, limiting the feasibility of large-scale industrial implementation. For example, methods like rapid cooling, flash sintering, and high-energy ball milling, which work well in small-scale settings, may face difficulties in terms of consistency and material properties when scaled up. The high energy requirements of these processes can further exacerbate the challenge of achieving cost-effectiveness in large-scale production.

Additionally, the production of HEMs often involves a broad range of raw materials, each with varying availability, cost, and sourcing logistics. The complexity of synthesis routes and the need for precise control over the elemental composition contribute to the overall cost of producing HEMs. Compared to conventional materials, which may have more straightforward production methods and lower material costs, HEMs can be significantly more expensive, raising concerns about their economic viability for large-scale industrial applications. To overcome these hurdles, innovations in cost-effective synthesis techniques, material sourcing, and process optimization are needed to make HEMs more accessible for commercial use.

Environment and Durability in Extreme Conditions. HEAs have shown great promise in extreme environments, such as high-temperature oxidation or corrosive atmospheres, due to their enhanced structural integrity and resistance to degradation. However, their long-term durability under these harsh conditions remains a subject of ongoing research. While they may initially perform well, the unique combination of elements in HEAs leads to complex interactions at the atomic and microstructural levels, which can result in phase changes, surface degradation, or other forms of material failure over time. For instance, the presence of multiple elements with differing oxidation rates could lead to selective corrosion or the formation of brittle phases, which would compromise the alloy's integrity.

To fully assess their suitability for industrial applications, it is crucial to investigate the behavior of HEAs under prolonged exposure to extreme conditions, including high temperatures and corrosive environments. This requires advanced experimental techniques to simulate real-world conditions and monitor the material's degradation over time. Furthermore, the development of predictive models that incorporate both the thermodynamic and kinetic effects of such environments on the alloy's structure and performance is essential for understanding their long-term stability. Addressing these challenges will be key to unlocking the full potential of HEAs for applications in industries such as aerospace, energy production, and chemical processing, where materials are subjected to extreme operating conditions.

10.2 Future Development of Functional HEA and Compounds

With a large number of elements in varying proportions, predicting the behavior, properties, and performance of HEMs is a daunting task. One of the primary challenges lies in the difficulty of efficiently screening for high-performance HEMs. The vast compositional space requires an approach that can quickly narrow down promising candidates, but current methods still struggle to keep pace with the rapidly expanding field. Despite the enormous potential, this technology remains in its early stages, and successful applications are still relatively rare. Moreover, a major hindrance to advancing HEMs is the lack of comprehensive databases that can be leveraged to predict material properties effectively. The absence of reliable and well-curated databases makes it challenging to use ML tools to their full potential. ML, despite its promise, is still limited by the availability and quality of the input data, and it is difficult to simultaneously simulate the actual material performance, the specific service environment, alloy composition, and the preparation technology in a single model. This complexity prevents the swift identification of high-performing HEAs, thus slowing down their widespread adoption.

Nevertheless, the combination of modern technologies such as materials science, gene engineering, data science, high-throughput experimentation, big data, and ML offers an exciting pathway forward. These innovative approaches can enable the design and optimization of high-functional HEMs by overcoming current limitations in screening and performance prediction. For instance, high-throughput experimentation can quickly generate large amounts of data, which, when paired with ML algorithms, can provide valuable insights into the relationship between alloy composition and performance, significantly accelerating material discovery. The intersection of artificial intelligence (AI) and materials science represents a new frontier in HEM development. AI, driven by the need to further enhance productivity, has already triggered a transformative revolution across multiple fields. ML, in particular, has emerged as a powerful tool for tackling the complexities of HEAs. Its applications in materials science are vast, encompassing tasks such as classification, regression, clustering, density estimation, and dimensionality reduction. ML

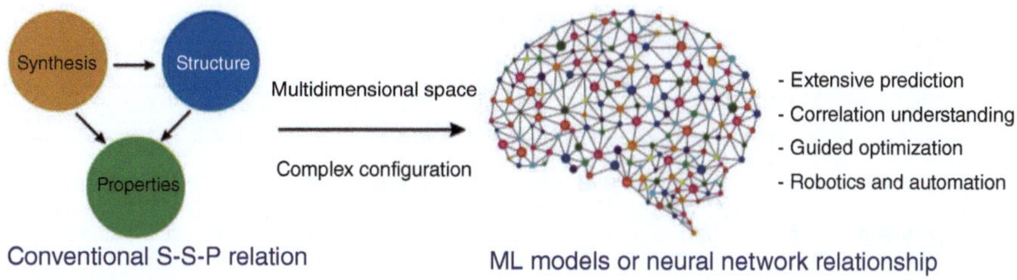

Figure 10.5 Data-driven approaches, powered by ML models, could replace traditional synthesis–structure–property relationships in material research, enabling faster predictions, deeper understanding, and even automated discovery [29] / with permission of The American Association for the Advancement of Science.

models can help identify patterns, predict material behaviors, and optimize synthesis parameters, facilitating the rapid discovery of new HEMs with desired properties (Figure 10.5).

10.2.1 High-Throughput Exploration of ML

One is based on ab initio calculations [30], such as Yoav et al. [31], which use ab initio calculations to predict the orderly and disordered transformation of alloy components by judging the formation capacity of solid solutions. Troparevsky et al. [32] calculated the enthalpy of the formation of a binary alloy subsystem ab initio and used this enthalpy to estimate the degree of stability of a multicomponent system. Based on high-throughput density-functional theory calculations, Chen et al. construct a chemical map of single-phase equimolar HEAs by investigating over 658,000 equimolar quinary alloys through a binary regular solid-solution model. We identify 30,201 potential single-phase equimolar alloys (5% of the possible combinations) forming mainly in body-centered cubic (BCC) structures [33] (Figure 10.6). The other is based on the calculation of phase diagram (CALPHAD) method; for example, Senkov et al. [34] used the high-throughput CALPHD method to predict the possible phases of the alloy and quickly evaluated more than 130,000 composite gold compositions. Abu-Odeh et al. [35] used the constraint satisfaction algorithm to reduce the traversal space and then used CALPHAD to verify the obtained results. This method frees researchers from tedious experiments, but it also has certain limitations, such as the accuracy depends on the accuracy of databases and simulations, and cannot establish a direct relationship with the experimental results.

10.2.2 Statistical Methods

Tancret et al. [36] used statistical methods to propose a statistical model based on thermodynamics and Gaussian processes, which uses nine parameters to identify a single solid-solution phase, and evaluated different thermodynamic databases. However, the thermodynamic database of HEAs is largely inherited from the database of nickel-based alloys, and the multicomponent database still needs to be

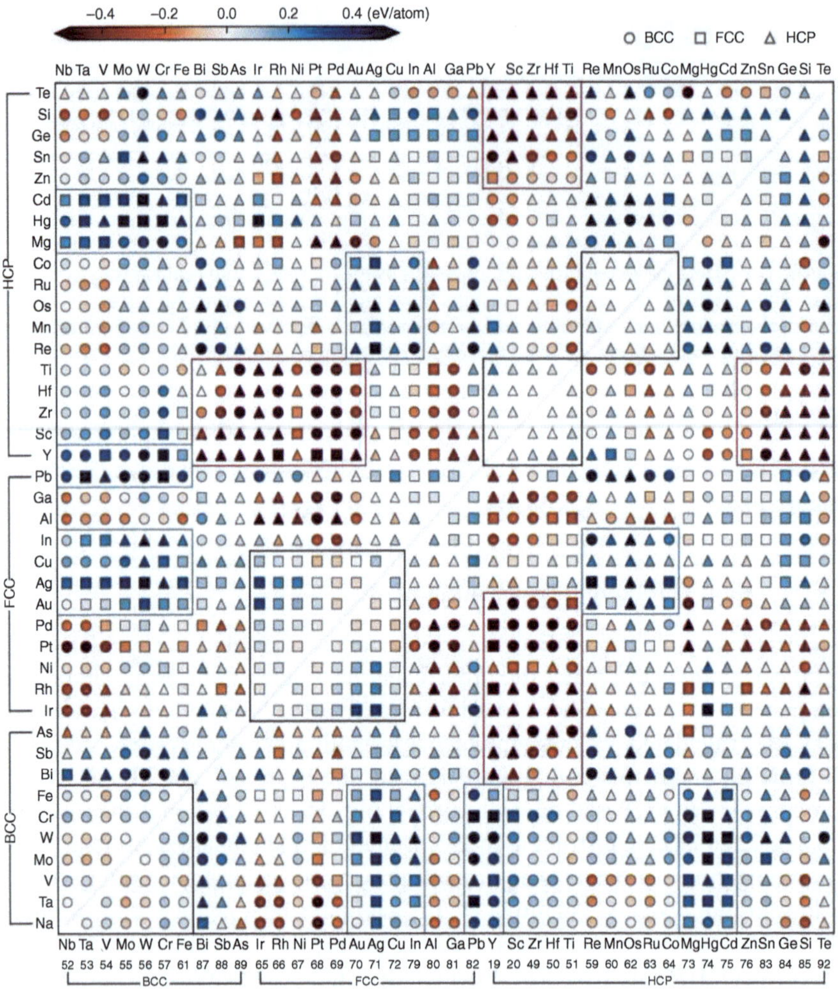

Figure 10.6 Map of formation enthalpy for binary solid solutions, as represented by SQS (Special Quasirandom Structure), obtained from DFT (Discrete Fourier Transform) calculations [33]. The formation enthalpy (ΔH^f) is determined with respect to the ground-state elemental phases. The 40 elements are grouped by their lowest energy structure at 0 K (BCC, face-centered cubic, or HCP [Hexagonal Close-Packed]) and are sorted according to Pettifor's Mendeleev numbers. Groups of elements mixing in the same crystal structure are shown by the black blocks. Groups of elements that strongly favor mixing ($\Delta H^f < -0.2$ eV/atom) are highlighted by the red blocks, whereas those strongly disfavoring mixing ($\Delta H^f > 0.2$ eV/atom) are highlighted by the blue blocks [33] / Springer Nature / CC BY 4.0.

further improved (Figure 10.7). Domínguez et al. [37] performed principal component analysis on a dataset of HEAs for the first time and made predictions for a series of alloys based on this. Authors extract useful information from raw data and use it to make predictions. The dataset used in the article is relatively small, with only 79 items. Among them, principal component analysis is a dimensionality reduction algorithm, which aims to reduce the dimensionality of high-dimensional

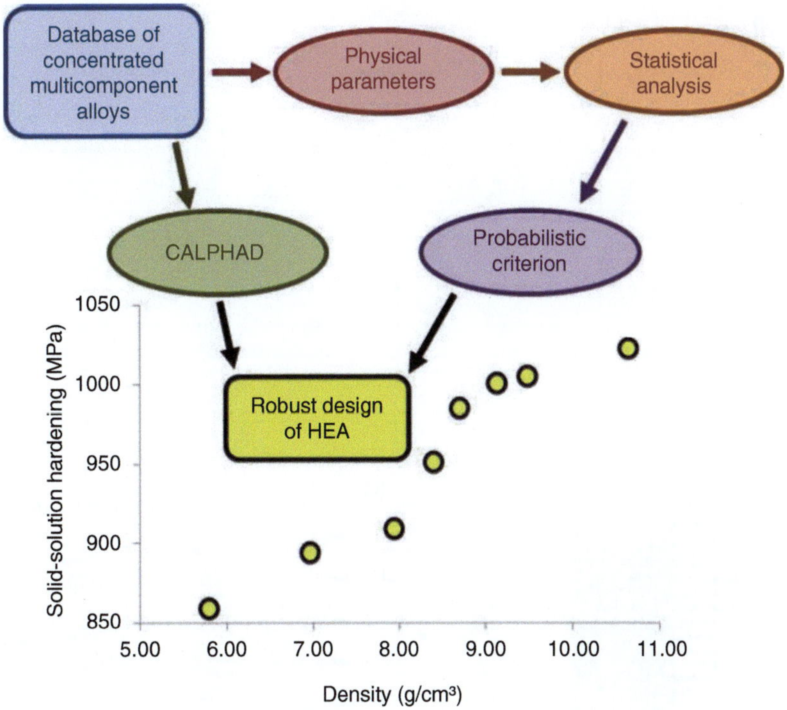

Figure 10.7 A guide to design new HEAs based on thermodynamics and Gaussian processes. [36] / with permission of Elsevier.

data, to retain the most important features and remove noise and unimportant features. This approach makes the dataset easier to use, reduces computational overhead, and makes the results easy to understand.

10.2.3 Artificial Neuron Networks

Islam et al. [38] used ML to classify HEA datasets accordingly. Five features were selected in the study, and the data were increased to 118. The original dataset was overfitted with an accuracy of 99%. Unevaluated prediction models lack meaning, and algorithms learn a lot of noise and lack generalization capabilities. The accuracy rate in the subsequent multifold training was only 86%. Multifold training is a commonly used method to reduce the error caused by the selection of the training set in a small dataset to the final result.

Cheng et al. [39] combined ML and experiments to find an HEA composition with higher hardness in the Al—Co—Cr—Cu—Fe—Ni system after two rounds of iterations. The database contains hardness data for 155 systems, including 22 quaternary alloys, 95 quinary alloys, and 38 six-membered alloys. Since the experimental data may come from different laboratories, the hardness data are likely to fluctuate, and the sample size of the database is relatively small, the high hardness data will have an anchoring effect on the prediction results of the algorithm.

Abhishek et al. [40] applied an adaptive neural fuzzy interface system (ANFIS) for AI to phase prediction of HEAs. The adaptive neural fuzzy interface is a hybrid

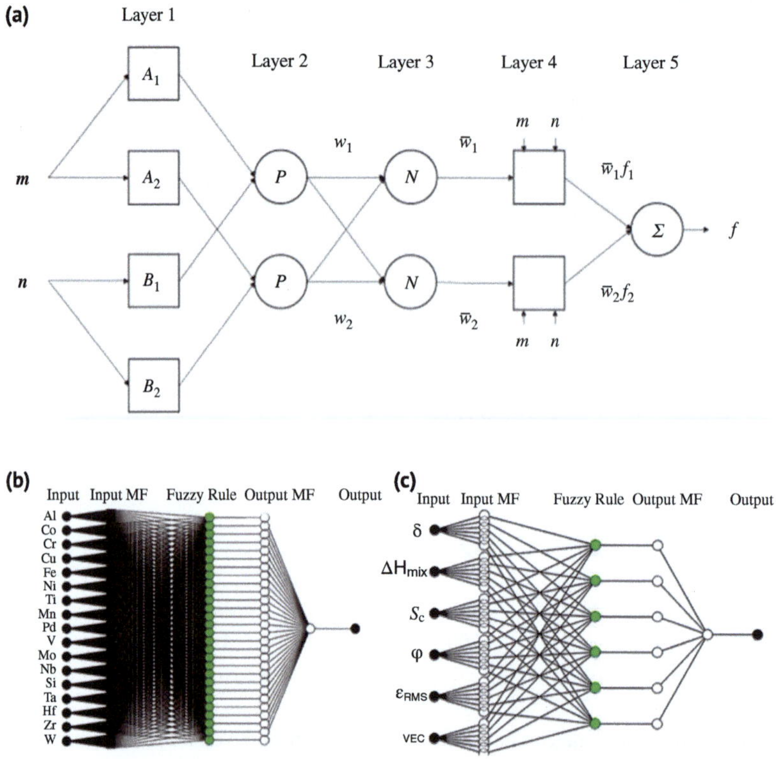

Figure 10.8 (a) Schematic diagram of working of ANFIS. (b and c) Framework of ANFIS: composition-based modeling (a), parameter-based modeling (b). *Source*: Reproduced with permission of Springer Nature.

intelligent system constructed by artificial neuron networks and fuzzy logic. Figure 10.8 shows the hybrid system algorithm framework they designed, changing the fuzzy logic can change the way knowledge is acquired, and the fuzzy rules can be optimized through the learning ability of the neural network. Figure 10.8 depicts a system with two inputs and one output: the inputs are m and n, and the output is f. The adaptive neural fuzzy inference system model is governed by a set of flexible rules called fuzzy if-then rules, where inputs are mapped to a series of outputs (also known as membership functions) based on their behavior. A membership function is a curve or function that defines how a membership is specified for each input parameter in the output. Membership ranges from 0 to 1, with membership 0 indicating that the input is not a member of the fuzzy set, 0.5 indicating partial membership, and 1 indicating full membership. For a model with 2 inputs, each input is mapped to 2 membership functions. Layer 1 is called the fuzzy layer or input layer because the input is mapped to the fuzzy range using membership functions. Layer 2 is called the product layer, labeled P, and it calculates the parameters from the fuzzy layer, which is sometimes referred to as the input member function layer. Layer 3 is called the fuzzy rule layer or normalization layer, labeled N, and normalizes the function weights by summing the output functions of layer 2 and all output functions of layer 2. Layer 4 is called the defuzzification layer or output membership function layer, and it ambiguates the values to give a clear output. Layer 5 is the total output layer, labeled R, which outputs the sum of all individual parameters obtained from the previous layer.

10.3 Concluding Regards: Materials Evolution from Simple to a Complex Yet Fruitful Future

Materials have always been central to human progress, evolving alongside societal advancements. From the Stone Age to the Metal Age, materials like bronze and iron enabled productivity and innovation, forming the foundation for modern civilization. The development of HEMs marks a significant leap in material complexity, offering superior strength and corrosion resistance for demanding applications in aerospace, biomedicine, electronics, and energy.

With the rapid advancements in AI and ML, materials development has entered a new era. The ability to leverage AI for materials discovery and optimization has revolutionized the process, shifting it away from traditional trial-and-error methods toward more efficient, low-cost, and rapid screening of new materials. High-throughput design, fabrication, and characterization techniques now enable accelerated material development by enabling the systematic exploration of vast compositional and structural spaces. This, in turn, facilitates the identification of novel HEMs with tailored properties that can meet the demanding requirements of various industries.

High-throughput experimentation plays a crucial role in advancing HEMs by generating the necessary experimental data to refine computational models and enhance material simulations. For instance, techniques like thin-film deposition, multivolume diffusion, jet-printing synthesis, and laser additive manufacturing enable the efficient production of material samples with varied compositions and structures. High-throughput characterization methods can quickly assess critical material properties, including microcomposition, crystal structure, microstructure, and mechanical properties. By generating large amounts of data, these techniques support the rapid identification of correlations between composition, structure, and performance, accelerating the research and development of HEMs for next-generation technologies.

The shift from simple materials to highly complex yet functional HEMs signifies a significant leap forward, promising not only better performance but also enhanced durability, scalability, and adaptability across various applications. As we continue to leverage the power of AI, big data, and high-throughput methods, the future of materials science—particularly in the realm of HEMs—looks poised for transformative growth.

References

1 Nellaiappan S, Katiyar N K, Kumar R, et al. High-entropy alloys as catalysts for the CO_2 and CO reduction reactions: experimental realization. *ACS Catalysis*, 2020, 10, 3658–3663. https://doi.org/10.1021/acscatal.9b04302

2 Wu S, Liu Y, Ren Y, et al. Microwave synthesis of single-phase nanoparticles made of multi-principal element alloys. *Nano Research*, 2021, 15: 4886–4892. https://doi.org/10.1007/s12274-021-3893-y

3 Qiao H, Saray M T, Wang X, et al. Scalable synthesis of high entropy alloy nanoparticles by microwave heating. *ACS Nano*, 2021. https://doi.org/10.1021/acsnano.1c05113

4 Yao Y, Huang Z, Xie P, et al. High temperature shockwave stabilized single atoms. *Nature Nanotechnoloy*, 2019, 14: 851–857. https://doi.org/10.1038/s41565-019-0518-7

5 Jiao M, Yao Y, Pastel G, et al. Fly-through synthesis of nanoparticles on textile and paper substrates. *Nanoscale*, 2019, 11: 6174–6181. https://doi.org/10.1039/c8nr10137j

6 Yao Y, Huang Z, Hughes L A, et al. Extreme mixing in nanoscale transition metal alloys. *Matter*, 2021, 4: 2340–2353. https://doi.org/10.1016/j.matt.2021.04.014

7 Feng J, Chen D, Pikhitsa P V, et al. Unconventional alloys confined in nanoparticles: building blocks for new matter. *Matter*, 2020, 3: 1646–1663. https://doi.org/10.1016/j.matt.2020.07.027

8 Wang X, Huang Z, Yao Y, et al. Continuous 2000 K droplet-to-particle synthesis. *Materials Today*, (2020), 35: 106–114. https://doi.org/10.1016/j.mattod.2019.11.004

9 Yao Y, Liu Z, Xie P, et al. Computationally aided, entropy-driven synthesis of highly efficient and durable multi-elemental alloy catalysts. *Science Advances*, 2020, 6: eaaz0510. https://doi.org/10.1126/sciadv.aaz0510

10 Liang, J., Cao, G., Zeng, M, et al. Controllable synthesis of high-entropy alloys. *Chemical Society Reviews*, 2024, 53: 6021–6041. https://doi.org/10.1039/D4CS00034J

11 Yang Y, Zhou J, Zhu F, et al. Determining the three-dimensional atomic structure of an amorphous solid. *Nature*, 2021, 592: 60–64. https://doi.org/10.1038/s41586-021-03354-0

12 Liu H, Zhang Y, Zhang L, et al. Unveiling atomic-scaled local chemical order of high-entropy intermetallic catalyst for alkyl-substitution-dependent alkyne semihydrogenation. *Journal of the American Chemical Society*, 2024, 146: 20193–20204. https://doi.org/10.1021/jacs.4c05295

13 Ding Q, Zhang Y, Chen X, et al. Tuning element distribution, structure and properties by composition in high-entropy alloys. *Nature*, 2019, 574: 223–227. https://doi.org/10.1038/s41586-019-1617-1

14 Chen X, Wang Q, Cheng Z, et al. Direct observation of chemical short-range order in a medium-entropy alloy. *Nature*, 2021, 592: 712–716. https://doi.org/10.1038/s41586-021-03428-z

15 Jiang B, Bridges C A, Unocic R R, et al. Probing the local site disorder and distortion in pyrochlore high-entropy oxides. *Journal of the American Chemical Society*, 2021, 143: 4193–4204. https://doi.org/10.1021/jacs.0c10739

16 Holder CF, Schaak RE. Tutorial on powder X-ray diffraction for characterizing nanoscale materials. *ACS Nano*, 2019, 13: 7359–7365. https://doi.org/10.1021/acsnano.9b05157

17 Song B, Yang Y, Rabbani M, et al. In situ oxidation studies of high-entropy alloy nanoparticles. *ACS Nano*, 2020, 14: 15131–15143. https://doi.org/10.1021/acsnano.0c05250

18 Zhao H, Zhu Y, Ye H, et al. Atomic-scale structure dynamics of nanocrystals revealed by in-situ and environmental transmission electron microscopy. *Advanced Materials*, 2022, 35: e2206911. https://doi.org/10.1002/adma.202206911

19 Cao G, Liang J, Guo Z, et al. Liquid metal for high-entropy alloy nanoparticles synthesis. *Nature*, 2023. https://doi.org/10.1038/s41586-023-06082-9

20 Sun X, Wu D, Zou L, et al. Dislocation-induced stop-and-go kinetics of interfacial transformations. *Nature*, 2022, 607: 708–713. https://doi.org/10.1038/s41586-022-04880-1

21 Lin Y, Zhou M, Tai X, et al. Analytical transmission electron microscopy for emerging advanced materials. *Matter*, 2021, 4: 2309–2339. https://doi.org/10.1016/j.matt.2021.05.005

22 Chen S, Aitken Z H, Pattamatta S, et al. Simultaneously enhancing the ultimate strength and ductility of high-entropy alloys via short-range ordering. *Nature Communications*, 2021, 12: 4953. https://doi.org/10.1038/s41467-021-25264-5

23 Li H, Han Y, Zhao H, et al. Fast site-to-site electron transfer of high-entropy alloy nanocatalyst driving redox electrocatalysis. *Nature Communications*, 2020, 11: 5437. https://doi.org/10.1038/s41467-020-19277-9

24 Wu D, Kusada K, Nanba Y, et al. Noble-metal high-entropy-alloy nanoparticles: atomic-level insight into the electronic structure. *Journal of the American Chemical Society*, 2022, 144: 3365–3369. https://doi.org/10.1021/jacs.1c13616

25 Feng G, Ning F, Song J, et al. Sub-2 nm ultrasmall high-entropy alloy nanoparticles for extremely superior electrocatalytic hydrogen evolution. *Journal of the American Chemical Society*, 2021, 143: 17117–17127. https://doi.org/10.1021/jacs.1c07643

26 Chun H, Lee E, Nam K, et al. First-principle-data-integrated machine-learning approach for high-throughput searching of ternary electrocatalyst toward oxygen reduction reaction. *Chem Catalysis*, 2021, 1: 855–869. https://doi.org/10.1016/j.checat.2021.06.001

27 Batchelor T A, Pedersen J K, Winther S H, et al. High-entropy alloys as a discovery platform for electrocatalysis. *Joule*, 2019, 3: 834–845. https://doi.org/10.1016/j.joule.2018.12.015

28 Lu Z, Chen Z W, Singh C V. Neural network-assisted development of high-entropy alloy catalysts: decoupling ligand and coordination effects. *Matter*, 2020, 3: 1318–1333. https://doi.org/10.1016/j.matt.2020.07.029

29 Yao Y, Dong Q, Brozena A, et al. High-entropy nanoparticles: synthesis-structure-property relationships and data-driven discovery. *Science*, 2022, 376: eabn3103. https://doi.org/10.1126/science.abn3103

30 Durandurdu M. Ab initio modeling of small diameter silicon nanowires. *Physica Status Solidi (B)*, 2006, 243: R7–R9. https://doi.org/10.1002/pssb.200541524

31 Lederer Y, Toher C, Vecchio K S, et al. The search for high entropy alloys: a high-throughput ab-initio approach. *Acta Materialia*, 2018, 159: 364–383. https://doi.org/10.1016/j.actamat.2018.07.042

32 Troparevsky M C, Morris J R, Kent P R C, et al. Criteria for predicting the formation of single-phase high-entropy alloys. *Physical Review X*, 2015, 5: 011041. https://doi.org/10.1103/PhysRevX.5.011041

33 Chen W, Hilhorst A, Bokas G, et al. A map of single-phase high-entropy alloys. *Nature Communications*, 2023, 14: 2856. https://doi.org/10.1038/s41467-023-38423-7

34 Senkov O N, Miller J D, Miracle D B, et al. Accelerated exploration of multi-principal element alloys with solid solution phases. *Nature Communications*, 2015, 6: 6529. https://doi.org/10.1038/ncomms7529

35 Abu-Odeh A, Galvan E, Kirk T. et al. Efficient exploration of the high entropy alloy composition-phase space. *Acta Materialia*, 2018, 152: 41–57. https://doi.org/10.1016/j.actamat.2018.04.012

36 Tancret F, Toda-Caraballo I, Menou E, et al. Designing high entropy alloys employing thermodynamics and Gaussian process statistical analysis. *Materials & Design*, 2017, 115: 486–497. https://doi.org/10.1016/j.matdes.2016.11.049

37 Domínguez L A, Goodall R, Todd I. Prediction and validation of quaternary high entropy alloys using statistical approaches. *Materials Science and Technology*, 2015, 31: 1201–1206. https://doi.org/10.1179/1743284715Y.0000000019

38 Islam N, Huang W, Zhuang H L. Machine learning for phase selection in multi-principal element alloys. *Computational Materials Science*, 2018, 150: 230–235. https://doi.org/10.1016/j.commatsci.2018.04.003

39 Wen C, Zhang Y, Wang C, et al. Machine learning assisted design of high entropy alloys with desired property. *Acta Materialia*, 2019, 170: 109–117. https://doi.org/10.1016/j.actamat.2019.03.010

40 Agarwal A, Prasada Rao A K. Artificial intelligence predicts body-centered-cubic and face-centered-cubic phases in high-entropy alloys. *JOM*, 2019, 71: 3424–3432. https://doi.org/10.1007/s11837-019-03712-4

Index

Note: Page numbers in *italics* refers to figures.

Functional High Entropy Alloys and Compounds, First Edition. Edited by Yong Zhang, Yonggang Yao and Haojie Liu.